W0246157

D. R. Axelrad

Stochastic Mechanics of Discrete Media

With 43 Figures

Springer-Verlag
Berlin Heidelberg New York
London Paris Tokyo
Hong Kong Barcelona Budapest

Prof. David Robert Axelrad

McGill University,
Institute of Micromechanics Research Laboratories,
Department of Mechanical Engineering,
817 Sherbrooke Street West, Montreal, PQ
Canada H3A 2K6

ISBN 978-3-642-51487-6   ISBN 978-3-642-51485-2 (eBook)
DOI 10.1007/978-3-642-51485-2

Library of Congress-in-Publication Data
Axelrad, D. R. Stochastic mechanics of discrete media / D.R. Axelrad. p. cm.
Includes bibliographical references and index.
1. Micromechanics. 2. Stochastic analysis. I. Title. QC176.8.M5A974 1993 530.4'12--dc20

© Springer-Verlag Berlin Heidelberg 1993
Softcover reprint of the hardcover 1st edition 1993

The use of general descriptive names, registered names, trademarks, etc. in this publication does not
imply, even in the absence of a specific statement, that such names are exempt from the relevant
protective laws and regulations and therefore free for general use.

Typesetting: Camera ready by author

61/3020 5 4 3 2 1 0 - Printed on acid-free paper

Preface

For the past three decades the mechanics of heterogeneous materials, frequently called "micro mechanics", has been recognized as an important new approach to the analysis of the behaviour of discrete media. The micro mechanics view originated from the condition that analytical models based on the principles of continuum mechanics could no longer be accepted for the interpretation of experimental results concerning discrete material. Even the introduction of higher order continuum theories could not provide satisfactory descriptions of the behaviour of real materials. Moreover, such theories relied upon the knowledge of a great number of parameters for whose determination no experimental procedure existed or could even be conceptualized.

It was therefore increasingly felt that an appropriate representation of discrete materials could only be achieved by including their respective "micro structures". This necessitated the adoption of some intermediate "length scale" in the analysis, since the information available for "single elements" of the structure could not be directly related to the required macroscopic relations. The difficulties arose also from earlier mathematical representations, which were influenced for nearly two centuries by continuum theory. It only became possible to free the mechanics analysis from the hegemony of the "continuum" by the introduction of set theory at the beginning of this century and by the subsequent establishment of measure theory and topology.

Thus in the characterization of material behaviour continuum theoretical concepts were gradually replaced by considerations of discontinuous statistics and the significance of "scales"; larger numbers of publications became concerned with "discretization" of continuumfields and the statistical treatment of the involved field quantities by recognizing their dependence on the existing microstructure of a given medium.

The stochastic mechanics approach introduced in 1966 by this author and pursued since both analytically and experimentally uses probabilistic concepts and statistical analysis for the description of the material behaviour. Thus field quantities are regarded as "random variables" or functions of such variables. Their axiomatic definitions on the basis of probability and measure theory were given in two preceding monographs. An important feature of this approach is the use of three "measuring scales", the micro, meso and macroscopic scales, with reference to a single structural element, an ensemble of the latter and finally a set consisting of a finite number of meso-domains forming the macroscopic material body. The stochastic mechanics approach further permits the inclusion of "interactions" between structural elements and applies to a large class of discrete media, which, for the most part, exhibit a certain randomness. This inherent characteristic also called "stochasticity" from the greek word: $\sigma\tau\omega\chi\omega\sigma$ is present in almost all discrete media. It leads in the mathematical representation of the behaviour to "stochastic differential equations" and hence the use of stochastic calculus.

The present volume seeks to review and extend the probabilistic formulation given in earlier work and to survey as well the application of stochastic mechanics to various discrete media of interest in practice. One of the main aims is to establish a link between the modern analysis and the mechanics of heterogeneous media. For the reader of this analysis who may not be familiar with the notion of sets, measures, topology and some aspects of the theory of stochastic processes a brief review of the fundamental concepts is given in Chapter 1. Chapter 2 is mainly concerned with the classification and phenomenology of discrete media, while Chapter 3 considers their "random evolution" and introduces the important concept of "geometric probability". Finally in Chapter 4 some applications of the stochastic mechanics approach are given together with a comparison of experimental results.

I would like to acknowledge the National Science and Engineering Council of Canada for support of the work carried out in the Micromechanics Laboratory of McGill University. It is also a pleasure to acknowledge Prof. O. Mahrenholtz, Institute of Mechanics, University of Hamburg–Harburg, Prof. Ch. Enz, Institute of Theoretical Physics, University of Geneva (Switzerland), Prof. H. Zorski, Institute of Fundamental Research, Polish Academy of Sciences (Warsaw) and Prof. R.K.T. Hsieh, Department of Mechanics, Royal Institute of Technology, Stockholm (Sweden) for their kind invitations at various times to deliver short courses and seminars in stochastic mechanics at their respective institutes.

I would also like to express sincerest thanks to Dr. W. Frydrychowicz, Senior Research Associate in Micromechanics and the Dept. of Chemical Engineering for undertaking the arduous task of proof reading the entire manuscript. Special thanks are due to Drs. J. Kalousek, K. Rezai and R. Peralta-Fabi for their careful performance of the experimental work associated with X-ray diffraction, stress-holographic interferometry and scanning electron microscopy concerning various materials. I would like to thank Ms. Lynda Corkum of the Dept. of Physics at McGill University for the conscientious preparation of the Computer script from the handwritten text; Ms. Patricia Butron-Guillen for supplying all diagrams and Dr. E. Turcott-Rios for providing the SEM-micrographs contained in this volume. Mr. W.Y. Hamad kindly furnished the subject index and aided in the final editing of the manuscript. Finally special acknowledgement is made to the Editorial staff of the publisher for their cooperation and efficient production of this volume.

MONTREAL, June 1993   D.R. AXELRAD

Contents

1 INTRODUCTION TO THE PROBABILISTIC ANALYSIS

1.1 Historical Remarks on Probability and its Application in Mechanics

(i) The notion of probability:

The main ideas and notions of probability, discontinuity and determinism discussed here, have also been considered in greater detail by historians and philosophers of science in various contemporary essays contained in two volumes on "the probabilistic revolution" [1,2]. The rather slow emergence of probabilism vs. determinism, may be ascribed to the arguments and discussions that were directed against the probabilistic view in the last century.

In this century an important change of this view occurred as well as a change in the development of the theory of probability itself. The definitions and principles due to Laplace [3] were no longer accepted and were replaced by the notions of the "frequentists" [4,5] in the foundations of applied probability (see also [5,6]). The definition of probability adopted in this study is based on Kolmogorov's axioms in probability theory [6]. It follows the frequency interpretation of probability whereby a random experiment, if carried out n-times under identical conditions, the probability of the "event E_i" to occur is postulated to hold almost everywhere, i.e.:

$$\mathcal{P}\{E_i\} \overset{a.e.}{=} \lim_{n \to \infty} n_i/n$$

This definition is based on the following axioms [6]:

A.1 $: \mathcal{P}\{E_i\} \geq 0$ (0 being an impossible event)

A.2 $: \mathcal{P}\{\Omega\} = 1$ (1 being always a certain event)

A.3   : For two mutually exclusive events E_i, E_j

$$\mathcal{P}\{E_i \cup E_j\} = \mathcal{P}\{E_i\} + \mathcal{P}\{E_j\}; E_k \cap E_j = \emptyset, i \neq j.$$

The notation $\mathcal{P}\{E\}$ or the "probability of occurrence" of the event E refers to the probability measure of E or simply the probability of E. In probability theory, in general, a "random experiment" is called a "trial". If such an experiment is carried out without any prior knowledge, the result is an "outcome ω".

The set of all possible outcomes $\{\omega\}$ is the "sample space Ω". Thus an event E is a subset of Ω such that $E = \{\omega \in \Omega; \; \omega$ satisfying certain properties$\}$. The statement that an event E occurs means that the outcome of the random experiment is an element of E or $\omega \in E$. The empty or null set $\emptyset$ contains no outcome and hence is an impossible event.

Around 1900 the position of probability theory, its methods and foundations were still controversial. The outstanding work of the mathematicians E. Borel, R. von Mises, Kolmogorov contributed considerably to its acceptance. E. Borel particularly stressed the importance of the application of probability to physical systems and denied the concepts of continuity and the continuum. He stated in his work (see for instance [7]) that "... the 19th century looked for unity in the field of mechanics and of differential equations, which without doubt will be found in the 20th century by studying the law of chance".

As early as 1900 D. Hilbert recognized the need for an axiomatic foundation of the theory of probability. In his work on mathematical problems [8] he states as the "sixth problem" the following: "... we have to treat by means of axioms those physical sciences in which mathematics already plays today an important part. These are above all probability theory and mechanics". In 1909 E. Borel published his famous paper on "denumerable probabilities", which reflected his conception of probability. He did not intend however to give a mathematical foundation to probability and was criticised for this by von Mises, who considered Borel's attempt as a pure formalism. Although at the beginning of 1925 Borel pursued his analysis in a purely abstract manner, he subsequently discussed the axiomatization of the theory [9]. Thus admitting that axiomatization, which has been stipulated earlier by D. Hilbert, removed some of the philosophical difficulties and permitted progress in this discipline. A satisfactory foundation of the theory of probability however was not achieved until the introduction of Kolmogorov's theory of probability spaces [10].

(ii) Continuity and discontinuity:

E. Borel's concerns of the relation between probability theory and the problem of continuity and discontinuity can be perceived from the aforementioned paper on denumerable probabilities [11]. This paper originated from his earlier work on set theory(1908–1909) in which every element of the continuum can be defined and where the set of all "effectively defined elements" is always denumerable. In his treatise "applications à l'arithmétique et à la théorie, vol. II", he discusses denumerable probabilities and continues with his considerations of probability in "sur les ensembles effectivement énumérables et sur les définitions effectives" in [11]. His conception of the relation between probability and the continuity-discontinuity problem is perhaps best illustrated by the following quote:

"... it is clear, that the set of analytic elements that can be actually defined and considered, can only be a denumerable set; I believe that this point of view will prevail more and more every day among mathematicians and that the continuum will prove to have been a transitory instrument whose present day utility is not negligible, but will come to be regarded only as a means of studying denumerable sets ...". As far back as 1912, Borel stated that the distinction between continuous and discontinuous probabilities corresponds to that between continuous and finite groups. For the intermediate stage corresponding to discontinuous groups he established the theory of denumerable probabilities (see also E. Knobloch in [7]).

(iii) Determinism and probabilism:

The introduction of probability in physics greatly affected two areas, namely the theory of measurement and the construction of analytical models characterizing various physical phenomena. The discovery of the "law of errors" and the subsequent development of the error theory (1790–1840) had the main purpose of establishing a more systematic evaluation of observations.

It was C.F. Gauss, who defined (1797) the normal or Gaussian "distribution of errors". On the basis of his arguments Legendre in 1805 introduced independently the method of "least squares". P.S. Laplace during 1809–11 was able to point out the probabilistic aspects of the error theory. He showed by the use of certain theorems that the sum of many independent errors is always normally distributed. In 1860 J.C. Maxwell [12] was the first to show that the velocities of molecules (gaseous) were distributed about a mean value, which depended on the temperature of the gas in accordance with the law of errors. His argument started from purely abstract probabilistic assumptions without consideration of the dynamics of motion. He only attempted later to establish the velocity distribution of the gas molecules by the "statistics of their collisions". L. Boltzmann employed this collision method to derive a mechanical equivalent to the entropy. However, he was unable to explain how the purely mechanical laws which interpreted all motions to be reversible in time, could lead to an "irreversible behaviour". In 1877 he changed his view and introduced an "intrinsic probability" of a given state of the system to be proportional to the number of "distinguishable microscopic realizations" of this state (see also [12]).

In order to relate the actual motion of the system to the distributions of properties of its constituents, Maxwell introduced the notion in (1878) (see also [13]), known as the "ergodic hypothesis". The latter states that a physical system in equilibrium will assume all conceivable states compatible with the conservation of energy. He pointed out, however, that this could only be true by virtue of suitable "random disturbances" of the system due to external sources. The probabilistic aspect of the ergodic properties of physical systems has been the topic of many analytical investigations and developed since about 1930 into the "mathematical ergodic" theory. Another important notion also due to Maxwell is that of "ensembles". It was W. Gibbs [13, 14] in 1902 who introduced the ensemble approach as another method to that of Boltzmann's collision theory, (see also [15]).

It should be pointed out that probabilistic methods have already been applied earlier by Krönig and Clausius (see for instance S.G. Brush [16]), and that the methods of Boltzmann and Maxwell differ mainly because of their disparate interpretation of probability. Thus, Boltzmann considers the "probability of state" and the average time under which the system is in this state. Whereas Maxwell assumes an infinity of equal systems (subsystems) with all possible initial states. One can say that Boltzmann advocated the "single system approach", while Maxwell proposed what has been called since Gibbs, the "ensemble approach". Both Boltzmann and Maxwell thought that their methods would lead to physically meaningful results. It was, however, only in 1930 that the single system approach gained importance mainly due to the new way of thinking in physics and mechanics brought about by the probabilistic view taken in quantum mechanics (see for instance Gudder [17]). Another contributing factor was the establishment of the modern ergodic theorems by von Neumann [18] and G.D. Birkhoff [19] (see also Khintchine [20]).

In early 1930 the ergodic problem assumed its present formulation with the development of the notion of an "abstract dynamical system". This framework is independent of classical physics although historically it evolved from it. It has been rigorously formulated by Khintchine. Contributing to this development were R. von Mises [4] views on probability, his criticism of classical dynamics and his introduction of a purely probabilistic scheme in the treatment of statistical physics. The aim of ergodic theory in classical statistical mechanics is the establishment of equilibrium values of the dynamic variables of a physical system.

Thus, if $X[q(t), p(t)]$ is a dynamic variable in terms of the generalized coordinates and momenta of the system belonging to the phase-space Γ, the equilibrium value of X is given by the "time-average":

$$\overline{X} = \lim_{t \to \infty} \frac{1}{t} \int_0^t X[q(\tau), p(\tau)] d\tau$$

One can define an invariant motion in $\Gamma(q, p)$ by: $H(q, p) = U$, where H is the Hamiltonian and U the energy of the system in its initial state, i.e. $U = H[q(0), p(0)]$. If the sample space $\Omega \subset \Gamma$ is a region of the phase-space that is compatible with the condition:

$$U - \Delta U < H(q, p) < U + \Delta U$$

then the "phase-average" is defined by:

$$< X > = \lim_{\Delta E \to 0} \frac{\int_\Omega X(q, p) dq dp}{\int_\Omega dq dp}$$

The classical ergodic theorem [20] asserts that:

$$\overline{X} = \, <X>$$

for all but a negligible class of initial states $[q(0), p(0)]$. The probabilistic form of this theorem can be readily obtained by introducing the modern concept of an "abstract dynamical system". The latter is characterized by a probability space X, a σ-algebra $\mathcal{F}$ of the events in X, a probability measure $\mathcal{P}$ and is designated by the triplet $[X, \mathcal{F}, \mathcal{P}]$. To obtain the probabilistic form of the ergodic theorem one can consider a measure preserving transformation "T_t" such that:

$$\mathcal{P}(T_t E) = \mathcal{P}(E) \text{ for all } t \in \mathbb{R}^+ \text{and } E \in \mathcal{F},$$

and for an arbitrary function $f(x) \in X$, the time-average is then given by:

$$\overline{f} = \lim_{t \to \infty} \frac{1}{t} \int_0^t f[T_t \cdot x] dt \quad , \quad t \in \mathbb{R}^+$$

and the space-average by:

$$<f> = \int_X f(x) d\mathcal{P} \text{ with the condition}: \mathcal{P}(X) = 1.$$

Hence the condition under which the "time-average" becomes equal to the "space-average" for an abstract dynamical system is that $\overline{f}$ has to be invariant under the transformation T_t. In this case one can define two subsets of X of non-zero measure associated with this invariance property. For a system to be ergodic the necessary and sufficient condition is that any invariant measurable set has the measure zero or one, i.e. that the system is indecomposable (see also Birkhoff [19], Yosida and Kakutani [21]). For an abstract dynamical system which is ergodic, the ergodic theorem simply states that:

$$\overline{f} \overset{a.e.}{=} <f>$$

It should be noted, however, that the ergodic theorems for an abstract dynamical system involves the convergence of some measure on the measure space $[X, \mathcal{F}, \mathcal{P}]$. This convergence is directly related to the motion of the system approaching equilibrium. The limitation of the classical ergodic theorem in its inability to distinguish between an equilibrium and non-equilibrium situation is therefore eliminated. The concept of an "abstract dynamical system" is fundamental in stochastic mechanics and will be repeatedly used in the representation and analysis of the behaviour of various discrete media in the following chapters.

1.2 Topological Spaces, Sets and Operators

It has been shown in previous work [22,23] that in the probabilistic mechanics of discrete media, two distinct features should be recognized. First, there exists a multitude of singular surfaces (interfaces) within a "finite volume" of the material body and second, the elements of the microstructure of actual materials exhibit random geometrical and physical characteristics. As a consequence, the relevant field quantities should be considered as random variables or functions of such variables together with their corresponding distribution functions which form a set of measures. Hence the formal structure of stochastic mechanics is based on the mathematical theory of probability and the axioms of measure theory. The field quantities used in the analysis of discrete media are scalars, but more often vector or tensor valued variables generating corresponding vector spaces. As mentioned in section 1.1, it is convenient to represent physical phenomena by an "abstract dynamical system" and hence a brief discussion on topological spaces, sets and measures is given in this and the following sections.

(i) Topological spaces and sets:

Evidently a topological vector space is a set, which makes the two structures, i.e. that of a vector space and of a topological space compatible so that the required algebraic operations are continuous.

A topological space X is a set with a structure which permits the definition of the neighbourhood of points and the continuity of functions. A system S of subsets of a set X defines a topology on X, if the system S contains:

(i) the null set $\emptyset$ and the set X itself.

(ii) the union of each of its subsystems.

(iii) the intersection of everyone of its finite subsystems.

The sets in S are open sets of the topological space X. If X is a non-empty set and S consists of $\emptyset$ and X only, the topology is called "trivial". If X is any non-empty set and the open sets consist of all subsets of X ($\emptyset$ and X included) the topology is called "discrete".

A subset $E \subseteq X$ is "open", iff it is contained in the topology of X. A set E is "closed", iff its complement E^c or $(X - E)$ is open. The null set $\emptyset$ and X are both closed.

It is important to consider the neighbourhood of a point $x \in X$ defined by a set $N(x)$ containing an open set that includes x. Similarly the neighbourhood of a subset $E \subset X$ is the neighbourhood of every point of E. A point $x \in X$ is called a "limit point" of $E \subseteq X$, if every neighbourhood $N(x)$ of x contains at least one point $e \in E$ different from x, i.e: $(N(x) - \{x\}) \cap E \neq \emptyset$ for all $N(x)$). A point x of E is called an "interior point" of E, if there exists an $N(x)$ such that $N(x) \subset E$.

The set of all interior points is the interior of E or $\overset{\circ}{E}$. The closure of a set E or $\overline{E}$ is the intersection of all sets closed relative to the topology S of X. The interior of a set E is the largest open set contained in E. The set E is said to be "dense" in X, if $\overline{E} \supset X$. It is nowhere dense, if $\overline{E}$ has an empty interior. A space $[X, S]$ is a topological space, if the system S of open sets is distinct with the previously given properties.

A "measurable topological space $[X, \mathcal{F}]$" is one in which the distinct σ-algebra $\mathcal{F}$ is generated by some basis of open sets of X. The sets $E \in \mathcal{F}$ are called Borel sets. This type of sets and their σ-algebra will be used extensively in the stochastic mechanics formulation. All countable unions and intersections of open and closed sets are Borel sets. There exist, however, other kinds of sets. Of particular interest here are "random sets", which will be considered later in conjunction with "geometric probabilities" in Chapter 3.

Among the basic topological spaces one can distinguish between a separated, a regular and normal space. A "separated" topological space is called Hausdorff, if for any two distinct points x, y of X, disjoint neighbourhoods $N(x)$ and $N(y)$ of x, y respectively, can be determined. The Hausdorff topological space has the characteristic that a sequence converges to one point at most. Hence in this space a set consisting of one point is "closed". A topological space is "regular", if for any closed set E and an arbitrary point $x \notin E$ there exist disjoint sets X_1 and X_2 such that $E \subseteq X_1$ and $x \subseteq X_2$. A topological space is "normal", if any pair of disjoint closed sets have disjoint open neighbourhoods. A topological space X is "separable", if there exists some countable basis of open sets, which is dense in X.

It is frequently necessary to use subspaces of a topological space as well as the product of spaces. Thus the product of spaces X_1 and X_2 is defined as the space whose points are all possible pairs (x_1, x_2) in which $x_1 \in X_1$ and $x_2 \in X_2$. It is written as $X = X_1 \otimes X_2$. In general a set E written as $E_1 \times E_2$ is called rectangular, i.e. it consists of all points $x = (x_1, x_2)$, where $x_1 \in E_1$ and $x_2 \in E_2$. For two systems of sets S_1, S_2 that belong to the spaces X_1, X_2, respectively and where each of them is a "semi-ring", the collection of all rectangular sets of the form $E = E_1 \times E_2$; $E_1 \in S_1, E_2 \in S_2$ is also a semi-ring. A system of sets S of X is called a "semi-ring", if together with arbitrary sets E_1 and E_2 in S, it also contains their intersections.

The semi-ring of sets S is a "ring R", if together with arbitrary sets E_1, E_2 of S, it contains their sum $E_1 \cup E_2$. If the ring contains the entire space X, it is called an algebra $\mathcal{F}$. The latter includes with any set E its complement E^c and with any sets $E_1, E_2 \ldots E_n$ in $\mathcal{F}$, it contains the union $E = \cup_{i=1}^{n} E_i$.

The algebra is a σ-algebra, if with every countable number of sets in $\mathcal{F}$, it also contains their union. In many instances use can be made of open sets for a given topology to construct simpler systems of sets on the basis of "compactness". This however involves the notion of "covering" of the space. Thus the system of open sets $\{U_i\}$ of a topological space is called a "covering $\mathcal{U}$", if each element in X belongs at least to one U_i, i.e. $\cup_i U_i = X$. The covering $\mathcal{U}$ is locally finite, if for every point x, there exists a neighbourhood $N(x)$ that has a non-empty intersection with only a finite number of elements of $\mathcal{U}$.

An important class of topological spaces are the "metric spaces". A metric space $[X, d]$ is a non-empty set of elements with a non-negative "distance function $d(x, y)$" defined for all pairs of elements or points $x, y \in X$ satisfying the conditions for all points $x, y, z \in X$:

$$\left. \begin{array}{ll} \text{(i)} & d(x,y) \geq 0; d(x,y) = 0, \text{ iff } x = y \\ \text{(ii)} & d(x,y) = d(y,x), \text{ (symmetry)} \\ \text{(iii)} & d(x,y) \leq d(x,z) + d(y,z), \text{ (triangle inequality)} \end{array} \right\} \quad (1.1)$$

The function $d(x, y)$ is called a metric. A neighbourhood $N_\delta(x)$ of a point $x \in X$, (δ being a positive number) is the set of all points y whose distance from X is smaller than δ. The metric space is Hausdorff. A subset E of the metric space X is open, if given a point $x \in E$, it contains some neighbourhood of this point. A function $f(x)$ on X is continuous, iff for every point $x \in X$ and a given positive number ϵ, a neighbourhood $N_\delta(x)$ of x is such that:

$$|f(x) - f(y)| \leq \epsilon \text{ for all } y \epsilon N_\delta(x) \quad (1.2)$$

For a "closed set C", where the distance function from a point x to C is given by:

$$d(x, C) = \inf d(x, y) \quad (1.3)$$

the function $f(x) = d(x, C)$ is continuous on X and $f(x) = 0$ for $x \in C$, and $f(x) > 0$ for $x \notin C$. Thus a point x of an arbitrary set E of a metric space $[X, d]$ is a "boundary point", if the distance from $x \rightarrow E$ and its complement is zero or if $d(x, E) = d(x, X \backslash E) = 0$. A metric space $[X, d]$ with a countable set of points $E = \{x_i\}; (i = 1, 2 \ldots)$ such that the closure $\overline{E} \supset X$ is called "separable".

It is often necessary to consider a sequence of points $x_1, x_2 \ldots$ of a topological space. Such a sequence is "convergent" to a point x, if for every open set $\mathcal{O} \ni x$, there exists an n such that $x_m \in \mathcal{O}$ for $m > n$.

A "Cauchy sequence" in $[X, d]$ of points $x_i, i = 1 \ldots$, is such that the $\lim d(x_i, x_j) = 0; i, j \rightarrow \infty$. The metric space is called "complete", if each Cauchy sequence $\{x_i\}$ has a limit point, $x \in X$ so that $\lim_{i \rightarrow \infty} d(x_i, x) = 0$.

Another significant notion in analysis is that of "compactness". Thus, if a separable topological space X is "compact", the distance $d(x_1, x_2)$ can be defined such that the neighbourhoods $N_\delta(x)$ form a basis of this space. A set E of X is compact, iff every sequence of points $x_i, (i = 1, 2 \ldots)$ of E contains a convergent "subsequence".

It is of interest to note that by Urysohn's lemma [24] for any given disjoint closed sets E_1, E_2 of a compact space X, there exists a continuous real function

$f(x)$ defined on this space with $0 \leq f(x) \leq 1$ for all $x \in X$ so that $f(x) = 0$ on E_1 and $f(x) = 1$ on E_2.

The terms mapping, function and transformation are synonymous in functional analysis. The notation $f : X \to Y$ means, if f is a single valued function whose domain $D(f)$ is X, its range $R(f)$ is contained in Y. A mapping of X into Y is also called a function on X with values in Y. The mapping f from X into Y associates with every element $x \in X$ a uniquely determined element $y \in Y$.

For any two mappings $f : X \to Y$ and $g : y \to Z$ one can define a "composite mapping" by $gf : X \to Z$, which is the mapping $g \circ f : X \to Z$ by $x \mapsto g(f(x))$. One can symbolically write $f(M)$ to designate the subset $\{f(x); x \in M\}$ of Y, where $f(M)$ is called the image of M under the mapping f. The notation $f^{-1}(M)$ is called the "inverse image of M" under the mapping f. If the mapping $f : X \to Y$ is such that for every $y \in R(f)$ there is only one $x \in X$ with $f(x) = y$, then the mapping f has an inverse f^{-1} or the mapping is one-one (injective). Hence, if f has an inverse it follows that:

$$\left.\begin{array}{l} f^{-1}(f(x)) = x \text{ for all } x \in D(f) \\ f(f^{-1}(y)) = y \text{ for all } y \in R(f) \end{array}\right\} \quad (1.4)$$

The function f maps X onto Y (surjective), if $f(X) = Y$ and into Y, if $f(X) \subseteq Y$. The mapping f is "bijective", if it is one-one and onto (see also [24–27]).

A function $f = f(x)$ on the topological space X with values in another space Y is a Borel function, if the inverse image $f^{-1}(B)$ of any open (or closed) set $B \subseteq Y$ is a Borel set of the space X. The function $f = f(x)$ on X is continuous, if the inverse image of any open (or closed) set $B \subseteq Y$ is open (or closed) in X. The existence of a "homeomorphism" between topological spaces is an equivalence relation. The term homeomorphism means a bijection f, which is bicontinuous, i.e. f and f^{-1} are continuous. Thus in a homeomorphism the images and inverse images of open sets are themselves open. A "topological invariant" is a property of a topological space, which is preserved under a homeomorphism. For a more detailed account of the concepts reviewed above the reader is referred to the bibliography of this volume, i.e. [24–35].

(ii) Vector spaces and convexity:

A vector space X is a set of objects that can be added and multiplied by real numbers to give, in either case, another element of the set. In particular X is linear, if for the elements $x \in X$, the operations are such that:

(i) for any arbitrary number α, β and $x_1, x_2 \in X$
$$x = \alpha x_1 + \beta x_2 \in X;$$
(ii) there is an element $0: x + 0 = x$ and $\alpha \cdot 0 = 0$;
(iii) for any $x \in X$, there exists an inverse element $-x$
so that: $x + (-x) = 0$
(iv) the following operations hold:
$$x_1 + x_2 = x_2 + x_1; 1 \cdot x = x; \alpha(\beta x) = (\alpha \beta)x$$
$$x_1 + (x_2 + x_3) = (x_1 + x_2) + x_3; (\alpha + \beta)x = dx + \beta x$$
$$\alpha(x_1 + x_2) = \alpha x_1 + \alpha x_2$$

$$\left. \right\} \quad (1.5)$$

If α is a complex number (i.e. $\bar{\alpha}$), X is called a "complex linear space".

As will be shown in the subsequent analysis, it is often convenient to consider subspaces of the vector space X. Thus a subspace of X is a subset of the vector space, which itself is a vector space with the set operations defined for X. A set $\{x_1, x_2 \dots x_n\}$ is called a "finite basis" for a vector space X, if it is linearly independent and it spans X, i.e. the subspace spanned by the set is X itself. A vector space is said to be "n-dimensional", if it has a finite basis consisting of n-elements. A vector space with no finite basis is infinite dimensional. A linear space X is called a "normed space", if for each element x a norm $\| x \|$ is defined as a function of x. The mapping $x \mapsto \| x \|$ of a vector space X on $\mathbb{R}$ (the field of real numbers) into $\mathbb{R}$ is a norm, if for $x \in X$ and any real number $\lambda \in \mathbb{R}$ the following properties hold:

(i) $\| x \| > 0$, if $x \neq 0$; $\| x \| = 0$, iff $x = 0$
(ii) $\| \lambda x \| = |\lambda| \cdot \| x \|$ for any $x \in X$ and any scalar λ
(iii) $\| x + y \| \leq \| x \| + \| y \|$ for any $x, y \in X$

$$\left. \right\} \quad (1.6)$$

The above norm induces a metric topology on X. However not every metric space is necessarily a normed space (see for instance Choquet [30]). The vector space X together with a norm topology is called a "normed topological vector space". A "complete normed vector space" is a "Banach space" or B-space.

It is often not possible to define a norm in accordance with(1.6) for the vector space X, but only a "semi-norm". The latter can be introduced by the notion of "locally convex" spaces, which are associated with convex sets and functionals. In general, for a topology S on the vector space to be compatible with the linear structure of X requires two mappings f_a and f_m to have the following properties:

$$f_a(x, y) = x + y, \text{ for all } x, y \in X$$
$$f_m(\alpha, x) = \alpha x, \text{ for all } \alpha \in \mathbb{R}, x \in X$$

$$\left. \right\} \quad (1.7)$$

in which the function f_a designates a vector addition and f_m a scalar multiplication. Thus vector spaces with compatible topologies and a base of convex neighbourhoods of the origin are "locally convex" topological vector spaces. Such spaces are defined by a system of semi-norms satisfying the axioms of separation [31, 32]. Semi-norms are fundamental in the theory of topological spaces and can be defined as follows:

Def. 1:

A real valued function $f(x)$ defined on a linear space X is called a "semi-norm on X", if the following conditions are satisfied:

$$\left.\begin{array}{l} \text{(i)} \ f(x) = 0, \text{if } x = 0 \\[4pt] \text{(ii)} \ f(\alpha x) = |\alpha| f(x); \\[4pt] \text{(iii)} \ f(x + y) \le f(x) + f(y); \ \text{(subadditivity)} \end{array}\right\} \qquad (1.8)$$

In addition a semi-norm has the following characteristics:

$$\text{(iv)} \ f(x_1 - x_2) \ge |f(x_1) - f(x_2)| \text{ and in particular } f(x) \ge 0 \qquad (1.9)$$

If $f(x)$ is a semi-norm on the topological space X and c an arbitrary positive number, the set $M = \{x \in X; f(x) \le c\}$ has the properties (see Yosida [24]):

$$\left.\begin{array}{l} \text{(i)} \ M \ni 0, M \text{ is convex} : x, y \in M \text{ and for } 0 < \alpha < 1 \\[4pt] \text{(ii)} \ \alpha x + (1 - \alpha)y \in M \\[4pt] \text{(iii)} \ x \in M \text{ and } |\alpha| \le 1 \text{ implies } \alpha x \in M \ (M \text{ is balanced}) \\[4pt] \text{(iv)} \text{ for any } x \in M, \text{ there exists } \alpha > 0 \text{ so that } \alpha^{-1} x \in M, \\[4pt] \qquad\qquad\qquad\qquad\qquad\quad (M \text{ is absorbing}) \\[4pt] \text{(v)} \ f(x) = \inf \alpha c; \ \alpha > 0, \alpha^{-1} x \in M. \end{array}\right\} \qquad (1.10)$$

The functional in property (v) is called the "Minkowski functional" of the set $M \subset X$. It will be denoted by $W_k^{(n)}(M)$ in Chapt. 3 (eqn. 3.182). It is a semi-norm on the space X.

In this context another definition is important, i.e. that of a locally convex linear space also referred to as a "Fréchet space", if its topology is given by just one "semi-norm" with the above properties.

Def. 2:

Using (1.10) the metric topology of such a space can be defined in terms of the distance function $d(x, y)$, i.e. $d(x, y) = \| x - y \|$, where $d(x, y)$ satisfies the axioms of a metric (1.1). It is important to note however, that condition (i) of (1.8) is not the same as condition (i) of (1.6), which refers to a norm.

It has been shown in ref. [24] that for a bounded sequence a "strong convergence limit" exists such that:

$$s\text{-}\lim_{n \to \infty} \alpha_n x_n = \alpha x, \text{ if } \lim_{n \to \infty} \alpha_n = \alpha \text{ and } s\text{-}\lim_{n \to \infty} x_n = x \qquad (1.11)$$

On this basis, a linear vector space X, can be referred to as a "quasi-normed" linear space, if for every $x \in X$, there is a number $\| x \|$ associated with the quasi-norm of the vector x, where:

(i) $\| x \| > 0$ and $\| x \| = 0$, iff $x = 0$,

(ii) $\| x + y \| \leq \| x \| + \| y \|$,

(iii) $\| -x \| = \| x \|$,

$$\lim_{\alpha_n \to 0} \| \alpha_n x_n \| = 0 \text{ and}$$

$$\lim_{\| x_n \| \to 0} \| \alpha_n x_n \| = 0$$

$$(1.12)$$

The "normed vector" space is a "Hilbert space", if a numerical function of two variables is defined such that the inner (scalar) product denoted by $< x_1, x_2 >$ satisfies the following conditions:

Def. 3:

(i) $< x, x > \geq 0; < x, x > = 0$, iff $x = 0$

(ii) $< x_1, x_2 > = < x_2, x_1 >$

(iii) $< x, \alpha x_1 + \beta x_2 > = \alpha < x, x_1 > + \beta < x, x_2 >$

$$(1.13)$$

for any α, β and elements $x_1, x_2 \in X$. In general the algebraic properties in (1.13) are the same as the rules concerning scalar products in ordinary vector algebra. However, it is not obvious that in a "complex linear space" the inner product is not linear, but rather "conjugate linear" (designated by a $\bar{}$ sign) with respect to the second factor, i.e. $< x_1, \alpha x_2 > = \bar{\alpha} < x_1, x_2 >$ for any scalar α and $x_1, x_2 \in X$. Hence a complex vector space C^n is an inner product space, if the product of the vectors $\mathbf{x}, \mathbf{y}$ (in the usual notation) is defined by $< \mathbf{x}, \mathbf{y} > = \sum_{i=1}^{n} x_i \bar{y}_i$, where x_i are the components of $\mathbf{x}$ and $\bar{y}_i$ (the complex conjugate components of $\mathbf{y}$). Any inner product space in accordance with the properties (1.12) can be defined by the norm:

$$\| x \| = < x, x >^{1/2} \qquad (1.14)$$

of an element $x \in X$ (see also refs. [30], [32]).

As mentioned previously, the notion of convexity is central in the theory of vector spaces. Thus, if X is a real vector space, a set $S \subset X$ is convex, if in accordance with (1.10):

$$\alpha x_1 + (1 - \alpha) x_2 \in S, \text{ whenever } x_1, x_2 \in S \text{ and } 0 \leq \alpha \leq 1 . \qquad (1.15)$$

All linear subspaces of X (including X) are convex and so is the empty set $\emptyset$. Any intersection of convex sets is convex, but the same is not true for any arbitrary union of convex sets.

Considering a set $S_1 \subset X$, the convex set of all finite linear combinations $\sum_i \lambda_i x_i$, $x_i \in S_1$ with $\sum_i \lambda_i = 1; (i = 1, 2 \ldots n)$ is a set called the "affine hull" of S_1. If $\lambda_i \geq 0, (i = 1, 2, \ldots n)$, the set is called the "convex hull" of S_1. It is the smallest convex subset of X, which contains S_1. A function $f : S \to \mathbb{R}$ is convex, if $\{(x, t) \in X \times \mathbb{R}^+ : x \in S \text{ and } t \in \mathbb{R}^+\}$ is convex. The function f is "concave", if $-f$ is convex; f is called affine, if f is convex and concave.

If X is a real vector space one says that $f : S \to X$ is affine, if it satisfies:

$$f(\alpha x + (1 - \alpha)y) = \alpha f(x) + (1 - \alpha)f(y) \tag{1.16}$$

whenever $x, y \in S$ and $0 \leq \alpha \leq 1$. If S is a convex open subset of $\mathbb{R}^n$, then every convex function on S is continuous. If $f(x)$ is a real function on the open interval $(a, b) \subset \mathbb{R}$ and $d^2 f(x)/dx^2 \geq 0$ for $x \in (a, b)$, then $f(x)$ is convex.

It is often necessary to consider the convexity of functionals on S. Thus a functional $f : S \to \mathbb{R}$ is defined to be convex on S, if for every $x, y \in S$ the condition (1.16) holds. In certain cases however a functional may assume infinite values, if for instance, $f(x) = -f(y)$ on the r.h.s. of (1.16). One can still consider convex functionals defined on $S \subset X$ in terms of an "indicator of S" (see for instance [24]). Thus, if $f : S \to \mathbb{R}$ is convex on S an extension $\overline{f}$ of f can be defined by: $\overline{f}(x) = f(x)$ for $x \in S$; $\overline{f}(x) = \infty$ for $x \notin S$ so that to every convex set $S \subset X$ one can associate a convex functional $\mathcal{J}_s : X \to \mathbb{R}$ given by:

$$\mathcal{J}_s(x) = \begin{cases} 0 & \text{for } x \in S \\ \infty & \text{for } x \notin S \end{cases} \tag{1.17}$$

where $\mathcal{J}_s$ is called the indicator of S. One can also define a set referred to as "epigraph of f", $(epi f)$ with respect to a convex functional $f : X \to \overline{\mathbb{R}}$. Thus the functional is convex, whenever the $epi f$ is a convex subset of $X \times \overline{\mathbb{R}}$. The standard definition of an "effective domain $\mathcal{D}(f)$" of the convex functional is given by:

$$\mathcal{D}(f) = \{x | x \in X; f(x) < \infty\} \tag{1.18}$$

The consideration of convex sets often requires "separation theorems". Thus two subsets S_1, S_2 of a topological vector space X are separated by a "closed hyperplane", if there is a continuous linear function $f : X \to \mathbb{R}$ and $c \in \mathbb{R}$ such that $f(x) \leq c$ for $x \in S_1$ and $f(x) \geq c$ for $x \in S_2$. In this case one says that S_1, S_2 are strictly separated. If S_1 and S_2 are disjoint convex sets, then:

(i) if S_1 is open, S_1 and S_2 are separated by a hyperplane

(ii) if S_1 and S_2 are open, then S_1 and S_2 are strictly separated by a hyperplane.

(iii) if X is locally convex, S_1 compact and S_2 closed, then S_1 and S_2 are strictly separated by a hyperplane.

It is to be noted that conditions (i), (ii) are forms of the Hahn-Banach theorems (see for instance Yosida [24]).

So far as functionals are concerned the notion of semi-continuity is important. This is particularly significant for considerations of Boolean random functions appearing later in dealing with geometric probability (see Chapt. 3).

Thus in accordance with (1.18) a functional $f: X \to \overline{\mathbb{R}}$ is said to be "lower semi-continuous (l.s.c.)" on X, if for every $\alpha \in \mathbb{R}$, the set $\{x | x \in X, f(x) \leq \alpha\}$ is closed in X. If the functional f is l.s.c., $-f$ is said to be "upper semi-continuous (u.s.c.)" on X and conversely. By this definition $f: X \to \overline{\mathbb{R}}$ is l.s.c. on X, iff:

$$\lim_{x_n \to x} \inf f(x_n) \geq f(x) \text{ for all } x \in X \tag{1.19}$$

it is u.s.c. on X, iff:

$$\lim_{x_n \to x} \sup f(x_n) \leq f(x) \text{ for all } x \in X \tag{1.20}$$

for every sequence $\{x_n\}$ converging to x in the topology of X. The above remarks concerning topological vector spaces and convexity are significant in the analysis of discrete media as shown later. The mappings of the elements of one vector space or subspace into the elements of another vector space is equally of importance in the subsequent analysis. Such mappings or functions are also termed "operators" on sets that have both an algebraic and a topological structure.

(iii) Linear operators and bilinear forms:

For a linear space X and a function $y = Tx$ on this space, $x \in X$, with values in another linear space Y, the "linear operator T" is defined by the relation $y = Tx$, i.e.:

$$T(\alpha x_1 + \beta x_2) = \alpha T x_1 + \beta T x_2 \tag{1.21}$$

for any $x_1, x_2 \in X$ and any number α, β. The operator T^{-1} defined in the elements of Y is called the "inverse operator", if $T^{-1}y = x$ for $y = Tx$. It is common usage to denote by $\mathcal{D}_T \subseteq X$ the domain of the operator and its range by $\mathcal{R}_T \subseteq Y: \{y \in Y; y = Tx; x \in X\}$. T is also called a linear map on X into Y. If the range of the operator $\mathcal{R}_T$ is contained in the scalar field $\mathbb{R}$, then T is a linear functional on $\mathcal{D}_T$. If T gives a one to one map of $\mathcal{D}_T$ to $\mathcal{R}_T$, the inverse map T^{-1} gives a linear operator with $\mathcal{R}_T$ on $\mathcal{D}_T$. Many important operators are linear and have the property (1.21).

If the spaces X, Y are normed vector spaces, there exist the norms $\| Tx \|$ and $\| x \|$ for $T : X \to Y$. The linear operator T is "bounded", if there exists a constant c such that:

$$\| \, Tx \, \| \leq c \, \| \, x \, \| \quad .$$ (1.22)

The smallest c for which this inequality holds is then $\| \, T \, \|$. If $x \neq 0$, it follows from the definition that:

$$\| \, T \, \| = \sup_{\substack{x \in \mathcal{D}_T \\ x \neq 0}} \frac{\| \, Tx \, \|}{\| \, x \, \|}$$ (1.23)

For bounded linear operators the following inequality holds:

$$\| \, Tx \, \| \leq \| \, T \, \| \| \, x \, \| \quad .$$ (1.24)

In certain applications the notion of the "projection" of a linear operator is used. A bounded linear operator T on a Banach space X is called a projection, if $T^2 = T$. The operator T also determines the subspaces X_1, X_2 of the Banach space X with $X_1 = \{x \in X; Tx = x\}$ and $X_2 = \{x \in X; Tx = 0\}$. For a given $x \in X$ one can write that $x = Tx + (x - Tx)$ with $Tx \in X_1$ and $(x - Tx) \in X_2$. Conventionally T is called a projection of X on X_1. A linear operator T is a "contraction operator", if $\| \, T \, \| \leq 1$. For two given bounded linear operators T_1, T_2 on the Banach spaces X_1 and X_2 with $X_1 \subset X_2$, the operator T_2 is called a "dilatation of T_1", if there is a projection operator T that projects X_2 on X_1 with $T_1 = TT_2T$. If T is a bounded linear operator on a Hilbert space H, there is also a bounded linear operator T^* on H such that:

$$(T^*x, y) = (x, Ty) \text{ for all } x, y \in H$$ (1.25)

in which T^* is the "adjoint of T" and $\| \, T \, \| = \| \, T^* \, \|$. If $T = T^*$, the operator T is called a "self-adjoint operator". Bounded linear operators will be further considered in connection with Banach space and Hilbert space-valued random variables. The continuity of linear operators can be considered also from the point of view of semi-norms. Thus from the definition given earlier (1.10) the following theorem [24] holds: "If X, Y are two locally convex spaces with semi-norms denoted by $\{p\}, \{p'\}$, respectively, which define the topologies of X, Y a linear operator T on $\mathcal{D}_T \subseteq X$ into Y is "continuous", iff for every semi-norm $p \in \{p\}$, there exists a semi-norm $p' \in \{p'\}$ and a positive number α such that:

$$p'(Tx) \leq \alpha p(x) \text{ for all } x \in \mathcal{D}_T$$ (1.26)

If T is a continuous linear operator on a normed linear space X into a normed linear space Y, one defines:

$$\| \, T \, \| = \inf_{\alpha \in X} \alpha; X = \{\alpha; \| \, Tx \, \| \leq \alpha \, \| \, x \, \| \text{ for all } x \in X\}$$ (1.27)

Invoking (1.26) or rather its corollary, i.e. that an inverse of the operator T exists, iff there is a positive constant β so that

$$\| \, T \, \| \geq \beta \, \| \, x \, \| \text{ for all } x \in \mathcal{D}_T$$ (1.28)

it is readily seen that:

$$\| T \| = \sup_{\|x\|=1} \| Tx \| \; . \tag{1.29}$$

Thus a continuous linear operator on a normed linear space X into Y is a "bounded linear operator". Linear operators on finite dimensional normed vector spaces are necessarily continuous, but this is not the case for infinite dimensional spaces.

In the subsequent analysis certain concepts of the theory of "semi-groups of operators" become rather significant (see also Butzer and Berens [39], [36]–[38]). Thus the proposition of Hille [38] concerning continuous linear operators on locally convex linear topological spaces is of interest. It may be stated as follows:

"If X is a Banach space and T_t a general linear bounded operator function for $t \geq 0$, the "one-parameter family of bounded linear operators" denoted by $L(X, X)$ satisfies the semi-group property, i.e.:

$$\left.\begin{array}{l} \text{(i) } T_{t+s} = T_t T_s \text{ for } t, s > 0 \\[2mm] \text{(ii) } T_0 = I \text{ (identity operator) .} \end{array}\right\} \tag{1.30}$$

By the one-parameter linear semi-group in a real or complex Banach space X, is meant the family $\{T_t, t \geq 0, t \in \mathbb{R}\}$ of operators in X satisfying the conditions (1.30).

Evidently, if the operator T_t is an automorphism one has: $T_{-t} = T_t^{-1}$ and the family is a "group". If every $T_t (t \geq 0)$ is a bounded operator in X satisfying (1.30), one can define a "strong limit" in the Banach space X, by:

$$s\text{-}\lim_{t \to t_0} T_t x = T_{t_0} x, \text{ for all } t_0 \geq 0 \text{ and all } x \in X \; . \tag{1.31}$$

In this case $\{T_t\}$ is called a semi-group of operators of the C_0-class (see also Yosida [24]). The one-parameter linear semi-group $\{T_t, t \geq 0\}$ is "contracting", if for all $t \geq 0$, one has:

$$\| T_t x \| \leq \| x \| \; ; \; x \in X \; , \tag{1.32}$$

The strong continuity condition given by (1.31) can be stated in an equivalent "weaker form", which is more closely related to the theory of Markov processes to be discussed later. It is given by:

$$w\text{-}\lim_{t \downarrow 0} T_t x = x, \text{ for all } x \in X \; , \tag{1.33}$$

or as the "weak limit" of the bounded operator T_t. A more general class of semi-groups on a locally convex topological space X is obtained, if $\{T_t, t \geq 0\}$ satisfies the conditions:

(i) $T_{t+s} = T_t T_s; T_0 = I$

(ii) $w\text{-}\lim\limits_{t \to t_0} T_t x = T_{t_0} x$ for any $t_0 \geq 0, x \in X$. $\left.\right\}$ (1.34)

and the family of mappings $\{T_t\}$ is "equi-continuous" in time, i.e. for any continuous semi-norm p on X, there exists a continuous semi-norm p' on X so that $p(T_t x) \leq p'(x)$ for all $t \geq 0$ and all $x \in X$. Every C_0 class semi-group $\{T_t\}$ on X has an "infinitesimal generator A", which is defined by:

$$Ax = \lim_{h \downarrow 0} h^{-1}(T_h x - x) \text{ or } \lim_{h \downarrow 0} h^{-1}(T_h - I)x. \qquad (1.35)$$

This definition arises from the assumption of an equi-continuous semi-group of class C_0, which is defined on a locally convex linear topological space X, to be "sequentially complete" (see also Yosida [24], and [39]).

The operator T is linear with the domain:

$$\mathcal{D}_T = \{x \in X; \lim_{h \downarrow 0} h^{-1}(T_h - I)x \text{ exists in } X\}, \qquad (1.36)$$

and where $\mathcal{D}_T$ is a non-empty set, i.e. it contains at least the vector zero. In the system approach to the behaviour of discrete media as shown later, the determination of extreme values of functionals is often required. For this purpose a certain analytical structure must be given to the functionals. Thus, considering a subset U of the vector space X. This set will be convex in accordance with earlier statements, if:

$u = \lambda u_1 + (1 - \lambda)u_2 \in U$ for all $u_1, u_2 \in X$,

and all $\lambda \in [0, 1]$. $\left.\right\}$ (1.37)

If $f(u) : U \subseteq X \to \mathbb{R}$ denotes a real functional defined on the convex subset U, then $f(u)$ is convex on U, if:

$f(\lambda u_1 + (1 - \lambda)u_2) \leq \lambda f(u_1) + (1 - \lambda)f(u_2)$

for all $u_1, u_2 \in U, \lambda \in [0, 1]$. $\left.\right\}$ (1.38)

The equality only holds, if $u_1 = u_2$. For this case $f(u)$ is "strictly convex". The functional $f(u)$ is "strictly concave", if $-f(u)$ is strictly convex. A particular attention must be given to functionals of the quadratic form. Thus, if U is an inner product space, $u \in U$, the functional $f(u) = <u, u>$ is strictly convex. The convexity also holds for the linear mapping $T : U \to X, u \in U \subseteq X$, where T is a linear operator and g any function in $X, g \in X$. The functional:

$$f(u) = <Tu, u> + <g, u> \qquad (1.39)$$

is convex (see Rockafeller [40] and others).

Since a functional on a real or complex normed space X is a map from X to the real or complex number, a "linear functional" is specified by:

$$f(\alpha x_1 + \beta x_2) = \alpha f(x_1) + \beta f(x_2) \tag{1.40}$$

for any real or complex number α, β and $x_1, x_2 \in X$.

If a linear functional is continuous at any one point, it is said to be "uniformly continuous". A linear functional can either be "discontinous" everywhere or uniformly continuous everywhere or simply continuous. A linear functional is "bounded", if there exists a real number $M \leq 0$ such that

$$|f(x)| \leq M \parallel x \parallel \text{ for all } x \in X. \tag{1.41}$$

The term bounded refers here to the ratio of $|f(x)|/ \parallel x \parallel$ to be bounded. A linear functional is bounded, iff it is continuous.

Denoting the set of all bounded linear functionals on a Banach space X by X^*, the latter is also a Banach space or the "dual of X". Since X^* is the dual of X, an element of X^*, i.e. $x^*(x)$ gives the value of x^* for an element of $x \in X$. Forming the set of all bounded linear functionals on X^* gives the second dual or "adjoint space of X" designated by X^{**}. If $X = X^{**}$ the Banach space X is said to be "reflexive". It is convenient to write $< x, x^* >$ instead of $x^*(x)$, where the notation indicates the duality existing between X^{**} and X as well as the action X on X^*. The following characteristics of duality can be given. Suppose B is a closed "unit sphere" of a normed space, then the norm of x^* is given by:

$$\parallel x^* \parallel = \sup\{| < x, x^* > | : x \in B\} \text{ for every } x^* \in X^* \tag{1.42}$$

and hence

    (i) this norm makes X^* a Banach space,

    (ii) if B^* is a closed unit sphere of X^*, then for every

        $x \in X \parallel x \parallel = \sup\{| < x, x^* > | : x^* \in B^*\}$ thus

        $x^* \rightarrow < x, x^* >$ is a bounded linear functional on

        X^* with the norm $\parallel x \parallel$,

    (iii) B^* is called weak-compact.

$$(1.43)$$

It is apparent that several topologies can be introduced in a Banach space each of which will be associated with a certain type of convergence of the variable. For instance, if the topology induced in X occurs by the metric (distance function) $d(x,y) = \parallel x - y \parallel$, it is called a metric norm or "strong topology of X". On the other hand a "weak topology of X" is induced by a sequence $\{x_n\}$ of elements in a Banach space converging to an element x so that, if:

    (i) the norms $\parallel x_n \parallel$ are uniformly bounded, i.e.

        $\parallel x_n \parallel \leq M$

    (ii) $\lim_{x_n \to x} x^*(x_n) = x^*(x)$ or $< x^*, x >$ for every $x^* \in X^*$.

$$(1.44)$$

The "weak topology" when introduced in the dual space X^* is called a weak-topology of X^*. The duality principle for Banach spaces also applies to Hilbert spaces.

Thus by the definition of a Hilbert space, if X is a vector space and $B(x, y)$ a Banach space, a symmetric "bilinear form" (inner product on $X \times X$), which is positive definite, gives together with the vector space X a real Hausdorff or "pre-Hilbert space" (see Choquet [30]), where $< x, x >^{1/2}$ is a norm on X. A pre-Hilbert space, which is complete, is a Hilbert space.

If the bilinear form is non-negative, i.e. $< x, x > \geq 0$ for all $x \in X$, the resulting space is not Hausdorff and $< x, x >^{1/2}$ will be a semi-norm on X. The norm $\| x \|$ on a normal space X is a Hilbert norm, if there exists a positive definite symmetric bilinear form $B(x, y)$ such that: $\| x \| = < x, x >^{1/2}$, satisfying:

$$(\| x \|^2 + \| y \|^2) = \frac{1}{2}(\| x + y \|^2 + \| x - y \|^2), \text{ for all } x, y \in X \qquad (1.45)$$

If U is a closed subspace of the H-space, then the so-called quotient space H/U is also a Hilbert space with the norm:

$$\| x \|_{X/U} = \inf_{y \in U} \| x - y \| . \qquad (1.46)$$

For an operator $T \in L(X, X^*)$ where X and X^* are Hilbert spaces, one can define a continuous bilinear form on X by:

$$B(x, y) = < Tx, y >, \text{ for all } x, y \in X. \qquad (1.47)$$

1.3 Probability and random variables

The brief review of fundamental notions and definitions of the functional and convex analysis will be extended in the following sections to their probabilistic counterparts that are required in the stochastic mechanics of discrete media.

(i) Probability:

The previously given definition of probability due to Kolmogorov (section 1.1) is adopted throughout the present analysis. Thus the probability of the occurence of an event E in the frequency interpretation is defined as before:

$$\mathcal{P}\{E_i\} \stackrel{a.e.}{=} \lim_{n \to \infty} n_i/n, \qquad (1.48)$$

in which n_i is the number of occurrences of the events E_i. The notation $\mathcal{P}\{E\}$ refers to the "probability measure of E" or simply the probability of E. A non-empty class of sets of events can be considered as a field, i.e. $\mathcal{F} = \{E_i, i = 1, 2 \ldots\}$ where $E_i \in \mathcal{F} \Rightarrow \cup_{i=1}^{\infty} E_i \in \mathcal{F}$. In particular a "$\sigma$-field" for the class of non-empty event sets E_i can be defined as follows:

$$\left.\begin{array}{ll} \text{(i)} & E_i \in \mathcal{F} \Rightarrow E_i^c \in \mathcal{F}; i = 1, 2 \ldots \\ \text{(ii)} & E_i \in \mathcal{F} \text{ and } E_j \in \mathcal{F} \Rightarrow E_i \cap E_j \in \mathcal{F} \\ \text{(iii)} & E_i \in \mathcal{F} \Rightarrow \cup_{i=1}^{\infty} E_i \in \mathcal{F} \end{array}\right\} \quad (1.49)$$

where $\mathcal{F}$ is closed under the operations of intersections, unions and complements. The probability measure "$\mathcal{P}$" is a set function since its argument is in general a set that may contain more than one point. Evidently, the probability associated with the set of all outcomes $\{\omega\}$ or Ω is unity. The space Ω containing all possible outcomes together with the σ-field $\mathcal{F}$ of events forms a measurable space $[\Omega, \mathcal{F}]$. The abstract dynamical system (section 1.1) or the triple $[\Omega, \mathcal{F}, \mathcal{P}]$ is a "probability space" consisting of $[\Omega, \mathcal{F}]$ and the measure $\mathcal{P}$ on $\mathcal{F}$ with $\mathcal{P}\{\Omega\} = 1$. A "random variable" is a measurable function from the probability space into a measurable "state-space". In the present study the probability space will be denoted by X and the state-space by Z (see also [23]). Since most of the random variables of interest in stochastic mechanics belong to topological vector spaces an appropriate topological structure must be specified for these spaces. The most frequently used quantities are Banach space-valued and Hilbert space-valued random variables that will be defined subsequently. If the state-space is a topological space, its measurable sets will be specified as "Borel sets" (event sets). The "distribution" of a real-valued random variable x is defined by the measure "μ" on the class of measurable sets of the state-space induced by x : $\mu(E) = \mathcal{P}\{x \in E\}$. A more detailed discussion on probability measures follows in section (1.4). Since the measure $\mathcal{P}$ on the probability space Ω or X is a countably additive set function on the σ-algebra $\mathcal{F}$ of the sets of $[X, \mathcal{F}]$, $\mathcal{P}\{E\}$ for any countable number of disjoint sets $E_1, E_2 \ldots \in \mathcal{F}, E = \cup_n E_n$, is defined by:

$$\mathcal{P}\{E\} = \sum_n \mathcal{P}\{E_n\} , \quad (1.50)$$

and is finite, if $\mathcal{P}\{X\} < \infty$ and σ-finite, if X is the union of countably many sets $E_n : \mathcal{P}\{E_n\} < \infty, (n = 1, 2 \ldots)$. The triple $[X, \mathcal{F}, \mathcal{P}]$ is then a measure space. Often "conditional probabilities" are required. Thus the conditional probability of an event E_1 with respect to another event E_2 is specified by the use of the definition (1.48) as follows:

$$\mathcal{P}\{E_1 | E_2\} = \frac{\mathcal{P}\{E_2 | E_1\} \mathcal{P}\{E_1\}}{\mathcal{P}\{E_2\}} \quad (1.51)$$

which is known as Baye's rule in probability [41]. For n mutually exclusive sets of events $E_i, i = 1, 2 \ldots n$, when $\sum_{i=1}^{n} E_i = X$ and if E is an arbitrary event in X, then:

$$P\{E\} = \sum_{i=1}^{n} P\{E|E_i\}P\{E_i\}. \tag{1.52}$$

One says that two events are pair-wise independent, if:

$$P\{E_i \cap E_j\} = P\{E_i\}P\{E_j\}; \ i \neq j \ . \tag{1.53}$$

It does not follow however, that pair-wise independence implies absolute independence. For two events E_1, E_2 equation (1.53) holds so that they are mutually independent, i.e.:

$$P\{E_1|E_2\} = P\{E_1\}; P\{E_2|E_1\} = P\{E_2\}$$

Generally events are said to be independent (see also Rényi [41]) if:

$$P\{\cap_{i=1}^{n} E_i^{(m_i)}\} = \prod_{i=1}^{n} P\{E_i^{(m_i)}\} \text{ for } m_i = 0,1, \ (i = 1 \ldots n) \qquad \left.\begin{array}{c} \\ \\ \end{array}\right\} \tag{1.54}$$

An infinite sequence of events $E_1, E_2 \ldots E_n \ldots$ with the probability that at least m of the events occur can be characterized by the probabilities $Q_m (m = 1, 2 \ldots)$ on the basis of the "Borel-Cantelli" theorem (see for instance [41], [42]). These probabilities are monotone decreasing and have a limit:

$$\lim_{m \to \infty} Q_m = Q.$$

If the sum $\sum_{i=1}^{n} P\{E_i\} < \infty$ then $Q = 0$.

(ii) Random variables:

The outcomes of physical phenomena when considered as random can be represented in probability theory by "random numbers". In accordance with the given probability definition and the event structure of the corresponding function space, a real finite valued function on Ω or X is called a real "random variable", if for every number ξ the inequality $x(\omega) \leq \xi$ defines a set $\{\omega\}$ or the event E, whose probability is defined.

In this context it is important to note, that in probabilistic mechanics [23], a random variable designated by $x = x(\omega)$ is considered in most cases to be "directly given", if each elementary outcome $\omega \in \Omega, x(\omega) = \xi$.

Hence, accordingly the "sample space Ω" will be identified with the probabilistic function space or the space of realization, i.e. $\Omega = X$. It follows that the corresponding σ-algebras "$\mathcal{B}$" and "$\mathcal{F}$" are also the same together with the associated measures. In the following as indicated earlier, the probability measure on the measurable topological space $[X, \mathcal{F}]$ will be denoted by "$\mathcal{P}$". The different types of probability measurs particularly those referring to "random sets" will be discussed later. There are two basic types of real-valued random variables, i.e. discrete and continuous ones. A discrete random variable $x = x(\omega)$ takes only a finite or denumerable number of different values with corresponding probabilities $P_x(\xi) = \mathcal{P}\{x = \xi\}$, where $\{x = \xi\}$ means that the event for the random variable x takes the value ξ or $\{\omega : x(\omega) = \xi\}$.

The probability of the event $\xi_1 \leq x \leq \xi_2$ is therefore expressed by:

$$\mathcal{P}\{\xi_1 \leq x \leq \xi_2\} = P_x(\xi) \tag{1.55}$$

in which the notation $P_x(\xi)$ reflects that it is a function for all possible values "ξ" of x. An important function of an arbitrary random variable x is its distribution function defined for all values of ξ on the real line $\mathbb{R}$, i.e.:

$$F_x(\xi) = \mathcal{P}\{x \leq \xi\}; \ (-\infty < \xi < \infty) , \tag{1.56}$$

and for any values ξ_1 and $\xi_2, (\xi_1 < \xi_2)$ one has:

$$F_x(\xi_2) - F_x(\xi_1) = \mathcal{P}\{\xi_1 \leq x \leq \xi_2\}. \tag{1.57}$$

The distribution function $F_x(\xi)$ of a discrete random variable is piece-wise constant. If the distribution function of x is absolutely continuous and differentiable, then:

$$p_x(\xi) = \frac{d}{d\xi} F_x(\xi) \tag{1.58}$$

is the "probability density" of the continuous distribution of x. It is a non-negative function such that for any ξ_1, ξ_2 with $\xi_1 < \xi_2$:

$$\mathcal{P}\{\xi_1 \leq x \leq \xi_2\} = \int_{\xi_1}^{\xi_2} p_x(\xi)d\xi . \tag{1.59}$$

The continous distribution function of x has the following properties:

   (i) $\lim_{\xi \to -\infty} F(\xi) = 0$; $\lim_{\xi \to \infty} F(\xi) = 1$

   (ii) if $\xi_1 \leq \xi_2 : F_x(\xi_1) \leq F_x(\xi_2)$; (monotone continuous function)

   (iii) $\lim_{\xi \uparrow \xi_0} F_x(\xi) = F(\xi_0)$; $F(\xi)$ is left continuous

or equivalently, if $\mathcal{P}\{x < \xi\} = F_x(\xi)$, it will be a right continuous non-decreasing function (see also Borovkow [43]). Experimental studies of discrete media frequently require the knowledge of two or more random variables. Thus if x_1, x_2 are two random variables the "joint distribution function" is expressed by:

$$F_{x_1 x_2}(\xi_1, \xi_2) = \mathcal{P}\{x_1 < \xi_1, x_2 < \xi_2\} , \tag{1.60}$$

where $F_{x_1 x_2}(.)$ is a non-decreasing right-continuous function with respect to each of the two dimensions, i.e.:

(i) $F_{x_1 x_2}(\xi_1, -\infty) = 0$; $F_{x_1 x_2}(-\infty, \xi_2) = 0$

(ii) $F_{x_1 x_2}(\infty, \infty) = 1$

(iii) $F_{x_1 x_2}(\xi_1, \infty) = F_{x_1}(\xi_1)$; $F_{x_1 x_2}(\infty, \xi_2) = F_{x_2}(\xi_2)$.

The joint probability density is defined by the mixed derivative of F_{x_1}, F_{x_2} given by the non-negative function:

$$p_{x_1 x_2}(\xi_1, \xi_2) = \frac{\partial^2}{\partial \xi_1 \partial \xi_2} F_{x_1 x_2}(\xi_1, \xi_2) , \tag{1.61}$$

writing the inverse form or:

$$F_{x_1 x_2}(\xi_1, \xi_2) = \int\limits_{-\infty}^{\xi_1} \int\limits_{-\infty}^{\xi_2} p_{x_1 x_2}(\xi_1', \xi_2') d\xi_1' d\xi_2' ,$$

and by letting one of the upper limits approach ∞, gives the distribution function of a single variable, i.e.:

$$\int\limits_{-\infty}^{\infty} d\xi_2' \int\limits_{-\infty}^{\xi_1} p_{x_1 x_2}(\xi_1', \xi_2') d\xi_1' = P\{x_1 < \xi_1\} = F_{x_1}(\xi_1) . \tag{1.62}$$

If both upper limits go to infinity, the interpretation will extend over the entire sample space in two dimensions so that:

$$\int\limits_{-\infty}^{\infty} \int\limits_{-\infty}^{\infty} p_{x_1 x_2}(\xi_1, \xi_2) d\xi_1 d\xi_2 = 1 . \tag{1.63}$$

In the stochastic mechanics of discrete media, frequently the field variables are vector or tensor-valued quantities and hence the above elementary forms must be extended to a multi-dimensional representation. Thus considering a more general real-valued random variable x on the space $[\Omega, \mathcal{B}]$ or $[X, \mathcal{F}]$ as a function from Ω onto the real line $\mathbb{R}$, an n-dimensional "random vector" variable $\mathbf{x} = \{x_1, x_2 \ldots x_n\}$ is a mapping from $[\Omega, \mathcal{B}]$ into $[\mathbb{R}^n, \mathcal{F}^n]$, where $\mathbb{R}^n$ is the n-dimensional Euclidean space and $\mathcal{F}^n$ the σ-algebra of Borel subsets of $\mathbb{R}^n$. For a given probability measure $\mathcal{P}$ on $[\Omega, \mathcal{B}]$ (Banach space), one can define another measure Q for every set $E \in \mathcal{F}^n$ generated by the random vector $\mathbf{x}$ so that:

$$Q\{E\} = \mathcal{P}\{\mathbf{x}^{-1}(E)\} . \tag{1.64}$$

A uniquely defined function $F(\xi_1, \xi_2 \ldots \xi_n)$ will then characterize the probability of the events: $(-\infty, \xi_1) \times (-\infty, \xi_2) \times \ldots x(-\infty, \xi_n)$ where the inequalities $x_1 < \xi_1, x_2 < \xi_2 \ldots x_n < \xi_n$ are simultaneously satisfied. Hence the "joint distribution function" of the random variables $x_1, x_2 \ldots x_n$ is then the distribution function of the random vector $\mathbf{x}$ and is a monotonically non-decreasing, left-continuous function in each variable (see also Rényi [41]). If the probability measure $Q(\cdot)$ in (1.64) is absolutely continuous with respect to the n-dimensional Lebesgue measure, the distribution function and probability density can be expressed by:

$$
\left.
\begin{aligned}
&F_{\mathbf{x}}(\xi_1, \xi_2 \ldots \xi_n) = \int_{-\infty}^{\xi_1} \cdots \int_{-\infty}^{\xi_n} p_x(\xi_1', \ldots \xi_n') d\xi_1', \ldots d\xi_n' \quad \text{(a)} \\
&\text{and} \\
&p_{\mathbf{x}}(\xi_1, \xi_2 \ldots \xi_n) = \frac{\partial^n F}{\partial \xi_1 \partial \xi_2 \ldots \partial \xi_n} \qquad\qquad\qquad \text{(b)}
\end{aligned}
\right\} \quad (1.65)
$$

where the derivatives exist almost everywhere in $\mathbb{R}^n$. As a consequence of the above statements one can define a "random vector" as follows:

"if $\mathbf{x} = (x_1, x_2 \ldots x_n)$ is an n-dimensional random vector in $\mathbb{R}^n$, the measure $Q(\cdot)$ defined for $E \in \mathcal{F}^n$ in (1.65) is the joint probability and $F_{\mathbf{x}}$ the joint distribution function. The function $p_{\mathbf{x}}(..)$ in (1.65b) is then the joint density function of the random variables $x_1, x_2 \ldots x_n$".

It is apparent, that one can also consider a fully "conditional probability space", which is generated by the σ-finite measure $\mathcal{P}$ on the σ-algebra $\mathcal{F}$ and define the corresponding "conditional distribution and density functions". Thus for two real-valued random variables x_1, x_2 the conditional distribution function becomes:

$$
\left.
\begin{aligned}
F_{x_1|x_2}(\xi_1|\xi_2) = F(\xi_1|\xi_2) = \mathcal{P}\{x_1 < \xi_1 | x_2 < \xi_2\}, \\
\text{for } \xi_1, \xi_2 \in \mathbb{R}
\end{aligned}
\right\} \quad (1.66)
$$

and the conditional density function:

$$
p_{x_1|x_2}(\xi_1|\xi_2) = p(\xi_1|\xi_2) = \frac{p(\xi_1, \xi_2)}{p(\xi_2)}. \qquad\qquad (1.67)
$$

It follows that for an n-dimensional "random vector $\mathbf{x} = \{x_1, x_2 \ldots x_n\}$" by the use of the chain rule of conditional probabilities one obtains:

$$
p(\xi_1, \xi_2 \ldots \xi_n) = p(\xi_n) \prod_{i=1}^{n-1} p(\xi_i | \xi_{i+1} \ldots \xi_n) \qquad\qquad (1.68)
$$

If the random vector $\mathbf{x}$ has a normal or "Gaussian distribution", the probability density then becomes:

$$
p(\xi_1, \xi_2 \ldots \xi_n) = \frac{\sqrt{|A|}}{(2\pi)^{n/2}} \cdot e^{-\frac{1}{2} B(\xi_1, \xi_2 \ldots \xi_n)} \qquad\qquad (1.69)
$$

in which $B = \sum\limits_{i,j=1}^{n} a_{ij}\xi_i\xi_j$ is a positive definite quadratic form and $|A|$ the determinant of the matrix $|a_{ij}|$. It can also be shown that:

$$\int_{\mathbb{R}^n} p(\xi_1,\xi_2\ldots\xi_n)d\xi_1 d\xi_2\ldots d\xi_n = 1. \qquad (1.70)$$

In the stochastic analysis of the discrete material behaviour, it is often necessary to consider random variables, whose values are in topological vector spaces such as a Banach space or a Hilbert space. Thus following Bharucha-Reid [44], one can give the following definitions:

Def. 1:

A mapping $x: X \to B$ (Banach space) is called a random variable with values in B, if the inverse image under x of every Borel set E belongs to $\mathcal{F}$; that is $x^{-1}(E) \in \mathcal{F}$ for all $E \in \mathcal{F}$ and where $[X, \mathcal{F}, \mathcal{P}]$ is the probability space and $[B, \mathcal{F}]$ the Banach space with the σ-algebra $\mathcal{F}$ of the sets E of X.

Similarly one can define a Hilbert space-valued random variable, if H is a Hilbert space with the inner product $< x, y >$ as follows:

Def. 2:

A mapping $x : X \to H$ is a Hibert space-valued random variable, if the inner product $< \cdot, \cdot >$ of H is a real (or complex) valued random variable for every $y \in H$.

Due to the importance of Gaussian random variables in the stochastic mechanics of discrete media one can also define such variables by:

Def. 3:

A random variable x with values in the Banach space X is said to be normal or Gaussian, if $x^* \in X^*$ is a scalar valued Gaussian random variable for every $x^* \in X^*$ where the latter is the dual space of X.

Considering now a discrete "directly given random variable" in the sense discussed previously, in the probability space $[X, \mathcal{F}, \mathcal{P}]$, that takes on only a finite number of distinct values, its probability distribution can be stated in terms of a sequence of "observable values", i.e.:

$$P_k = \mathcal{P}\{x = \xi_k\}; (k = 1, 2\ldots n). \qquad (1.71)$$

Heuristically, one can expect the arithmetic mean or "average value" of the observed values of x to be very near to the expected value, i.e.:

$$E\{x\} = p_1\xi_1 + p_2\xi_2 + \ldots p_n\xi_n.$$

which, according to the given frequency interpretation of probability, would be obtained for n-random experiments in which each ξ_k is equal to n_k. A more rigorous defintion of the "mean or expected value of x" is however (see also Doob [45]) given by:

Def. 4:

The expectation of a real-valued directly gives random variable x on $[X, \mathcal{F}, \mathcal{P}]$ is defined as the integral of x over X with respect to the measure $\mathcal{P}$, where:

$$E\{x\} = \int_X x \, d\mathcal{P} , \tag{1.72}$$

if this Lebesgue integral exists. This conforms with the assumption, that for a large number of observations of a random variable, the "mean" of the observed values would be $E\{x\}$ or near the value of this integral. It follows immediately from the properties of the integral in (1.72) that:

(i) the expectation is a linear operation, i.e. for two random variables on the same probability space one has:

$$E\{\alpha x + \beta y\} = \alpha E\{x\} + \beta E\{y\}, \text{ where } \alpha, \beta \text{ are real constants,}$$

(ii) if x is a non-negative random variable, $E\{x\} \geq 0$

(iii) if x is bounded, then $E\{x\}$ exists.

The expectation of a discrete random variable that takes on infinitely distinct values ξ_n with probabilities $p_n = \mathcal{P}\{x = \xi_n\}, n = 1, 2 \ldots$ is given by:

$$E\{x\} = \sum_{n=1}^{\infty} p_n \xi_n , \tag{1.73}$$

if this series converges absolutely. If it converges only conditionally $E\{x\}$ is not defined (see also Rényi [41]). This is easily recognized by the fact that only a random variable which is constant with probability one ($P\{x = c\} = 1 \Rightarrow E\{x\} = c$), has a "centred distribution".

In general there are positive and negative deviations from the mean so that one uses the notion of "variance" for the determination of the distribution. The variance is defined by:

$$D^2(x) = E\{(x - E\{x\})^2\} , \tag{1.74}$$

if this quantity exists. If it does not exist, x is said to have an infinitely large variance. The square root of $D^2(x)$ is known as the "standard deviation of x". The two parameters, i.e. the expectation and the variance are the principal characteristics of the distribution of x. In certain cases this information about the distribution may not be sufficient and thus "higher moments" or the expectation of higher power of a random variable must be used. These are defined as follows:

$$M_n(x) = E\{x^n\} = \int_{-\infty}^{\infty} x^n dF(\xi); (n = 1, 2 \dots). \tag{1.75}$$

A related definition to the above (1.75) is that of a "central moment" of order n, i.e.:

$$m_n(x) = E\{[x - E(x)]^n\}. \tag{1.76}$$

It is apparent that $M_1(x) = E\{x\}, m_1(x) = 0, m_2(x) = D^2(x)$ and so on. For a bounded random variable all moments exist and the distribution is uniquely determined. By using the Fourier-Stieltjes transform of the distribution function $F(\xi)$ or:

$$\varphi_x(t) = \int_{-\infty}^{\infty} e^{i\xi t} dF(\xi); \; (-\infty < t < \infty) \tag{1.77}$$

one obtains the "characteristic function" of the random variable. If the distribution is continuous with the probability density function $p(\xi)$ then:

$$\varphi_x(t) = \int_{-\infty}^{\infty} e^{i\xi t} p(\xi) d\xi. \tag{1.78}$$

Since by definition the Lebesgue integral of any real Borel measurable function $g(x)$ of the real-valued variable x is given by:

$$E\{g(x)\} = \int_{-\infty}^{\infty} g(\xi) dF(\xi)$$

one can express (1.77) also by:

$$\varphi_x(t) = E\{e^{ixt}\}. \tag{1.79}$$

The moments sometimes required for a more complete description of the distribution of a bounded random variable can be obtained from the characteristic function by taking the derivative of the function at $t = 0$. Thus

$$M_n(x) = i^{-n} \varphi_x^{(n)}(0); (n = 1, 2 \dots). \tag{1.80}$$

Assuming that all moments of a certain distribution are given and by taking the Laplace transform of the distribution $F(\xi)$, i.e.:

$$\psi_x(s) = \int_{-\infty}^{\infty} e^{\zeta s} dF(\xi) = \int_{-\infty}^{\infty} e^{\zeta s} p(\xi) d\xi = 1 + \sum_{n=1}^{\infty} \frac{M_n s^n}{n!} \tag{1.81}$$

in which for a bounded random variable the series in (1.81) converges. The function $\psi_x(s)$ is called the "moment generating function" of the random variable x. The notion of a characteristic function also applies to random vectors. Thus, if $\mathbf{x} = (x_1, x_2 \ldots x_n)$ is an n-dimensional random vector, its characteristic function $\varphi_{\mathbf{x}}(t_1, t_2 \ldots t_n)$ for any real values $t_1, t_2 \ldots t_n$ is given by:

$$\varphi_{\mathbf{x}}(t) = E\{e^{i(\mathbf{x_1 t})} F_{x_1, x_2 \ldots x_n}(d\xi_1, d\xi_2 \ldots d\xi_n)\} \tag{1.82}$$

where $\mathbf{t}$ denotes the vector $(t_1, t_2 \ldots t_n) \in \mathbb{R}^n$ and $(\mathbf{x}, \mathbf{t})$ is the scalar product of the vectors $\mathbf{x}$, and $\mathbf{t}$, i.e. $< \mathbf{x}, \mathbf{t} > = \sum_{k=1}^{n} x_n t_k$. A more comprehensive discussion on distribution, characteristic functions, etc. is given amongst others in Rényi [41], Neveu [46].

It has been stated earlier that the probability measure $\mathcal{P}$ on the space $[\Omega, \mathcal{B}, \mathcal{P}]$ is a countably additive set function. For a finite number of random variables $x_1(\omega), x_2(\omega) \ldots x_n(\omega)$ on that space, the distribution function in terms of the real values $\xi_i, (i = 1 \ldots n)$ of the random variables can be expressed by:

$$F\{\xi_1, \xi_2 \ldots \xi_n\} = \mathcal{P}\{x_1(\omega) \leq \xi_1, x_2(\omega) \leq \xi_2 \ldots x_n(\omega) \leq \xi_n\} \tag{1.83}$$

This distribution determines the probability measure on the n-dimensional Borel sets, where one loosely refers to this measure as the distribution function F. Assuming a real-valued function f, which is measurable with respect to the n-dimensional Borel sets and if:

$$\int (|f(x_1(\omega) \ldots x_n(\omega)|)|dP < \infty, \text{ then}$$

$$\int f[x_1(\omega) \ldots x_n(\omega)]dP = \int f(x_1, x_2 \ldots x_n)dF, \tag{1.84}$$

where these integrals are understood as Lebesgue integrals. In this context an important theorem should be given here, i.e. the "Radon-Nikodym theorem" [42, 44], which can be stated as follows:

T.1.:

"If $\mathcal{P}$ is a probability measure on the Borel field $\mathcal{B}$, a finite completely additive set function on $\mathcal{B}$ is absolutely continuous with respect to $\mathcal{P}$, iff it is an integral of an integrable $\mathcal{B}$-measurable function on $\mathcal{B}$."

Hence, if the random variable $x(\omega)$ is integrable w.r.t. $\mathcal{P}$, then the function:

$$\varphi(B) = \int_{B \in \mathcal{B}} x(\omega)\mathcal{P}(d\omega) \tag{1.85}$$

is a completely additive and absolutely continuous function w.r.t. $\mathcal{P}$, i.e. given any Borel set $B \in \mathcal{B}$ for which $\mathcal{P}(B) = 0$, then $\varphi(B) = 0$. This function can also be thought of as the "derivative of φ" w.r.t. $\mathcal{P}$ relative to the Borel field $\mathcal{B}$ and is usually denoted by:

$\varphi(d\omega)/\mathcal{P}(d\omega)$. (Radon-Nikodym derivative).

It is to be noted that this theorem remains valid, if the measure $\mathcal{P}$ is σ-finite. As shown in earlier work [22, 23], it is often necessary to use instead of the Borel field $\mathcal{B}$ a somewhat smaller field $\mathcal{A}$, ie, $\mathcal{A} \subseteq \mathcal{B}$ in the representation of the events of a random experiment. This is usually imposed by experimental constraints corresponding to a "cruder experiment" than that first visualized. One may also attempt to find a "conditional mean value" of the random variable x that is associated with this experiment. This suggests to look for a function φ and measure $\mathcal{P}$ with regard to the sets $A \in \mathcal{A}$. In this case, if the random variable $x(\omega)$ on $[\Omega, \mathcal{A}, \mathcal{P}]$ is integrable w.r.t. $\mathcal{P}$ and $\mathcal{A} \subset \mathcal{B}$, then there is an $\mathcal{A}$-measurable function integrable w.r.t. $\mathcal{P}$, if:

$$\int_{A \subset \mathcal{A}} x(\omega)\mathcal{P}(d\omega) = \int_{A \subset \mathcal{A}} f(\omega)\mathcal{P}(d\omega) \text{ for all } A \in \mathcal{A}. \tag{1.86}$$

The above statement is a corollary of the Radon-Nikodym theorem (see also [24, 37, 47]). The $\mathcal{A}$-measurability of the function f determines it uniquely almost everywhere, that is up to a set of the $\mathcal{P}$-measure zero. The function f or the Radon-Nikodym derivative is also referred to as the "conditional expectation of x" given the Borel subfield $\mathcal{A}$, i.e.:

$$f(\omega) = E\{x|\mathcal{A}(\omega)\}, \tag{1.87}$$

with the certain characteristics given in refs. [47, 50]. Another theorem of interest in the subsequent analysis of discrete media concerns the second moment of a random variable (see also Elliott [48], and others). It can be stated as follows:

T.2.:

> "if x is a random variable having a finite second moment $E\{x^2\} < \infty$ and if $\mathcal{A}$ is a subfield of the Borel field $\mathcal{B}$, the best $\mathcal{A}$-measurable estimate of x in the sense of "minimizing" a mean-square error approximation is given by $E\{x|\mathcal{A}\}$."

This theorem and the characteristics of the conditional expectation given in (1.87) become important in dealing with the "transient behaviour" of microstructures, when considered from a system theoretical point of view (see also [49]). In the formulation of "evolution relations" of discrete media as discussed in Chapter 3 of this volume, it is often convenient to consider a class of real random processes known as "martingales". Hence by considering a sequence of random variables $\{x_n, n = \ldots - 1, 0, 1 \ldots\}$ on a probability space $[\Omega, \mathcal{B}, \mathcal{P}]$ with an increasing family of Borel fields $\mathcal{B}_n$ such that the sequence $\{x_n\}$ is $\mathcal{B}_n$-measurable, $(\mathcal{B}_n = \mathcal{B}(x_n, k \leq n)$; if $E\{|x_n|\} < \infty$ for all n and

$$E\{x_n|\mathcal{B}_m\} = x_m; m < n,$$ (1.88)

then the sequence of random variables is called a "martingale".

If the sequence and equality in (1.88) is such that:

$$E\{x_n|\mathcal{B}_m\} \geq x_m; m < n,$$ (1.89)

the sequence is referred to as a "submartingale" (see for instance Doob [50]). One frequently encounters in stochastic mechanics sequences of independent, identically distributed (i.i.d.) random variables $\{x_1, x_2 \ldots\}$ with $E\{x_i\} = 0$. A simple example of a martingale is then given by the "partial sum" of this sequence, i.e.:

$$S_n = \sum_{j=1}^{n} x_j \quad \text{with} \quad B_j = \mathcal{B}(x_k, k \leq j) ,$$ (1.90)

since $E\{S_n|\mathcal{B}_m\} = S_n$, $m < n$. Such partial sums will be further considered in connection with the "central limit theorem" and the dependence of random variables (section 1.5). A more detailed treatment of real valued martingales can be found among others in Doob [50], Neveu [46], Brémaud [47], Elliott [48].

Another concept of significance in stochastic mechanics associated with the definition of martingales is the following:

Def. 5:

If $[\Omega, \mathcal{F}, \mathcal{P}]$ is a probability measure space and T an interval either on the real line $\mathbb{R}$ or the extended line of integers $\mathbb{Z}$ in the continuous and discrete parameter case, respectively, and $\mathcal{F} = \{\mathcal{F}_t, t \in T\}$ an increasing family of sub σ-algebra of $\mathcal{F}$, a real valued function $x(t, \omega)$, which is "adapted to the family $\mathcal{F}$", i.e.: $x(t, \omega)$ is $\mathcal{F}_t$-measurable, is called a martingale, if

(i) $E\{|x(t, \omega)|\} < \infty$ for every $t \in T$,
(ii) for every $s, t \in T : x(t, \omega) = E\{x(t, \omega)|\mathcal{F}_s\}$.
$\left.\begin{array}{r}\\\\\end{array}\right\}$ (1.91)

This definition also applies to Banach space-valued random functions, which are adapted to $\{\mathcal{F}_t\}$ in the Banach space X. (see also Hanš [51], Umegaki and Bharucha-Reid [52]).

Another notion, which is significant in the stochastic mechanics of discrete media is that of a "filtered probability space". By formulating the material behaviour in probabilistic terms, the identification of an appropriate outcome space Ω or X may be regarded as a first step in the probabilistic modelling. The abstract dynamical systems representation by means of the triple $[\Omega, \mathcal{F}, \mathcal{P}]$ in which $\mathcal{F}$ specifies the set of all events to which one can assign probability numbers, reflects then the "likelihood" of the various events.

The second step in the development of a stochastic theory consists of specifying these events. It is apparent that the Borel field $\mathcal{B}$ or for σ-finite measures, the σ-algebra $\mathcal{F}$ plays a distinctive role in the abstract dynamical system representation. One usually considers in "system theory" the triple $[\Omega, \mathcal{F}, \mathcal{P}]$ as a primitive probability space and a family $\mathbb{F} = \{\mathcal{F}_t, t \geq 0\}$ of σ-algebras on Ω in such a manner, that:

$$\left.\begin{array}{l} \text{(i)} \ \ \mathcal{F}_t \subseteq \mathcal{F} \text{ for all } t \geq 0, \\[2mm] \text{(ii)} \ \ \mathcal{F}_s \subseteq \mathcal{F}_t, \text{ if } s \leq t \end{array}\right\} \quad (1.92)$$

In order to express (i) and (ii) above, one speaks of $\mathbb{F}$ as an "increasing family" of sub-σ-algebras or a "filtration of the measurable space $[\Omega, \mathcal{F}]$". The sub σ-algebra $\mathcal{F}_t$ is interpreted as the set of all events, whose occurrence or non-occurrence is determinable at time t. Hence $\mathbb{F}$ shows how the information from the events arises and the "uncertainty" is resolved with time. Conventionally, one may denote by $\mathcal{F}_\infty$ the "smallest σ-algebra on Ω" containing all events in $\mathcal{F}_t$ for all $t \geq 0$ and assume without loss of generality, that the "ambient" σ-algebra $\mathcal{F} = \mathcal{F}_\infty$ (see also [47]). In this sense, whenever the triple $[\Omega, \mathcal{F}, \mathcal{P}]$ is used as a "filtered probability space", it is understood that the abstract dynamical system is characterized by a probability space Ω or X and that $\mathbb{F} = \{\mathcal{F}_t, t \geq 0\}$ is a filtration of $[\Omega, \mathcal{F}]$ with $\mathcal{F} = \mathcal{F}_\infty$.

Associated with filtered probabilty spaces is the notion of a "stopping time τ". The latter is a measurable function $\tau(\omega)$ from $[\Omega, \mathcal{F}] \to [0, \infty]$ such that $\{\omega \in \Omega : \tau(\omega) \leq t\} \in \mathcal{F}_t$ for all $t \geq 0$. Hence τ is a stopping time with respect to $\mathbb{F} = \{\mathcal{F}_t, t \geq 0\}$. By letting $\mathbb{F}_\tau$ to consist of all events $E \in \mathcal{F}$, where:

$$\{\omega \in \Omega : \omega \in E \text{ and } \tau(\omega) \leq t\} \in \mathcal{F}_t \quad \text{for all} \quad t \geq 0 ,$$

it is seen that this condition means $E \cap \{\tau \leq t\} \in \mathcal{F}_t$. The above concepts are useful in the consideration of stochastic processes from the point of view of system theory as discussed later in the text.

1.4 Probability measures.

In this section some general remarks on measures well-known from functional analysis are made and, then, probability measures on the relevant topological spaces will briefly be given. "Random measures" required in the consideration of geometric probabilities and the application of stereology to the behaviour of various discrete media will be discussed in Chapter 3.

(i) General remarks on measures:

It has been mentioned previously that a measure μ on a topological space X is a countably additive set function $\mu(A)$ on the σ-algebra $\mathcal{A}$ of the measure space $[X, \mathcal{A}]$. This measure is defined for any countable number of disjoint sets, by:

$$\mu(A) = \sum_n \mu(A_n). \tag{1.93}$$

It is "finite", if $\mu(X) < \infty$ and "σ-finite", if X can be represented by the union of countably many sets $A_n : \mu(A_n) < \infty, (n = 1, 2 \dots)$. Thus the probability space $[X, \mathcal{A}, \mu]$ with $\mu(X) = 1$ is a measure space. A "distribution" is a non-negative finite function $\mu = \mu(A)$ on a "semi-ring S", if for any set $A \in S$, the above definition of the measure holds. If the semi-ring S is a ring, then the function μ is a distribution, if it is "finitely additive" (see also Halmos [53]).

One speaks of a "weak distribution" for $\mu(A)$ on S, (semi-ring of sets in X), if for any set $A \subseteq S$ as the union of a finite number of disjoint sets, the definition of $\mu(A)$ holds. Every weak distribution on S can be extended to the ring of all sets $A \subseteq X$ that are the unions of a countable number of disjoint sets of S. Considering a "bounded" measure $\mu(A)$ of the sets S in X with $\mu(X) < \infty$, then the "exterior measure" of a set $Z \subseteq X$ is defined by the number:

$$\mu^e(A) = \inf \sum_i \mu(A_i) , \tag{1.94}$$

where the infinum is taken over all countable sets $A_1, A_2 \dots \in S$ covering A. If $A_1^*, A_2^* \dots \in S$ is a family of sets contained in A then:

$$\mu^*(A) = \sup \sum_i \mu(A_i^*) , \tag{1.95}$$

is called the "interior measure" of this set. If these two measures are equal, i.e. $\mu^e = \mu^*$, the measure is called "regular".

Equivalent definitions can be found, amongst others, in refs. [24], [50]. Often "product measures" must be considered. Thus for two σ-finite measure spaces $[X, \mathcal{A}, \mu]$ and $[X^*, \mathcal{A}^*, \mu^*]$ one can define the σ-field $\mathcal{A} \otimes \mathcal{A}^*$ of subsets of the direct product $X \otimes X^*$ as the smallest σ-field containing all sets from $A \times A^*$, $A \in \mathcal{A}$, $A^* \in \mathcal{A}^*$.

It can be shown (see also [50]) that there exists a unique measure $\mu \otimes \mu^*$ defined on $\mathcal{A} \otimes \mathcal{A}^*$ such that for all $A \in \mathcal{A}$, $A^* \in \mathcal{A}^*$ one has:

$$(\mu \otimes \mu^*)(A \times A^*) = \mu(A)\mu^*(A^*) \tag{1.96}$$

with the product measure space $[X \otimes X^*, \mathcal{A} \otimes \mathcal{A}^*, \mu \otimes \mu^*]$.

The term "measurable almost everywhere or a.e." is a characteristic to hold for all points $x \in X$, except perhaps for points of a set $A \subseteq X$ of measure

$\mu(A) = 0$. One can extend the class of measurable sets so that every subset of a set with measure zero is measurable and has the measure zero. This results in a "complete measure" with respect to the original one. The positive measure on Borel sets of a locally compact topological space (Hausdorff) is called a "Borel measure". Another measure defined on $\mathbb{R}$ is the "Lebesgue measure". It is a regular Borel measure on $\mathbb{R}$ and is invariant by translation. The definition of regularity of a finite measure $\mu(A)$ in the σ-algebra $\mathcal{A}$, is given by:

$$\mu(A) = \sup \mu(K) \tag{1.97}$$

in which the supremum is taken over all compact sets $K \subseteq A$. The Lebesgue measure on $\mathbb{R}^n$ is the product measure "$L \otimes L \otimes \ldots \otimes L$" and is the only regular Borel measure invariant by translation and rotation (see for instance Choquet [30], Halmos [53]).

It is well-known in functional analysis that a real-valued simple func-. tion $f(x)$ on X is "integrable" with respect to the measure μ, if the series $\sum_i |y_i| \mu(A_i)$ converges. One can then express the integral (see also relation 1.86) by:

$$\int_X f(x)\mu(dx) = \sum_i y_i \mu(A_i) \tag{1.98}$$

An arbitrary real function $f(x)$ on X is integrable, if it is the limit of a uniformly converging sequence of simple integrable functions, i.e.:

$$f(x) = \lim_{n\to\infty} f_n(x); \quad \int_X f(x)\mu(dx) = \lim_{n\to\infty} \int_X f_n(x)\mu(dx) \ . \tag{1.99}$$

The sequence of measurable functions on $[X, \mathcal{A}, \mu]$, which are finite a.e. is said to "converge in measure" or $f_n(x) \overset{\mu}{\leadsto} f(x)$ to the measurable function $f(x)$, if:

$$\lim_{n\to\infty} \mu(\{x : |f_n(x) - f(x)| > \epsilon\}) = 0 \text{ for every } \epsilon > 0 \tag{1.100}$$

It is seen that a function $f(x)$ is integrable, if it is measurable and its absolute value $|f(x)|$ is integrable. It has been shown by Yosida [24] using Egorov's theorem, that if $f_n(x)$ converges point-wise a.e. in a finite measure μ on X to $f(x)$, then there exists a subset $A : \mu(X \backslash A) \leq \epsilon$ for every $\epsilon > 0$ and that the convergence is uniform on A. Evidently there are other measures in topological spaces.

For instance, one can define a "generalized measure ν" of a set A, that involves the notion of the bounded variation as a measure on $\mathcal{A}$ (see Prohorov and Rozanov [42]). Thus given any function $f(x)$, which is integrable with respect to the measure μ, then the integral:

$$I(A) = \int_A f(x)\mu(dx) , \quad A \in \mathcal{A} \tag{1.101}$$

can be regarded as a "generalized measure" on the σ-algebra $\mathcal{A}$.

It is absolutely continuous with respect to $\mu(A)$ and the variation is bounded, i.e.:

$$\mathrm{Var} I(A) = \int_A |f(x)|\mu(dx) ; \quad \mathrm{Var} I(X) = \int_X |f(x)|\mu(dx) < \infty.$$

Hence

$$\nu(A) = \int_A f(x)\mu(dx) ; \quad A \in \mathcal{A} \tag{1.02}$$

is a "generalized measure" on $\mathcal{A}$.

If X is a compact space with $\mu(A)$ on $A \subset X$, any real continuous function $f(x)$ on X is measurable, bounded and integrable. In the normed space X of all real continuous functions $f(x)$ on X with the norm:

$$\| f \| = \sup_x |f(x)| , \tag{1.103}$$

the integral:

$$< f, I >= \int_X f(x)\mu(dx) , \tag{1.104}$$

represents a "linear continuous functional". This functional is positive, i.e. $< f, I >\geq 0$ for $f(x) \geq 0$. In this context the "convergence in measure" as stated in (1.100) is equivalent to the convergence in the metric space of all measurable functions. In this case the metric can be defined by:

$$d[f_1, f_2] = \int_X \frac{|f_1(x) - f_2(x)|}{1 + |f_1(x) - f_2(x)|}\mu(dx) \tag{1.105}$$

where $f_1 = f_2$ means that for almost all x : $f_1(x) = f_2(x)$.

A sequence of integrable functions $f_1(x) f_2(x) \ldots$ is said to "converge in mean", if

$$\left.\begin{array}{l} \lim_{n \to \infty} \int_X |f_m(x) - f(x)|\mu(dx) = 0 , \\[2mm] \text{implying that} \\[2mm] \lim_{n \to \infty} \int_X f_n(x)\mu(dx) = \int_X f(x)\mu(dx). \end{array}\right\} \tag{1.106}$$

In the normed space X mean convergence is the convergence of all integrable functions $f(x)$, where the norm is given by:

$$\| f(x) \| = \int_X |f(x)| \mu(dx) \tag{1.107}$$

and where $f_1 = f_2$ means that $f_1(x) = f_2(x)$ at almost all $x \in X$. Often the functions $[f_1(x)]^2$ and $[f_2(x)]^2$, if they are integrable are of interest. Then the product $f_1(x)f_2(x)$ and the sum $f_1(x) + f_2(x)$ are also integrable. The $L^2(X)$ space of all functions that are square integrable is then a "complete Hilbert space" with the scalar product:

$$< f_1, f_2 > = \int_X f_1(x)f_2(x)\mu(dx) . \tag{1.108}$$

For a more detailed discussion of measures on topological spaces, reference is made among other to Halmos [53], Saks [54].

(ii) Probability measures.

In view of the later application in the stochastic mechanics of discrete media some probability measures are briefly reviewed in this section. Of particular interest here are the measures pertaining to Banach and Hilbert spaces (see also Bharucha-Reid [44], Gel'fand and Vilenkin [55], Mourier [56] and others). It is important to note that the following notation concerning topological spaces, the σ-algebras and probability measure will be used in the subsequent chapters. Thus a "probability measure $\mathcal{P}$" on a measurable space $[X, \mathcal{F}]$, where X is a metric space and $\mathcal{F}$ the σ-algebra (Borel) is said to be "regular", if for every set $E \in \mathcal{F}$ and $\epsilon > 0$, there exists a closed set F and an open set G such that $F \subseteq E \subseteq G$ and $\mathcal{P}(G - F) < \epsilon$.

Related to the weak convergence of probability measures is the smaller class of such measures known as "tight measures". Thus the probability measure $\mathcal{P}$ on $[X, \mathcal{F}]$, (X being a metric space) is said to be tight, if for every $\epsilon > 0$, there exists a compact set. $K \subseteq X$ such that $\mathcal{P}(X - K) < \epsilon$.

For separable Banach spaces every probability measure on $[X, \mathcal{F}]$ is tight. A measure $\mathcal{P}$ on $[X, \mathcal{F}]$, X being a metric space is "perfect", if for any $\mathcal{F}$-measurable real valued function and any set E on $\mathbb{R}$, $f^{-1}(E) \in \mathcal{F}$, there are Borel sets F_1, F_2 on $\mathbb{R}$ such that $F_1 \subseteq E \subseteq F_2$ and $\mathcal{P}(f^{-1}(F_2 - F_1)) = 0$. If the measure $\mathcal{P}$ is perfect, the probability space $[X, \mathcal{F}, \mathcal{P}]$ is called a "perfect measure space".

One can distinguish three types of convergence for sequences of probability measures on $[X, \mathcal{F}]$, i.e., a strong, uniform and a weak convergence. The most significant in the present context is the "weak convergence". Thus a sequence of probability measures $\{\mathcal{P}_n\}$ is said to converge to the measure $\mathcal{P}$, if:

$$\int_X f(x)d(\mathcal{P}_n(x)) \rightsquigarrow \int_X f(x)d(\mathcal{P}(x)) \qquad (1.109)$$

for every bounded and continous function f on X, i.e. for every $f \in C(X)$. A necessary and sufficient condition for "weak convergence" can also be given as follows:

If $\{\mathcal{P}_n\}$ is a sequence of probability measures on $[X, \mathcal{F}]$, then $\{\mathcal{P}_n\}$ converges weakly to $\mathcal{P}$, if

$$\lim_{\substack{n\to\infty \\ f\in E}} \sup |\int f d\mathcal{P}_n - \int f d\mathcal{P}| = 0 \qquad (1.110)$$

where $E \in F$ is a family of open sets $\subseteq C(X)$, which is equincontinuous for all $x \in X$ and uniformly bounded.

For the consideration of probability measures on Banach and Hilbert spaces, it is often convenient to use the concept of characteristic functions and functionals. Thus, if x is a random variable with values in an arbitrary Banach space and x^* any real bounded linear functional on X, where $x^*(x)$ is the value of x^* for $x \in X$, then the characteristic functional $\varphi(x^*)$ by virtue of eq. (1.79) is given by:

$$\varphi(x^*) = E\{e^{ix^*(x)}\} = \int_X e^{ix^*(x)} d\mathcal{P}(x) \qquad (1.111)$$

and is defined for all real $x^* \in X^*$, (X^* is the dual of X). This functional uniquely determines a probability measure $\mathcal{P}$ on the minimal σ-algebra $\mathcal{F}_m$ relative to which all linear functionals are measurable. If the space X is separable then $\mathcal{F}_m$ and the σ-algebra $\mathcal{F}$ of all Borel sets of X, coincide. The characteristic functional $\varphi(x^*)$ is positive-definite and a uniformly continous function of x^* in the strong and weak topology of X^*. If $\{\mathcal{P}_n\}$ is a sequence of probability measures in the measure space X with the characteristic functionals $\varphi_n(x^*)$, it converges weakly to a measure $\mathcal{P}$ with the characteristic functional $\varphi(x^*)$, i.e. $\varphi_n(x^*) \rightsquigarrow \varphi(x^*)$ for all $x^* \in X^*$.

If the measurable space $[H, \mathcal{F}]$ is a real separable Hilbert space with the inner produce $< \cdot, \cdot >$ and the σ-algebra $\mathcal{F}$ of Borel sets of H, then the probability measure on H for an H-valued random variable is defined for all $y \in H$ by the characteristic functional:

$$\varphi(y) = \int_H e^{i<x,y>} d\mathcal{P}(x). \qquad (1.112)$$

The expected value of the measure $\mathcal{P} \in H$ is an element $m \in H$ and is defined for all $y \in H$, by:

$$< m, y >= \int_H < x, y > d\mathcal{P}(x), \tag{1.113}$$

which may not exist. In the bounded case however, if $\int_H \parallel X \parallel d\mathcal{P}(x) < \infty$, m exists and is equal to $\int_H x d\mathcal{P}(x)$ and $\parallel m \parallel \leq \int_H \parallel x \parallel d\mathcal{P}(x)$.

In the study of probability measures in Hilbert spaces an operator formalism is often employed. In this context the "Hilbert-Schmidt" or S-operator is used to represent the corresponding characteristic functionals. Theorems concerning the latter and the so called S-topology are given, amongst others, by Bharucha-Reid [44], Prohovov [57], Kappos [58].

Finally, in the development of evolution equations as discussed later, the $C[0, 1]$ space of continuous functions is often of interest. Thus the measurable space denoted by $[C, \mathcal{C}]$, where $\mathcal{C}$ is the σ-algebra of Borel sets on C, the following theorem due to Kolmogorov can be given as follows:

T.1.:

"If $x_t(\omega), t \in [0, 1]$ is a random function and $F_{t_1} \ldots t_n$ the probability distribution on $(\mathbb{R}^n, \mathcal{C}^n)$ of the random vector $[x_{t_1}(\omega) \ldots x_{t_n}(\omega)]$ and if there are positive constants α, β, M such that:

$$\{E|x_t(\omega) - x_{t_2}(\omega)|^\alpha\} \leq M|t_2 - t_1|^{1+\beta}$$

for all $t_1, t_2 \in [0, 1]$, then there exists a unique probability measure on C, where:

$$F_{t_1} \ldots t_n = \mathcal{P}_{t_1 \ldots t_n} \text{ for all } n \text{ and } (t_1 \ldots t_n) \in [0, 1]."$$

Another theorem concerning the so called "Wiener measure" on $C[0, 1]$ is often required. It relates to a Wiener process or the Brownian motion process to be considered later in this text. The existence of such a measure is given by the following theorem (Doob [50]):

T.2.:

"There exists a unique probability measure $\mathcal{P}_W$ on $[C, \mathcal{C}]$ called a "Wiener measure" with the following characteristics:

(i) $\mathcal{P}_W\{x : x_0(\omega) = 0\} = 1$

(ii) If $0 \leq t_1 < t_2 < \ldots t_n \leq 1$, the random variables in the C-space: $x_{t_1}(\omega), x_{t_k}(\omega) - x_{t_{k-1}}(\omega), 1 < k \leq n$ are independent.

(iii) If $0 \leq s \leq t \leq 1$, the random variable $x_t(\omega) - x_s(\omega)$ is Gaussian with the expectation equal to zero and the variance $(t - s)$.

The above measures as well as other types of probability measures will be used later in this text (see also Billingsley [59], Itô and McKean [60], Nelson [60] and others).

1.5 Dependence of random variables:

It is usually assumed in probability theory that the random variables considered in the modelling of physical phenomena are "independent" and that
under certain conditions this assumption continues to hold even when the variables become "dependent". A dependence often occurs when discrete or multi-
component media undergo "phase-transitions". Hence, it is natural to ask how
much dependency can still be allowed for the "classical limit theorem (C.L.T.)"
of probability to be valid. It may be indicated therefore to briefly review the
various assumptions underlying the rather complex relationship between the
various dependence relations and in particular the limiting procedures available to distinguish between "independent" and "dependent" random variables.
A sequence of random variables may exhibit long-range dependence or too
much dependence for the conventional C.L.T. to hold. It is known that the
latter holds when summands of sequences are only "weakly dependent". Such
a dependence also occurs when the summands satisfy a suitable "mixing condition" or satisfy other dependence structures. Thus for the C.L.T. or the more
modern notion of a "functional central limit theorem (F.C.L.T.)" to hold, the
mixing coefficients between the past and distant future have to decay rapidly.
Also "moment conditions" of the involved random variables must be satisfied.
A more rigorous discussion on these topics can be found in ref. [62].

(i) Independent random variables:

The random variables $x_1, x_2 \ldots x_n$ are called stochastically independent, if

$$\mathcal{P}\{x_1 \in B_1 \ldots x_n \in B_n\} = \mathcal{P}\{x_1 \in B_1\} \ldots \mathcal{P}\{x_n \in B_n\} \qquad (1.114)$$

where $B_1, \ldots B_n$ are arbitrary Borel sets of the real line $\mathbb{R}$. Thus, if F_k denotes
the distribution function of $x_k (k = 1, 2 \ldots n)$ the "joint distribution function
of the random variables $x_1, x_2 \ldots x_n$ is given by:

$$F(x_1, x_2 \ldots, x_n) = F_1(x_1) F_2(x_2) \ldots F_n(x_n) \qquad (1.115)$$

If this distribution is aboslutely continuous with the density function $p_k(\xi), (k = 1, 2 \ldots n)$ then the density function of the joint distribution of the random variable is:

$$p(\xi_1, \xi_2, \ldots \xi_n) = p_1(\xi_1) p_2(\xi_2) \ldots p_n(\xi_n). \qquad (1.116)$$

These statements can be extended to vector-valued random variables. Thus, let
$x_1, x_2 \ldots x_n$ be independent real or vector-valued random variables and $g_1(\xi)$,
$g_2(\xi) \ldots g_n(\xi)$ Borel-measurable real or vector-valued functions such that, if
x_j is an r_j-dimensional vector-valued random variables $(r_j \geq 1)$, then g_j is a
function of the r_j variables. The random variables $g_1(x_1), g_2(x_2) \ldots g_n(x_n)$ are

then independent. In the case of only two random variables, i.e., x, y having finite expectations, one has:

$$E\{x, y\} = E\{x\}E\{y\} \ . \tag{1.117}$$

The two random variables for which $E\{x\}, E\{y\}$ and $E\{x, y\}$ exist are then said to be "uncorrelated". In terms of a "correlation coefficient $c(x, y)$" defined by:

$$c(x, y) = \frac{E\{x, y\} - E\{x\}E\{y\}}{D\{x\}D\{y\}} \ , \tag{1.118}$$

where D denotes the variances, the two random variables are uncorrelated, if $c(x, y) = 0$ (see also Rényi [41], Borowkow [43] and others). An obvious generalization of (1.117) is that for a sequence of independent random variables $x_1, x_2 \ldots x_n$ having finite expectations, one has:

$$E\{x_1, x_2 \ldots x_n\} = \prod_{k=1}^{n} E\{x_k\}. \tag{1.119}$$

The notion of independence is also related to the independence of σ-algebras. Thus, considering the event structure in a probability space $[X, \mathcal{F}, \mathcal{P}]$ and two classes of events $\mathcal{E}_1, \mathcal{E}_2$ contained in $\mathcal{F}$ of the space X, the two classes are called independent, if for all events $E_1 \in \mathcal{E}_1, E_2 \in \mathcal{E}_2$ the following relation holds:

$$\mathcal{P}\{E_1 E_2\} = \mathcal{P}\{E_1\}\mathcal{P}\{E_2\}. \tag{1.120}$$

In a sequence of classes of events $\{\mathcal{E}_n\}_{n=1}^{\infty}$, the events will be independent, if for any choice of the integer n:

$$\mathcal{P}\{\cap_{i=1}^{k} E_{n_i}\} = \prod_{i=1}^{k} \mathcal{P}\{E_{n_i}\}, \tag{1.121}$$

for arbitrary $E_{n_i} \in \mathcal{E}_i$. Thus, the sub-algebras of independent algebras of events are also independent. In this context an approximation theorem by Borowkow [43] in relation to directly given random variables mentioned in section (1.3(ii)) states that:

"if $[X, \mathcal{F}, \mathcal{P}]$ is a probability space and $\mathcal{E}$ the σ-algebra of a certain class of events, then there exists for each $E \subset \mathcal{F}$ a sequence $E_n \in \mathcal{E}$ such that:

$$\lim_{n \to \infty} \mathcal{P}\{E_n E' \cup E_n' E\} = 0 \ ,$$

or equivalently:

$$\lim_{n \to \infty} \mathcal{P}\{E \backslash E_n\} = \lim_{n \to \infty} \mathcal{P}\{E_n \backslash E\} = 0 \ ",$$

from which it follows that:

$$P\{E\} = \lim_{n \to \infty} P\{E_n\}. \tag{1.122}$$

(See also Halmos [53]). Hence, if $\sigma(x)$ is the smallest sub-algebra of $\mathcal{F}$ with respect to which the mapping of x is still measurable, the sequence of the random variables $x_1, x_2 \ldots x_n$ will be independent, if the sub-algebras $\sigma(x_1), \sigma(x_2) \ldots \sigma(x_n)$ are independent. Analogously the independence of random vectors is based on:

$$P\{x_1 \in B_1, x_2 \in B_2 \ldots x_n \in B_n\} = \prod P\{x_j \in B_j\}, \tag{1.123}$$

where the B_j's represent the Borel sets in the subspaces of corresponding dimensions. Hence, it is always possible to define a finite sequence of independent random variables by considering the random quantities $x_1, x_2 \ldots x_n$ in the space of elementary events $\mathbb{R} \times \mathbb{R} \ldots n$-times$= \mathbb{R}^n$ and the corresponding σ-algebras $\mathcal{B}_1 \times \mathcal{B}_2 \times \ldots \times \mathcal{B}_n = \mathcal{B}^n$ induced by the Borel sets $B_1 \times B_2 \times \ldots \times B_n \subset \mathbb{R}^n$. As mentioned earlier, for the distinction between independent and dependent random variables it is important to consider the notion of convergence of a sequence of random variables $x_1, x_2 \ldots x_n$. The following definitions can be given (see also Fréchet [63], Feller [64] and others):

(i) "Convergence everywhere" (e-convergence):

$$x_n(\omega) \to x(\omega), \text{ if } \lim_{n \to \infty} x_n(\omega) = x(\omega) \text{ for } \forall \omega \in \Omega, \tag{1.124}$$

(ii) "Convergence almost everywhere" (a.e.-convergence):
$(x_1, \ldots x_n, \ldots)$ converges almost everywhere (a.e.) to x as $n \to \infty$, if the probability of every sequence of "realized values of x_n" converging to x is one, i.e.:

$$P\{\lim_{n \to \infty} x_n = x\} = 1; \lim_{n \to \infty} x_n = x \text{ (a.e.)}, \tag{1.125}$$

(iii) "Convergence in probability" (p-convergence):
x_n converges in probability to x as $n \to \infty$, if for every $\epsilon > 0$

$$\lim_{n \to \infty} P\{|x_n - x| > \epsilon\} = 0, \forall \epsilon > 0, \tag{1.126}$$

(iv) "Convergence in quadratic mean" (mean square convergence, m.s.)
x_n converges to x in the m.s. as $n \to \infty$, if $\lim E\{|x_n - x|^2\} = 0$ or

$$\lim_{n \to \infty} x_n = x \text{ by definition, if } \lim_{n \to \infty} E\{|x_n - x|^2\} = 0, \tag{1.127}$$

(v) "Convergence in distribution":

$$x_n \text{ converges to } x \text{ as } n \to \infty, \text{ if } \lim_{n \to \infty} F(x_n) = F(x), \tag{1.128}$$

at every point of the continuous distribution function $F(x)$.

It is to be noted, that the above definitions do not always hold, e.g. it is possible for a sequence of random variables to converge almost everywhere, but not in the mean square. The above defintions can be extended to vector valued random variables and stochastic functions. They are significant for the consideration of "sums of independent identically distributed" random variables or vectors.

Thus considering first the one-dimensional case, the "normalized sum Z_n" of a sequence $\{x_n\}$ is:

$$Z_n = \frac{s_n - a_n}{b_n} = \frac{x_1 + x_2 + \ldots x_n - a_n}{b_n},$$

it will converge to the "normal distribution $\Phi(x)$" [42], if:

$$\lim_{x \to \infty} \frac{x^2 \int\limits_{|x| > x} dF(x)}{\int\limits_{|x| < x} x^2 dF(x)} = 0 , \tag{1.129}$$

for some constants $a_n, b_n > 0$. This condition is necessary and sufficient. It is to be noted that the normalizing constant b_n can increase in a manner, of $1/2$ with n or differ from this value by a slowly varying factor. The constant b_n will be approximately $n^{1/2}$, if the terms in the sequence have a finite variance. To find limit distributions of such normalized sums it would be necessary to consider all possible limit distributions, which is rather complicated. Hence it is more convenient to use the concept of "stable distributions" [42]. The latter can be defined for a sequence of random variables, if for arbitrary constants a_1, a_2 (positive) and b_1, b_2 one has:

$$F_1(a_1 x + b_1) * F_2(a_2 x + b_2) = F(ax + b) \text{ for all } x \in \{x_n\} \text{ in } Z_n \tag{1.130}$$

in which * stands for the convolution of the two distributions F_1, F_2.

Hence by a theorem due to Gnedenko [65] concerning a class of "stable distributions", (if the distribution function $F_x(\xi) = \mathcal{P}\{x_n < \xi\}$) of the random variable x is to be a "limit distribution" for sums of i.d.r.v., a necessary and sufficient condition is that it is "stable".

In the multi-dimensional case the normalization of a sequence of sums of identically distributed vectors will occur by subtracting the expectations and dividing by $n^{1/2}$. For non-identical distribution terms, the normalization can then be carried out in form of a linear transformation (see also [42]).

It is of interest to note, that there is a connection between the concept of "infinitely divisible distributions" and "limit theorems". Thus a distribution on $\mathbb{R}$ with the distribution function $F(x)$ and a characteristic function $\varphi(t)$ is said to be 'infinitely divisible", if for any n, it can be represented as a convolution of n-identical distributions:

$$\varphi(t) = [\varphi_n(t)]^n ,$$

where φ_n is then a characteristic functional. A random variable x on $[X, \mathcal{F}, \mathcal{P}]$ is called infinitely divisible, if for any n, it can be represented as a "sum of n identically distributed independent random variables (see also Gnedenko [65]).

(ii) Dependent random variables:

A significant aspect of the stochastic mechanics of discrete media is the "critical behaviour" of the latter. It involves the theory of limit distributions for sums of "strongly dependent variables". As mentioned earlier the "critical behaviour" of various media is associated with "phase-transitions". However, there exists also a large class of phenomena where "structural changes" occur during which the involved field variables do not attain their critical values. Such changes will be considered later in this text as "transients" and can be generally referred to as "subcritical". A "strong dependence" of field variables when considered as random variables implies that the C.L.T. no longer holds. Thus, if the field variables of a stochastic system are denoted by X_i with reference to a subspace of the probabilistic function space X, like the configuration, velocity, deformation space etc. the sum of these random variables with expectation zero can be written as:

$$Z_n = \frac{\sum\limits_{1}^{n} X_i}{b_n}, \tag{1.131}$$

if b_n is properly chosen, i.e. to have a smooth distribution for large n. Evidently, this holds for homogeneous macroscopic material domains for which Z_n can be regarded as a "collective variable". However, if the forces or interactions on the "microlevel" are taken into account and are short-ranged, the variables X_i as $n \to \infty$ will tend to become independent of each other and the sum Z_n over a certain domain will fluctuate. In this sense the C.L.T. will still apply. By letting $b_n \sim n^{1/2}$ one obtains by means of the square-root law of fluctuations, a distribution of the form:

$$F_n(a) = \mathcal{P}\{s_n < a\} . \tag{1.132}$$

There is unfortunately no unique way to delineate "strong and weak" dependency. However, it can be stated that the C.L.T. holds, when the summands of a sequence of random variables are "weakly dependent". A weak dependence occurs when the summands satisfy a suitable "mixing condition". In general "mixing sequences" are sequences for which the past and distant future are asymptotically independent. There are various definitions for the so-called "mixing coefficients" characterizing the dependency of the random variables. These coefficients can be defined in terms of "measures of dependency" between two σ-algebras based on pairs of events or on correlations (see for instance Peligrad [66], Bradley [67]). Thus, if $\{x_n\}$ is a sequence of random variables on a probability space $[X, \mathcal{F}, \mathcal{P}]$ one has the σ-algebras:

$$\mathcal{F}_n^m \triangleq \sigma(x_k; n < k < m) \, , \tag{1.133}$$

for two real numbers p, r and the positive sequence $a_n(p, r)$ approaching zero as $n \to \infty$ is a measure of dependency, where:

$$|E\{xy\} - E\{x\}E\{y\}| \le a_n(p, r) \parallel x_p \parallel \parallel y_r \parallel \tag{1.134}$$

for every $x \in L_p(\mathcal{F}_1^m) \subset X$ and $y \in L_r(\mathcal{F}_{n+m}^\infty) \subset X$ for every m.

More specifically "strong mixing" occurs, if the sum of these numbers $(\frac{1}{p} + \frac{1}{r}) < 1, a_n(p, r) \to 0$ or equivalently $\alpha_n \to 0$ where the latter is defined by:

$$\alpha_n := \sup\{|P(E_1 \cap E_2) - P(E_1)P(E_2)| : E_1 \in \mathcal{F}_1^k; E_2 \in \mathcal{F}_{k+n}^\infty, k \ge 1\} \tag{1.135}$$

(see also [66]). If the sum $(\frac{1}{p} + \frac{1}{r}) = 1, p \ne 1, r \ne 1$, one says that the sequence is "p-mixing" (Ibragimov [69])or equivalently if $\lambda_n \to 0$, where λ_n is defined by:

$$\left. \begin{aligned} \lambda_n := \sup\Big\{ & \frac{P(E_1 \cap E_2) - P(E_1)P(E_2)}{[P(E_1)P(E_2)]^{1/2}}; \\ & E_1 \in \mathcal{F}_1^k, E_2 \in \mathcal{F}_{k+n}^\infty, \ P(E_1)P(E_2) \ne 0, k \ge 1 \Big\} \, . \end{aligned} \right\} \tag{1.136}$$

This type of mixing has been studied earlier by Kolmogorov and Rozanov. There are three more defintions of "mixing sequences" as follows:

(i) the so-called "ϕ-mixing", when in the above sum of real numbers is formed for
$p = 1, r = \infty$.

(ii) the "ϕ^r-mixing", when $p = \infty$, $r = 1$, which is then also valid for the reversed sequence, and

(iii) the "ψ-mixing", when $p = 1$ and $r = 1$ with the corresponding mixing coefficient (see also [62]).

The partial sum $Z_n = \sum_{i=1}^n X_i; \sigma_n^z = \text{Var } Z_n$ is said to be "attracted" (or partially attracted) to the distribution F as $n \to \infty$, if there exist sequences of real numbers $\{a_n\}, \{b_n\}$ with $b_n > 0, b_n \to \infty$ such that:

$$\frac{s_n - a_n}{b_n} \to F \text{ in distributions as } n \to \infty$$

(along a "subsequence" of positive integers).

The "limiting behaviour" of partial sums Z_n, if $\{x_n\}$ is a strictly "stationary mixing sequence" (strong mixing) is characterized by a "stable law" of the form:

$$a_n = n^{1/2}h(n) \, ; \ h(n) \text{ a slowly varying function} \, , \tag{1.137}$$

and Z_n can only be attracted to this law (Ibragimov [68, 69]). The sequence $\{x_n\}$ will satisfy the C.L.T., if $\left\{\frac{s_n - a_n}{b_n}\right\}$ is attracted to the normal distribution $\Phi(x_n)$. More recent studies on the C.L.T. use the notion of the "invariance principle in distribution" or (IPID) due to Erdös and Kac [70, 71]. It is also referred to as the "functional central limit theorem" or F.C.L.T. It is regarded as synonomous to an approximation theorem of partial sums, empirical processes, extremal processes etc. Such processes can be approximated in "distribution", "probability" or "almost sure (a.s.)" in terms of a canonical process such as the Brownian motion process to be discussed later. By considering an open set C of the real-valued continuous function space $\mathcal{C}(0, 1)$ and a sequence of i.i.d. random variables $\{x_k, k \geq 1\}$ with mean zero and the variance equal to one, the partial sum Z_k can be written as:

$$f_n(t) = n^{-1/2} Z_k, \text{ if } t = \frac{k}{n}, \ 0 \leq t \leq 1 , \tag{1.138}$$

which is linear on $(\frac{k-1}{n}, \frac{k}{n})$, $(1 \leq k \leq n)$ and defines a sequence of random variables with values in the C-space. Denoting by P_n the distribution of f_n, i.e. $P_n(E) = \mathcal{P}\{f_n(t) \in E\}, E \in \mathcal{C}$, the latter being the σ-algebra of the open set in $\mathcal{C}$, then the distribution of the Brownian motion: $\{B(t), 0 \leq t \leq 1\}$ converges weakly to the "Wiener measure W" or $\mathcal{P}_W\{\mathcal{C}(0, 1), \mathcal{C}\}$ so that:

$$\begin{array}{rl} P_n \to W \ ; \ (n \to \infty) & \text{(a)} \\ \text{and for any bounded continuous function } h : C \to \mathbb{R} : & \\ h(f_n) \to h(B) \ ; \ (n \to \infty) & \text{(b)} \end{array} \right\} \tag{1.139}$$

(see also Donsker [72]). This is referred to as the "invariance principle in distribution" mentioned above and was established by Erdös and Kac [70, 71]. They noted that the "limit distribution, or $\mathcal{L}(h(B))$" does not depend on the common distribution function F of the sequence $\{x_k, k \geq 1\}$, but remained invariant under a change of F as long as:

$$\int x \, dF(x) = 0 \ ; \ \int x^2 \, dF(x) = 1 \tag{1.140}$$

A stronger form of Donsker's theorem is the "invariance principle in probability" or (IPIP). For a more detailed discussion on these concepts, i.e. invariance principles, mixing conditions of sequences, etc. the reader is referred to the above references and Phillip [73].

1.6 Stochastic Processes:

(i) Characteristics of stochastic processes:

In the formulation of the behaviour of discrete media certain concepts of the theory of stochastic processes and stochastic differential equations are required. A stochastic process can be generally regarded as a family of scalar or vector valued random variables (functions) that are indexed by a parameter set $\mathcal{J}$. The latter is usually identified with the time set T. If T contains a countable number of values $T = \{0, 1, 2 \ldots\}$ or $T = \{0, \pm 1, \pm 2 \ldots\}$, the stochastic process is a "discrete parameter" or "discrete time" process with a corresponding discrete state-space $Z \subset X$. If however the random variables $x(t)$ are continuous, i.e. $T = \{t; t \geq 0\}$ or within a closed time interval of the set $T : T = \{t : a \leq t \leq b\}$, the process is said to be a "continuous parameter" or "continuous time" process with a state-space Z. The notation used subsequently for a stochastic process $\{x_t, t \in T\}$ is an abbreviated form of $\{x(t, \omega), t \in T, \omega \in \Omega\}$, where the random function x_t for a specific ω is referred to as a "realization or sample function $x_t(\omega)$". In accordance with an earlier definition (section 1.3(ii)) concerned with directly given random variables, the function $x(t, \omega)$ or $x_t(\omega), t \in T$ having values in $[X, \mathcal{F}]$ for each outcome $\omega \in \Omega$ is called a "realization of the random process $x(t)$", if the parameter set T is continuous. If the set is discrete, one refers to a "sample sequence". A scalar or vector valued stochastic process $\{x_t, t \in T\}$ is characterized by the finite dimensional distribution function $F(x_{t_1} \ldots x_{t_n})$ for any finite set $\{t_k\} \in T$. For the specification of a random process the "probability densities" are fundamental. Thus, the first density function $p(x_t)$ is called the "first order density" and the density $p(x_t, x_s)$ the "second-order density" of the process. One can write for $p(x_t)$ also $p(x, t)$ to indicate that for each t, $p(\cdot)$ is the density of x_t. Similarly, the second-order density for each (t, s) can be expressed by the joint density $p_{x_t x_s}(\cdot, \cdot)$ of the random variables (scalar or vector), i.e.:

$$p(x_t, x_s) = p(x, y; t, s) , \tag{1.141}$$

in which x, y are to be considered as dummy variables. In terms of "conditional densities", one can also write:

$$p(x_t | x_s) = p(x, t; y, s) . \tag{1.142}$$

There are statistical parameters associated with the above densities such as the "mean value function" of the scalar process $\{x_t, t \in T\}$, i.e.:

$$m_x(t) \overset{\Delta}{=} E\{x(t)\}; \ t \in T , \tag{1.143}$$

and correspondingly a (t, s)-function known as the second-order moment function or the "correlation function" of the process $x(t)$:

$$E\{x_t, x_s\} \stackrel{\Delta}{=} R_x(t, s); \ (t, s) \in T. \tag{1.144}$$

It exists, if $E\{|x(t)|^2\} < \infty$ for all $t \in T$ since by Schwartz's inequality (see also [42]):

$$E\{x(t)x(s)\} \le [E\{|x(t)|^2\}E\{|x(s)|^2\}]^{1/2}. \tag{1.145}$$

Any random process satisfying $E\{|x(t)|^2\} < \infty$ for all $t \in T$ is said to be a finite variance process. The two functions, i.e. the mean and correlation functions can be used to describe the variance of the value of the process at any given time t. Evidently one can define the covariance between the values at any two points t, s of $\{t_k\} \in T$ as follows:

$$R_x(t, t) = E\{|x(t)|^2\} \to \sigma_{x(t)}^2 = R_x(t, t) - m_{x(t)}^2$$

so that the covariance function becomes:

$$C\{x(t)x(s)\} = R_x(t, s) - m_x(t)m_x(s) \tag{1.146}$$

in which $\sigma_x^2(t), \ t \in T$ is the variance function of the process.

In the usual terminology of stochastic processes, the function $R_x(t, s)$ is also called "auto-correlation" function to indicate that the two random variables $x(t)x(s)$ belong to the same process or $\{x_t, t \in T\}$.

If two different processes are analyzed the term "cross-correlation function" is employed, then:

$$R_{xy}(t, s) \stackrel{\Delta}{=} E\{x(t)x(s)\} \ ,$$

in which $x(t)$ and $x(s)$ belong to different (perhaps related) random processes. It is apparent that for a vector-valued process the mean value function of (1.144) will be defined by:

$$m_{\mathbf{x}(t)} \stackrel{\Delta}{=} E\{\mathbf{x}(t)\} \ , \tag{a}$$

and the correlation function is then a matrix or:

$$\underline{R}_{\mathbf{x}}(t, s) = E\{\mathbf{x}_t, \mathbf{x}_s^T\} \tag{b}$$

$$\left.\begin{array}{c}\\\\\\\\\end{array}\right\} (1.147)$$

similarly the covariance matrix is given by:

$$C_{\mathbf{x}}(t, s) = E\{[\mathbf{x}_t - m_{\mathbf{x}}(t)][\mathbf{x}_s - m_{\mathbf{x}}(x)^T]\}. \tag{1.148}$$

(ii) Regularity and continuity:

Similar to the limit conditions for "sequences" of random variables considered
before, i.e. convergence in probability, convergence in distribution etc., one can
also give such conditions for random functions or a family of such functions
that comprises a stochastic process. Thus for instance:

$$\lim_{t \to t_0} x(t) = x(t_0), \text{ if } \lim_{t \to t_0} E\{|x(t) - x(t_0)|^2\} = 0. \tag{1.149}$$

Other characteristics of random functions of importance are "regularity and
continuity". Random functions that have a finite mean square are called regular
random functions. At points or set of points, where this criterion does not hold,
one speaks of m.s. singularities of the function $x(t)$. A random function $x(t)$
with a realization at $x(t)$ and $x(t + s)$, $s, t \in T$ is said to be continuous in
probability at point t, if

$$\lim_{s \to 0} \mathcal{P}\{|x(t + s) - x(t)| \leq \epsilon\} = 1 \text{ for } \forall \epsilon > 0, \qquad \text{(a)}$$

or equivalently, if

$$\lim_{s \to 0} \mathcal{P}\{|x(t + s) - x(t)| > \epsilon\} = 0 \text{ for } \forall \epsilon > 0, \qquad \text{(b)}$$

$$\left.\begin{array}{c}\\\\\\\end{array}\right\} \tag{1.150}$$

A random function is continuous in the m.s. sense, if:

$$\text{l.i.m.} E\{|x(t + s) - x(t)|^2\} = 0 . \tag{1.151}$$

Analogously, a stochastic process $\{\mathbf{x}(t), t \in T\}$ is continous in the m.s. sense
at t, if:

$$\text{l.i.m.}_{\epsilon \to 0} \mathbf{x}(t - \epsilon) = \mathbf{x}(t) , \qquad \text{(a)}$$

or equivalently, if

$$\text{l.i.m.}_{\epsilon \to 0} E\{|(\mathbf{x}(t + \epsilon) - \mathbf{x}(t)|^2\} = 0 . \qquad \text{(b)}$$

$$\left.\begin{array}{c}\\\\\\\end{array}\right\} \tag{1.152}$$

In the subsequent stochastic mechanics of "steady-state" phenomena, the notion of "stationarity" of a stochastic process becomes significant. A stochastic
process $\{x_t, t \in T\}$ is said to be "stationary in the strict sense", if it has the
same probability density as the process $\{x_{t+s}, t \in T\}$ for any $s \in T$, i.e., if:

$$p(x_{t_1} \cdots x_{t_n}) = p(x_{t_1+s}, \cdots x_{t_n+s}). \tag{1.153}$$

The process x_t is strictly stationary of order k, if the above relation holds for
$n \leq k$ only. Evidently, if (1.153) holds for $n = k$, it will also hold for $n \leq k$. In
the special case where $n = 1$, (1.153) becomes:

$$p(x_t) = p(x_{t+s}) \text{ for all } s \in T ,$$

indicating that the first-order density is independent of t and thus the mean
value function $m_x(t) =$const. If $n = 2$, the relation (1.153) will only depend on
$(t - s)$ so that the correlation function $R_x(t, s)$ is given in the following form:

$$R_x(t+s,t) = E\{x_{t+s}, x_t\} = R_x(s) \ ,$$

and the covariance function: $C_x(t+s,t) = C(s)$.

A weaker notion of stationarity (see also [74]) states that the process $\{x_t, t \in T\}$ is "weakly stationary" or "stationary in the wide sense", if it has finite second moments or if $E\{|x_t|^2\} < \infty$ for all $t \in T$ and the conditions given for $m_x(t)$ and $R_x(t+s,t)$ hold. The first and second order statistical properties of a stochastic process become important when dealing with a process in which higher characteristics can be established from the knowledge of the first and second order ones. This is the case for "Gaussian processes" frequently used in the description of physical phenomena. Moreover, by employing the well-known Chebyshev inequality [44] pertaining to random variables, i.e.:

$$P\{|x - m_x| \ge \epsilon\sigma_x\} \le \frac{1}{\epsilon^2}; \ \epsilon > 0 \ , \tag{1.154}$$

an upper bound for the probability of the event $|x(t) - m_x(t)| \ge \epsilon$ at any point $t \in T$ can be formed from the mean and variance functions of the process $x(t)$. It is equally possible to assess the probability of such an event for any t in the closed time interval $(a \le t \le b) \in T$ (see for instance Parzen [75]). It is of interest to note that for an arbitrary function $f(t)$ the auto-correlation function satisfies the following inequality:

$$R_x(t_i, t_j)f(t_i)f^*(t_j) \ge 0 \tag{1.155}$$

where repeated indices mean summation and i, j is the range from 1 to any finite integer n. The asterisk indicates the complex conjugate of the function. This relation also holds for complex-valued random processes in which the auto-correlation function is defined by:

$$R_x(t_1, t_2) \overset{\Delta}{=} E\{x(t_1)x^*(t_2)\}. \tag{1.156}$$

The symmetry property in this case is Hermitian, i.e. $R_x(t_1, t_2) = R_x^*(t_2, t_1)$. The concept of stationarity is important in the molecular dynamics of simple and multi-phase fluid structures to be considered later. In this context a theorem by Bochner [76] should be mentioned here. It asserts that every non-negative definite function satisfying (1.156) has a non-negative Fourier transform, which for $R_x(\tau)$, is given by:

$$S(\omega) = \frac{1}{\sqrt{2\pi}} \int\limits_{-\infty}^{\infty} R_x(\tau) \exp[-i\omega\tau]d\tau \ge 0 \ . \tag{1.157}$$

This function in the theory of stationary random processes is called the "mean square spectral density" or briefly "spectral density" of the process $x(t)$. By the Fourier inversion theorem one has:

$$R(\tau) = \frac{1}{\sqrt{2\pi}} \int\limits_{-\infty}^{\infty} e^{i\omega t} S(\omega) d\omega \ . \tag{1.158}$$

Relations (1.157, 1.158) are known as the Wiener-Khintchine theorem for stochastic processes and forms a basis for a generalized harmonic analysis (Wiener [77]).

In the theory of stochastic processes as a wider class of functions has to be frequently considered than the ordinary time-functions. These functions are then regarded as operators that include time derivatives and integrals of the relevant random variables and must be determined in accordance with the rules of stochastic calculus. Thus, so far as the differentiation of a random function $x(t)$ is concerned, the derivative or product of derivatives is given by:

$$E\{\frac{dx(t)}{dt}\} = \frac{d}{dt} E\{x(t)\} \ , \tag{a}$$

and

$$E\{x(t)\frac{dx(s)}{ds}\} = \frac{\partial}{\partial s} E\{x(t)x(s)\} \ , \tag{b}$$

$$\left.\right\} (1.159)$$

where the derivatives are "mean square" differential operators, i.e., $x'(t) = \frac{d}{dt} x(t)$, if the following limit exists:

$$\lim_{\Delta t \to 0} E\{|\frac{x(t + \Delta t) - x(t)}{\Delta t} - x'(t)|^2\} = 0 \ . \tag{1.160}$$

A necessary and sufficient condition that $x(t)$ has a m.s. derivative at a point $t \in T$ is then:

$$\frac{\partial^2}{\partial t \partial s} E\{x(t)x(s)\} \text{ exists at } (t, s) = (t, t).$$

Using the product moment of $x(t)$, i.e.: $m_{1,2...,n}(t_1, t_2 ... t_n) = E\{x_1, x_2 ... x_n\}$, relation (1.159(b)) can be generalized to:

$$E\{x_1^{(k_1)}, x_2^{(k_2)} ... x_n^{(k_n)}\} = (\frac{\partial}{\partial t_1})^{k_1} (\frac{\partial}{\partial t_2})^{k_2} ... (\frac{\partial}{\partial t_n})^{k_n} E\{x_1, x_2 ..., x_n\} \ , \tag{1.161}$$

in which $x_i^{(k_i)} = \frac{\partial^{k_i} x(t_i)}{\partial t_i^{(k_i)}}$.

"Stochastic integrals" in the mean-square sense of a random function $x(t)$ can be defined in the following manner:

(i) Considering the stochastic function $x(t)$ on an interval $(a, b) \subset T$ subdividing it into arbitrary time intervals Δt_k and taking the sum $\sum_{x=1}^{n} x_k \Delta t_k$, where x_k is the random variable associated with an arbitrary point in Δt_k and if:

$$\text{l.i.m.} \sum_{k=1}^{n} x_k \Delta t_k = \int_a^b x(t)dt \qquad (1.162)$$

exists, the function $x(t)$ is said to have a stochastic "Rieman integral" on (a, b).

(ii) Defining a suitable measure $\mathcal{P}(E)$ on the time-parameter space, where $\mathcal{P}(E)$ is a non-negative set function defined for all Borel sets of the T-space, then $x(t)$ can be considered on the set E of finite Lebesgue measures. Dividing E into n-disjoint sets E_k and forming $\sum_{k=1}^{n} x_k \mathcal{P}\{E_k\}$, ($x_k$ is here the random variable associated with a point in E_k), and if

$$\text{l.i.m.} \sum_{k=1}^{n} x_k \mathcal{P}\{E_k\} = \int_E x(t)d\mathcal{P}(t) \qquad (1.163)$$

exists, then $x(t)$ is said to have a stochastic "Lebesgue integral" on E.

(iii) Similarly the stochastic "Stieltjes integral" of the function $x(t)$ with respect to a set of random functions on E_k or $S(E_k)$ for all Borel sets E_k of the T-space is defined by:

$$\underset{\substack{n \to \infty \\ S(E_k) \to 0}}{\text{l.i.m.}} \sum_{k=1}^{n} x_k S(E_k) = \int_{E_k} x(t)dS(t) \ . \qquad (1.164)$$

For a family of random variables $\mathbf{x}(t)$ or the stochastic process $\{\mathbf{x}_t, t \in T\}$ the stochastic integral in accordance with (1.162) is given by:

$$\underset{\substack{k \to \infty \\ \epsilon \to 0}}{\text{l.i.m.}} \sum_{k=1}^{n} \mathbf{x}(t_k)[t_x - t_{k-1}] = \int_a^b \mathbf{x}(t)dt \ , \qquad (1.165)$$

where ϵ designates $[t_k - t_{k-1}]_{max}$, $a = t_0 < t_1 \ldots t_n = b$, $k = 1 \ldots n$, if the limit exists. Such a limit also implies that for $(t, s) \in T$ the correlation function $R_x(t, s)$ to be Rieman integrable over $a \le t, s \le b$ for a time-continuous process with finite second moments and variance.

(iii) Some basic stochastic processes:

As mentioned before a large number of physical phenomena can be modelled by Gaussian random processes. They have been briefly mentioned in section (1.3 ii). A Gaussian (or normal) process is one, where all distribution functions are joint normal distributions. In terms of the nth order probability density function the process is described by:

$$p_n(x_1, x_2 \ldots x_n; t_1, t_2 \ldots t_n)$$

$$= \frac{\Delta^{1/2}}{(2\pi)^{1/2}} \exp[-\frac{1}{2} \sum_{j,k} \Delta_{jk}(x_j - m_j)(x_k - m_k)] , \quad (1.166)$$

in which $m_j = m(t_j)$ are the means, $\Delta = |C_{jk}|$ the determinant of the auto-covariance function C_{jk} and Δ_{jk} the cofactor of C_{jk} in the determinant. By using a "joint characteristic functional $\phi_x(t)$" of the jointly distributed Gaussian random variables $x(t_1), x(t_2) \ldots x(t_n)$ of the form:

$$\phi_x(t) = \exp[i \sum_{j=1}^{n} \varphi_j m_x(t_j) - \frac{1}{2} \sum_{j,k=1}^{n} C_{xx}(t_j, t_k)\varphi_j\varphi_k] , \quad (1.167)$$

it is seen that the normal random function is completely specified by its mean $m(t_j)$ and the auto-covariance $C_{jk}(t_j, t_k)$, since from these functions the distribution at n-points of the process can be obtained. An assignment of different parametric values to a Gaussian process yields again Gaussian random variables. By the consistency theorem of probability (see for instance Loève [78]), if for any n, $x(t_1), x(t_2) \ldots x(t_n)$ are Gaussian random variables, the process $\mathbf{x}(t)$ must also be a Gaussian process. A linear transformation of a set of Gaussian random variables leads to a new set of such variables. Analogously, any linear operation on a Gaussian process results in another Gaussian process. Hence, one can generalize such processes by using a linear functional on a linear space U, i.e. $f = < u, v >$ and a Gaussian process has then a characteristic functional of the form:

$$\phi_x(u) = \exp[iE(< u, v >) - \frac{1}{2}C(u, v)] , \quad (1.168)$$

where $E(< u, v >)$ is the expectation of the generalized process and $C(u, v)$ the scalar product of $u, v \in U$ or the correlation functional (see also Prohorov and Rozanov [42]).

Another class of stochastic processes $x(t), t \in T$ is that of "independent increments", if for all $t_1 < t_2 < \ldots < t_k < t_{k+1} \in T$ and every $k = 1, 2 \ldots$, the random variables $x(t_2) - x(t_1), x(t_3) - x(t_2) \ldots, x(t_{k+1}) - x(t_k)$ are mutually independent (see also Feller [64], Doob [50]). Since there exists a simple relationship between $x(t_1), \ldots x(t_k)$ and their increments, the process will be specified by the density of an increment $x(t) - x(t')$ for all $t' < t$ and the first order density $p_0(x)$ at some $t_0 \in T$.

An important representative of this group is the "Poisson process". Thus a scalar random process with independent increments satisfying:

$$P\{x(t) - x(t') = k\} = \frac{[\lambda(t - t')]^k}{k!}e^{-\lambda(t-t')} , \quad (1.169)$$

for any $t, t' \in T, t' < t$ and $\lambda > 0$ is called a "simple Poisson process" having a discrete state-space and a continuous time-parameter space. On the other hand, if $x(t)$ has a continuous state and time-parameter space with the following properties:

(i) $\{x(t), t \geq 0\}$ has stationary increments, i.e. for any $t_1, t_2 \in T$ and $t_1 < t_2$, the distribution of $[x(t_2) - x(t_1)]$ is the same as $[x(t_2 + \tau) - x(t_1 + \tau)]$ for any $\tau > 0$. For any intervals $[t_1, t_2]$ and $[t_3, t_4]$ with $t_1 < t_2 < t_3 < t_4$, the random variables $[x(t_2) - x(t_1)]$, $[x(t_4) - x(t_3)]$ are independent.

(ii) For any given interval $[t_1, t_2]$, the difference $[x(t_2) - x(t_1)]$ has a normal distribution with zero mean and a variance $\sigma^2(t_2 - t_1)$, the process is called a "Wiener or Brownian motion" process. These processes and in particular their vector valued forms will be used in Chapter 4. It should also be pointed out that Poisson processes and the class known as Poisson-Point processes are important in considerations of the stochastic geometry of various discrete materials particularly for the simulation of their response behaviour as discussed later in the text.

Random phenomena in general in the continuous time parameter space are characterized by a family of random variables $\{x(t), t \geq 0\}$ or a stochastic process $\mathbf{x}_t$ on the basis of an abstract dynamical system $[X, \mathcal{F}, \mathcal{P}]$ and on the assumption of the continuous observability of the relevant variables. It is apparent that the process $\mathbf{x}_t$ has for each $t \geq 0$ a "sub-σ-field" of $\mathcal{F}$, that can be designated by:

$$\mathcal{F}_t^x = \sigma\{x_s, s \in (0, t)\} , \tag{1.170}$$

which is generated by the family of random variables $\{x_s, s \in (0, t)\}$. The family $\mathbb{F} = \{\mathcal{F}_t^x, t \geq 0\}$ can be referred to as the "internal history" of the process $\mathbf{x}_t$. However, it is often necessary to consider another random variable Y with values in the space R^k, but which is $\mathcal{F}_t^x$-measurable for a fixed time instant $t \geq 0$. Thus by using a result on measurability (see also Brémaud [47]) this variable can be expressed by:

$$Y = f(x_s, s \in S); S \subset (0, t) , \tag{1.171}$$

in which S is a countable subset of $(0, t)$, f a Borelian function from $R^{n\overline{\overline{S}}}$ into R^k and $\overline{\overline{S}}$ the cardinality of the set S (see also Mostowski [79]). Similarly, if Y_t is an R^k-valued process such that for each $t \geq 0, Y_t$ is $\mathcal{F}_t^x$-measurable, then:

$$Y_t = g_t(x_0^t) , \tag{1.172}$$

for some function g_t, indicating that the process Y_t depends "causally" on the process x_t. As discussed earlier, if the family of sub-σ-fields increases, i.e. $\mathcal{F}_t \supseteq \mathcal{F}_t^x$ for all $t \geq 0$, then $\mathcal{F}_t$ is called a "history of the process $\mathbf{x}_t$". The latter is then said to be adapted to $\mathcal{F}_t$. The above concepts become important in the formulation of "structural changes" in heterogeneous solids for example, during a given time interval $(0, t) \in T$ and will be dicsussed in greater detail in Chapt. 3.

In most practical applications the histories $\mathcal{F}_t$ can be represented by $\mathcal{F}_t = \mathcal{F}_t^z$, where the superscript refers to a random process z_t in the state-space $Z \subset X$ of the material. In particular in the analysis of structural changes an "incomplete observed" Markov process $z_t \in Z$ can be employed. From a system theory point of view, the history $\mathcal{F}_t$ can be regarded as an "increasing information pattern" in which $\mathcal{F}_t^z$ is the overall information related to the observations of two processes, e.g. an "observable" and an "unobservable" one that are components of the overall process z_t.

Due to the importance of Markov processes in stochastic mechanics and in the development of general probabilistic theories concerning the deformation and flow of discrete media such processes will briefly be reviewed here. A more detailed study will be given in later chapters. Thus considering conditional probability functions, a discrete (one-dimensional) random process $x(t)$ is said to be Markovian, if:

$$\left. \begin{aligned} &P\{\xi_n, t_n | \xi_{n-1}, t_{k-1}; \ldots; \xi_2, t_2; \xi_1, t_1\} \\ &= P\{\xi_n, t_n | \xi_{n-1}, t_{n-1}\}; t_n > t_{n-1} \ldots > t_2 > t_1 , \end{aligned} \right\} \quad (1.173)$$

indicating that the probability of the process $x(t)$ to have value ξ_n in $\mathbb{R}^n$ at times $t = t_n$ under the condition that its value at some earlier times are known, depends only on the most recent value $x(t) = \xi_{n-1}$ at time $t = t_{n-1}$. The conditional probability function is referred to as the "transition probability" of the Markov process. Hence a discrete Markov process is completely defined by its first probability function $P(\xi, t)$ and the transition probabilities. The Markov property of the process x_t can also be expressed by "conditional density" functions of the form:

$$\left. \begin{aligned} &p(\xi_n, t_n | \xi_{n-1}, t_{n-1}; \ldots; \xi_2, t_2; \xi_2, t_2; \xi_1, t_1) \\ &= p(\xi_n, t_n | \xi_{n-1}, t_{n-1}); t_n > t_{n-1} > \ldots > t_2 > t_1 \end{aligned} \right\} \quad (1.174)$$

in which the quantity on the right-hand side of (1.174) is called the transition probability density or briefly "transition density". Denoting it by $p_{t|t'}(\xi|\xi')$, $t' < t, \xi, \xi' \in X$, the Markov process is completely specified by these transition densities and the first-order density $p_{t_0}(\xi), \xi \in X$ for some fixed instant of time $t_0 \in T$. Although (1.174) remains valid for any parametric values of $t_1, t_2 \ldots t_n$ so long as $t_1 < t_2 < \ldots < t_n$, the transition densities are not completely arbitrary, e.g. they must satisfy the well-known Chapman-Kolmogorov functional relation shown subsequently.

The theory of Markov processes is due to A.A. Markov [80], who first gave an analytical formulation of the Brownian motion of particles in statistical fluid mechanics. This motion was characterized by the transition probability or the probability, that a particle starting from a given point ξ is in a certain event E at time t. Subsequently a more general theory of Markov processes was developed by A.N. Kolmogorov [6], W. Feller [64], Doob [50] and others.

A comprehensive study of Markov processes is due to Dynkin [81] (see also Blumenthal and Getoor [82], Rosenblatt [83], Revuz [84]).

The most advanced theory of Markov processes is concerned with the motion $x(t)$ and its path in terms of transition probability measures. Thus, if X or Z is the state-space, i.e. the space of points $\omega = \{\xi_0, \xi_1 \ldots\}$ representing possible observations at fixed instants of time, the possible events for which a probability is well defined will be the elements of the Borel field $\mathcal{F}$ of subsets of X. A "stable generating mechanism" for a Markov process through time is then characterized by its transition probability function $P(\xi, E)$ which is assumed to be $\mathcal{F}$-measurable as a function of ξ for each event $E \in \mathcal{F}$ and a probability measure on the Borel field $\mathcal{F}$ for each ξ in X.

Hence for a simple Markov process, if a phenomenon is observed at time instants $t = 0, 1, 2 \ldots$, the function $P(\xi, E)$ represents the probability that an observation at time $(t + 1)$ falls within the set E given that at time t the observation ξ was made. The fundamental condition relating the random process $x(t, \omega)$ in X or equivalently $\{x(t_i, \omega) = \xi_i; t_i = 0, 1, 2 \ldots\}$ to the measure $\mathcal{P}\{x(t) \in E | x(s) = \xi\}$ is the Markov principle. By using an initial probability measure $\mathcal{P}_0$ at time $t = 0$ on $\mathcal{F}$, a discrete time-parameter process can be formulated from the distribution P and a stationary transition probability function. Evidently for any finite collection of event sets $E_0, E_1 \ldots E_n \in \mathcal{F}$ the following relation will hold:

$$
\left.
\begin{aligned}
\mathcal{P}\{E_0 \times E_1 \times \ldots \times E_n\} &= \mathcal{P}\{\xi_0 \in E_0, \xi_1 \in E_1, \ldots \xi_n \in E_n\} \\
&= \int_{E_0} P_0 d\xi_0 \int_{E_1} P(\xi_1, d\xi_1) \ldots \int_{E_{n-1}} P(\xi_{n-2}, d\xi_{n-2}, d\xi_{n-1}) P(\xi_{n-1}, E_n)
\end{aligned}
\right\} \quad (1.175)
$$

in which the above set functions can be extended on the basis of the Jonescu-Tulcea theorem (see also Doob [50]) to a measure $\mathcal{P}_\infty$ on the Borel field $\mathcal{F}_\infty$. However, it restricts the Markov process to a discrete time-parameter process. Another extension theorem due to Kolmogorov (see Dynkin [81]) is based on measure theoretical and topological arguments, which are equally applicable to Markovian and non-Markovian processes, becomes significant for certain applications in stochastic mechanics.

Evidently the probability function in the manner described above, represents the transition of a material system from one state to another in one-step only. However, often higher step transition probabilities may be required. Thus designating such probabilities by $P^{(n)}$, they can be generated by a recursive procedure such that:

$$
P^{(n+1)}(\xi, E) = \int_X P(\xi, d\eta) P^{(n)}(\eta, E); \quad (n = 1, 2 \ldots), \quad (1.176)
$$

where the following relation can be obtained:

$$P^{(n+m)}(\xi, E) = \int\limits_{X} P^{(n)}(\xi, d\eta) P^{(m)}(\eta, E); (n, m = 1, 2 \ldots). \qquad (1.177)$$

The above represents the well-known Chapman-Kolmogorov equation which is important in stochastic mechanics since it connects the probabilistic theory with the functional analytic formulation. Considering the transition after a certain unit of time has elapsed from a point $\xi \in X$ in more than one step into the set $E \in \mathcal{F}$, then $P^{(n)}(\xi, E)$ will be given as follows:

$$\left. \begin{array}{l} P^{(n)}(\xi, E) = \displaystyle\int\limits_{X} P^{(n-1)}(\xi, d\eta) P^{(1)}(\eta, E); \ (n = 1, 2 \ldots) \\[4pt] \text{with} \\[2pt] \qquad P^{(1)}(\xi, E) \equiv P(\xi, E); \ P^{(1)}(\xi, E) \geq 0 \\[4pt] \text{and} \\[2pt] \qquad P^{(n)}(\xi, X) = 1. \end{array} \right\} \qquad (1.178)$$

Although there exist many types of Markov processes that are useful in the mathematical modelling of the behaviour of discrete media, they can be approximately grouped into those that have a non-denumerable number of states and those which have a denumerable number of states. The latter group is commonly referred to as "Markov chains". The most important feature of a Markov chain is its reference to the discretness of a set of time-parameter since a denumerable "state-space" can be regarded as a special case of a more general state-space (see also Chapt. 3 on the state-space representation). However, discrete times are not a particular case of continuous time and, hence, by the definition of a Markov chain due to Revuz [84], the latter is discrete in time, whilst a Markov process as such is one in which the time parameters are continuous. Thus, if the stochastic process $x(t)$ representing a discrete system or parts of it at time s, which is in the "state i" and goes to a "state j" at a later time, the probability pertaining to $x(t)$ is given by:

$$P\{x_t = j | x_s = i\} = P_{ij}, i, j = 1, 2 \ldots t, s \in T \subset \mathbb{R}^{+} . \qquad (1.179)$$

If the behaviour of the system or the Markov chain x_t is referred to an initial time t_0, the corresponding initial probability distribution of some field quantity, which may be experimentally accessible is therefore expressed by:

$$\left. \begin{array}{l} \qquad P\{x(t_0) = i_0\} = \overset{\circ}{p_i}; \ (i = 1, 2 \ldots) , \\[4pt] \text{so that} \\[2pt] \qquad P\{x(t_0) = i_0, x(t_1) = i_1 \ldots x(t_n) = i_n\} \\[4pt] \qquad\qquad = \overset{\circ}{p_i} p_{i_0 i_1}(t_0, t_1) \ldots p_{i_{n-1} i_n}(t_{n-1}, t_n) , \end{array} \right\} \qquad (1.180)$$

for any $i_0, i_1 \ldots i_n$ and $t_0 \leq t_1 \leq \ldots \leq t_n$. A Markov chain is called homogeneous, if the transition probability $p_{ij}(t)$ depends only on the time difference, i.e.:

$$p_{ij}(s,t) = p_{ij}(t-s); (i,j = 1, 2 \ldots), \ s < t, \ (s,t) \in T \subset \mathbb{R}^+. \tag{1.181}$$

By using the notation introduced in (1.179 and 1.180), it is apparent that for a Markov chain the following relation holds:

$$p_{ij}(s,t) = \sum_k p_{ik}(s,u) p_{kj}(u,t); \ (i,j = 1, 2 \ldots), \ s \leq u \leq t. \tag{1.182}$$

If the process $x(t)$ is homogeneous and the time instants given by $t = nh$; $(n = 0, 1, 2 \ldots; h > 0)$, the probabilities of a transition in n-steps $p_{ij}(nh)$ are uniquely determined by the probabilities $p_{ij} = p_{ij}(h)$ of a "one-step transition", where:

$$\left. \begin{array}{c} p_{ij}(nh) = \sum_k p_{ik} p_{kj}[(n-1)/h] = \sum_k p_{ik}[(n-1)/h] p_{kj}, \\[2mm] (i,j = 1,2 \ldots) \text{ for all } n = 1, 2 \ldots \end{array} \right\} \tag{1.183}$$

If the time-parameter is continuous with $p_{ij}(0)$ specified by:

$$\left. p_{ij}(0) = \lim_{h \to 0} p_{ij}(h) = \begin{cases} 1 & \text{for } j = i \\ 0 & \text{for } j \neq i \end{cases} \right\} \tag{1.184}$$

and on the assumption that the Markov chain has this property, the probabilities will be continuously differentiable for $t > 0$. Moreover, if the following limit exists:

$$\lim_{h \to 0} \frac{p_{ij}(h) - p_{ij}(0)}{h} = q_{ij}, (i,j = 1, 2 \ldots), \tag{1.185}$$

where $0 \leq q_{ij} < \infty$ for $i \neq j$ and $q_{ij} = q_{ii} = -q_i$ for $i = j$, the coefficients q_{ij} are called "transition density" from states i to j. It is of interest to note that frequently in modelling the morphology of discrete media by Markov processes or Markov chains, it becomes necessary to introduce "Markov times". Thus, if τ_i denotes the instant of time when $x(t)$ leaves the state i for the first time and by τ_{ij}, when $x(t)$ reaches the state j for the first time, then:

$$\tau_i = \sup_{x^{(i)}(t) = i} t; \tau_{ij} = \inf_{x^{(i)}(t) = j} t, \tag{1.186}$$

in which $x^{(i)}(t)$ designates the process when $x(0) = i$. Hence the probability that the Markov chain $x(t)$, when leaving the state i and first moves to the state j, is given by:

$$P(\tau_{ij} = \tau_i | x(0) = i) = P\{x^{(i)}(t) = j\} = \frac{q_{ij}}{q_i}; i \neq j, \tag{1.187}$$

since $x^{(i)}(t)$ is a step function. Assuming that the above transition densities always satisfy the inequality:

$$\sum_{i \neq j} q_{ij} \leq q_i; \ (i = 1, 2 \dots) \,,$$

the transition probabilities $p_{ij}(t)$ under these conditions will satisfy the so-called "backward Kolmogorov" equation. Both the "backward" and "forward" Kolmogorov equation will be considered in more detail in Chapt. 3 and 4. The above basic characteristics of Markov chains are significant not only for the classification of "states", i.e., whether a chain is irreducible, recurrent or transient etc., but also for the "state-space" representation of discrete media to be discussed in Chapt. 3. An interpretation of the transition probabilities is also possible in terms of "graph theory".

Thus one can identify the "states" of a Markov chain with the "vertices of a graph" and the possible one-step transition probabilities with the "edges or paths" connecting these vertices. Evidently the path (i, j) between the states i and j exists, if the transition probability from i to j is greater than zero. Such a graph is then referred to as a Markov or stochastic graph (see for instance Kindermann and Snell [85] among others). Thus by considering a state vector $\mathbf{z} \in Z \subset X$, ($Z$ being the state-space), its components: $\mathbf{z} = \{z_1, z_2 \dots z_n\}$ correspond to a finite set of vertices and the "directional paths" between the vertices form then a set $\mathbf{e} = \{e_i\}$ so that the graph $\mathbf{G} = \{\mathbf{z}, \mathbf{e}\}$ represents the states of the material. In general a graph is said to be "strongly connected", if for any two distinct vertices z_i, z_j there is at least one path going from $z_i \rightarrow z_j$. A "maximal subgraph" of $\mathbf{G}$ is a strong subgraph of $\mathbf{G}$, which is not contained in any other strong subgraph of $\mathbf{G}$. Such a subgraph is called a "strong component of the graph $\mathbf{G}$".

Since in stochastic mechanics probability measures and distribution can only be assessed from step-wise experimental observations that correspond to some states in "dynamic equilibrium" of the discrete media, a somewhat wider concept of the graph structure is required. This will be further discussed in Chapt. 3 that is concerned with the state-space analysis of discrete media and where interaction effects between elements of the material structure can be related to the graph structure.

(iv) Random fields:

Although most evolutions of the behaviour of discrete media can be represented by stochastic processes and in particular by Markov processes there are certain phenomena which require a generalization of such processes. Hence to finalize this introductory chapter, a few remarks on the generalization are given below. It may be recalled that, if $[X, \mathcal{F}, \mathcal{P}]$ is a probabilistic function space, the stochastic process $x_t(\omega)$ is a map $x_t \colon \mathbb{R} \times X \to \mathbb{R}$, $t \in \mathbb{R}, \omega \in X$, that can be considered as a random variable characterizing the random motion of an element of the microstructure. In order to generalize this basic notion consider a real or complex inner product space U which is assumed to have the norm topology determined by the inner product. Denoting the set of complex-valued random variables on $[X, \mathcal{F}, \mathcal{P}]$ by $R(X)$, the map $\Phi: U \to R(X)$ is called a "random functional". It is a stochastic process indexed by U. A random functional Φ is linear, if:

$$\Phi(\alpha u + \beta u) = \alpha \Phi(u) + \beta \Phi(v) \; ; \; u, v \in U \text{ and for all } \alpha, \beta \in \mathbb{R}$$
$$\text{or } C \text{ (complex space).}$$
$$\left. \right\} \quad (1.188)$$

The random functional is continuous, if $u_i \to u$ in U implying that $\Phi(u_i) \to \Phi(u)$ in probability.

A continuous linear random functional is called in general a "random field". If the physical domain in the Euclidean space $\mathbb{R}^n$ is characterized by the stochastic process $\mathbf{x}_t$, it corresponds then precisely to measurements with reference to a point $X_0 \in \mathbb{R}^n$ (for instance the C.M. of a micro element in the configuration space $\mathcal{C} \subset X$). However measurements due to experimental constraints can only be carried out over a specific range of accuracy (see subsequent definitions in Chapt. 2) and hence can only be interpreted by the mean value or a weighting function on $\mathbb{R}^n$ with a random variable. By physical considerations such a functional should be linear and continuous and hence represents a random field. In accordance with earlier definitions of probabilistic quantities [22], [23], one can define the random functional $\Phi : U \to R(X)$, its expectation or mean value (see for instance [17]) by:

$$\Phi_m = E\{\Phi(u)\} = \int \Phi(u) d\mathcal{P} \qquad (1.189)$$

If Φ_m exists and Φ is a random field, then Φ_m will be linear, but not necessarily continuous in U. One can also define a covariance functional Φ_c such that:

$$\Phi_c : U \times U \to C$$

where

$$\Phi_c(u, v) = E\{[\Phi(u) - \Phi_m(u)][\Phi(v) - \Phi_m(v)^*]\}. \qquad (1.190)$$

It is to be noted that $\Phi_c(u, v)$ need not exist for all $u, v \in U$, but if it does and Φ is a random field on U, Φ_c will be a positive semi-definite bilinear form on U, which may not be bounded. Similarly, the variance functional $\Phi_V : U \to \mathbb{R}^+$ is defined by:

$$\Phi_V(u) = \Phi_c(u, u) , \qquad (1.191)$$

and the correlation functional $\Phi_C : U \times U \to C$ by:

$$\Phi_C(u, v) = \Phi_c(u, v) + \Phi_m(u)\Phi_m(v)^* = E\{\Phi(u)\Phi(v)^*\}. \tag{1.192}$$

One says that a random functional Φ is of "second order", if $\Phi_C(u, u) < \infty$ for all $u \in U$ and if Φ is bounded there exists a Φ_C such that:

$$\Phi_C(u, u)^{\frac{1}{2}} \le \Phi_C \| u \| \text{ for all } u \in U . \tag{1.193}$$

Another definition often required is that of the "characteristic random functional L_Φ" or:

$$L_{\Phi(u)} = E\{e^{i\Phi(u)}\} = \int e^{i\Phi(u)} d\mathcal{P}. \tag{1.194}$$

It exists for all $u \in U$ and is continuous, if Φ is continuous. Recalling that, if x is a random variable in $[X, \mathcal{F}, \mathcal{P}]$, the characteristic function $\varphi_x(t)$ is given by:

$$\varphi_x(t) = E\{e^{itx}\} = \int e^{itx(\omega)} d\mathcal{P}(\omega) , \tag{1.195}$$

and for a sequence of random variables $x_1, x_2 \ldots x_n$, the joint characteristic unction $\varphi_{x_1 \ldots x_n}$ (see also Doob [45] and Loève [78]) can be expressed by:

$$\varphi_{x_1 \ldots x_n}(t_1, \ldots t_n) = E\{\exp(i \sum_{k=1}^{n} t_k x_k)\}. \tag{1.196}$$

Since the joint distribution function of the sequence $x_1, x_2 \ldots x_n$ is uniquely determined by $\varphi_{x_1 \ldots x_n}$, it follows that if Φ is a random field, L_Φ will determine Φ uniquely within an equivalence. Thus for a random field Φ, L_Φ is a continuous positive definite functional with $L_\Phi(0) = 1$. It can be shown that for a "second-order" random field, the functional is proper, if and only if Φ_c is an inner product on U (dim. $U \ge 2$). If Φ is proper and uncorrelated, it is bounded. If U is a Hilbert space and Φ is bounded, there exists a unique vector $\mathbf{u}_\Phi \in U$ and two unique operators T_Φ^1, T_Φ^2 on U such that:

$$\left.\begin{array}{ll} \text{(i)} & \Phi_m(u) = < u, u_\Phi > \\ \text{(ii)} & \Phi_c(u, v) = < T_\Phi^1 u, v > \\ \text{(iii)} & \Phi_C(u, v) = < T_\Phi^2 u, v > \text{ for all } u, v \in U. \end{array}\right\} \tag{1.197}$$

These quantities are referred to as the "mean vector", the "covariance and correlation operators" respectively, for the functional Φ, (for proof see for instance Gudder [17]).

A certain group of random fields is sometimes important, which occurs when Φ has strongly independent values, if $< u, v >= 0$, which implies that:

$$E\{\Phi(u)\Phi(v)^*\} = E\{\Phi(u)\}E\{\Phi(v)^*\} . \tag{1.198}$$

Hence, $\Phi(u)$ and $\Phi(v)$ are stochastically independent.

Apart from the above generalization of stationary random processes to random fields another group is often of importance, i.e. Markov random fields and their equivalence to Gibbsian random fields. This equivalency has already been pointed out in previous work [23]. For the present purposes the generalization of a stochastic function $x_t(\omega, t), t \in T$ to a Markov random field can be achieved by the use of conditional probabilities and the corresponding distributions. This approach has been taken by Dobrushin [86], Averintsev [87] and Rozanov [88]. Thus following Dobrushin, consider x_t to be a random field, i.e. a random function parametrized by $t \in T^r$ taking values ξ_t in a finite set X. One can regard T^r as a finite dimensional "lattice" on which a domain $\mathcal{D}$ is a subset of T^r. Denoting this subset by $\mathcal{D} = \{t_1 \ldots t_n\} \subset T^r$, the required probability distribution can be written as:

$$P\{x_{t_1}(\omega) = \xi_{t_1}, \ldots x_{t_n}(\omega) = \xi_{t_n}\} = P_{\mathcal{D}}(\xi_{t_1} \ldots \xi_{t_n}) \,. \tag{1.199}$$

The distributions $P_{\mathcal{D}}(\cdot, \cdot)$ form a system corresponding to all finite subsets $\mathcal{D} \subset T^r$, satisfying the consistency conditions. This distribution is called by Dobrushin the distribution of the random field. If such a field is homogeneous, it can be characterized by the invariance of the distributions $P_{\mathcal{D}}(\cdot, \cdot)$ with respect to the group of translations in T^r. This means that for any shift vector $\tau \in T^r$ one has:

$$P_{\mathcal{D}}(\xi_{t_1}, \ldots \xi_{t_n}) = P_{\mathcal{D}+\tau}(\xi_{t_1+\tau}, \ldots \xi_{t_n+\tau}). \tag{1.200}$$

Introducing conditional probabilities and denoting the corresponding set of probability distributions by $Q = \{q_{\mathcal{D}, x(t)}(\xi_{t_1} \ldots x_{t_n}); \xi_{t_n} \in X, i = 1 \ldots n\}$ for each subset $\mathcal{D} \subset T^r$ and each function $x(t) = \xi_t, t \in T^r \setminus \mathcal{D}$ with values in X results in:

$$\left. \begin{aligned} &P\{x(t_1) = \xi_{t_1} \ldots \xi(t_n) = \xi_{t_n} | x(t), \mathbf{t} \in T^r \setminus \mathcal{D}\} \\ &= q_{\mathcal{D}, x(t)}(\xi_{t_1} \ldots \xi_{t_n}). \end{aligned} \right\} \tag{1.201}$$

Thus the system Q includes the conditional probabilities for the value of the process $x(t)$ on $\mathcal{D}$ under the condition that the values of the random field outside of this domain are known. If $\mathcal{D}_d$ denotes the set of points $(T^r \setminus D)$, d being an integer, the random field is called a d-Markovian field $(d > 0)$ in which:

$$q_{\mathcal{D}, x(t)}(\xi_{t_1} \ldots \xi_{t_n}) = q_{\mathcal{D}, x'(t)}(\xi_{t_1} \ldots \xi_{t_n}) \text{ for } x(t) = x'(t), \ t \in \mathcal{D}_d \,. \tag{1.202}$$

It can be shown that a complete chain of order d is a Markov random field with $r = 1$ (see Dobrushin [86]).

It is always assumed that for any fixed set $\mathcal{D}$ and $(\xi_{t_1} \ldots \xi_{t_n})$ the variable $q_{\mathcal{D}, x(t)}(\cdot, \cdot)$ is a measurable function of $x(t)$, if in the space of all functions $\{x(t), \ t \in T^r \setminus \mathcal{D}\}$, the σ-algebra $\mathcal{F}_{T^r \setminus \mathcal{D}}$ is used, which is generated by the following sets:

$$\left. \begin{array}{l} \{x(t) : \xi_{t_1} = \xi_1 \ldots, \xi_{n-1} \ ; \ \xi(t_n) = \xi_n \} \ , \\ \text{and} \\ \{t_1 \ldots, t_n \subset T^r \setminus \mathcal{D}\} \ ; \ \xi_1 \in X, \ \xi_n \in X \ . \end{array} \right\} \quad (1.203)$$

The above representation of a Markov random field is based on the notion of a regular lattice structure. There are however many problems in the stochastic mechanics of discrete media for which a more general representation may be required. Such a formulation should then be based on the random geometry in the physical domain and the notion of nearest neighbour interactions. Another aspect for the purpose of the stochastic mechanics formulation is the consideration of a multi-dimensional time space or the notion of distinguishing between a "macro" and "micro-time" scale at which the actual macroscopic and microscopic phenomena take place.

2 PHENOMENOLOGY OF DISCRETE MEDIA

2.1 Classification of Materials

(i) Introduction:

One of the main objectives of the stochastic mechanics of discrete media is the analysis of the behaviour of various classes of materials and the establishment of macroscopic relations that include microstructural effects. Conventionally, the description of the material behaviour is based on "continuum mathematical" models that generally refer to homogeneous media, ignoring thereby the non-deterministic influences of the microstructure on the overall behaviour of real materials. The realization of this shortcoming has led to the development of the probabilistic mechanics theory of discrete media presented in previous work (see [22, 23]). Thus, it is important to recognize that the probabilistic theory permits the formulation of the deformational or flow behaviour of discrete media without the use of additional physical parameters, which in general are unidentifiable, but nevertheless are introduced in the phenomenological models of materials. Another significant aspect of the continuum theoretical approach is the claimed "computational efficiency" of the models. This efficiency however depends largely on the assumption of "smoothness" and "continuity" of the computed fields. This view stems directly from the persistent belief that only continuum theory can adequately describe the material behaviour.

It is well-known that real materials have "defects", which induce a certain disorder in an otherwise ordered structure. The notion of "disorder" is however a larger one, since it is also brought about by the possible interactions between elements of the structure and the "spatial variation" of the local material characteristics. In the continuum mechanics approach, it is common usage to define a "representative volume element or RVE", that is small enough compared with the dimensions of a material sample, but much larger than the "length scale of disorder". Hence on the assumption that the material is "statistically homogeneous", i.e. that the local material properties are constant when averaged over an RVE, one can replace the actual material by a homogeneous one in which the local characteristics are identical to the averages over the RVE's.

This identification constitutes the "homogenization process", which is used to calculate the "macroscopic material properties".

By contrast the stochastic mechanics of discrete media considers from the outset, in order to take account of the disorder effects as well as the inherent random nature of the physical and configurational characteristics, the field quantities as random variables or functions of such variables. The importance of this approach, as recent developments indicate is mainly due to the possibility of measuring relevant microstructural parameters and thus obtaining quantitative data on the structure to advance further the determination of microstructural effects on the overall behaviour of discrete media. Another perspective of the stochastic mechanics is the significance of connecting the statistical analysis to the quantitative description of the microstructures. In this context two main morphological parameters become important, i.e. higher order correlation functions and moments. These parameters can be established from experimental observations and/or "image analysis". They can also be assessed by means of appropriate "random models" of the structures. This will be discussed further in other sections of this text. Image analysis has become recently an extremely useful tool in the "simulation" of the behaviour of various classes of structured media. Another class of such materials to be considered later in this volume is that of "structured fluids". There is a great variety of substances that are designated as "complex fluids", which differ remarkably in their response to the application of a stress and for which the "viscosity" is a quantitative measure of their response. In general structured fluids have distinctive properties, which originate from the large "polyatomic structure" that is many times larger than that of a "simple molecule". Condensed-matter science has focused in the past on small-molecules of solids and liquids. Structured fluids may respond like solids or liquids depending on the time-scale or how strongly they are perturbed. Their properties generally do not comply with the conventional notion of solidity or liquidity. Many of their characteristics can be attributed to large scale statistical features of the constituents of the polyatomic structure. These properties can often be better understood by means of a stochastic theory.

Most of the concepts required in the stochastic mechanics approach have already been given previously [23] as well as in Chapt. 1. In the development of a more extensive analysis, it is necessary however to introduce some additional notions and definitions, which will be considered in this chapter.

(ii) Classification of Microstructures:

The great variety of discrete media makes a classification rather difficult, if not impossible. An attempt has been made in [22, 23] to group certain types of materials and hence in the following analysis certain representatives of these "classes of discrete materials" will be dealt with. It is further apparent that such an analysis must be restricted due to the desired size of this volume. In

order to illustrate the discreteness and random nature of some real materials
a few examples are given below. Thus a crystalline solid is represented by the
transmission-electron micrograph of a silicon-steel (magnification 5700X) that
has been subjected to a compression load at $\sim 800°$ C. It indicates clearly the

Fig. 1. TEM-micrograph of Silicon-Steel (magnification 5700X)

existing elements of the structure (grains) separated by the grain boundaries
(Fig. 1). A compound material consisting of Tungsten-Carbide particles embedded in a Cobalt-matrix is shown by the TEM-micrograph (magnification
1500X) in Fig. 2. The material reveals the randomness of the particles (average

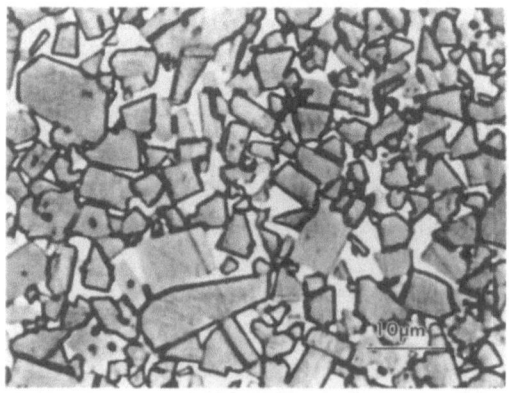

Fig. 2. TEM-micrograph of Tungsten-Cobalt compound (magnification 1500X)

size $\sim$ 2–4 μm) in the matrix. This composite belongs to a wider class of discrete solids known as "high-temperture" materials and includes Boron-Nitride (BN), Silicon-Nitride ($Si_3 N_4$), Silicon-Carbide, Zirconium-alloy etc. (Fig. 3).

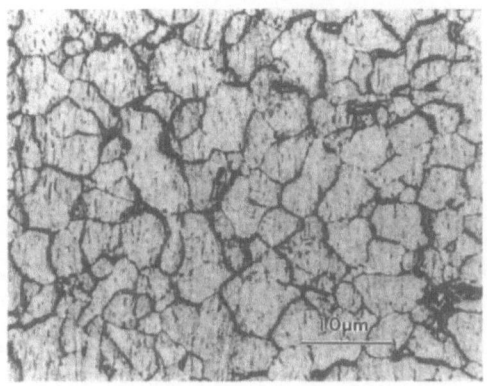

Fig. 3. Scanning-Electron (SEM)-micrograph of Zirconium-Alloy (magn. 1700X)

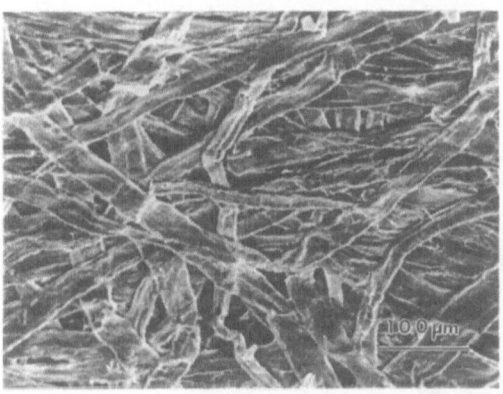

Fig. 4. SEM-micrograph of a fibrous network (magn. 170X), (sulphite-paper)

A typical Scanning-Electron micrograph of a cellulose structure is shown in Fig. 4. It represents a strip of "beaten sulphite" paper, bleached and dried as it occurs in the paper manufacturing. Fibrous materials of this type have been studied extensively in this Laboratory particularly with regards to the effects of the microstructure on the overall behaviour of cellulose and polymeric networks (see also Chapt. 4).

Fig. 5. Microstructure of a Kaolin-Water compound (soil) SEM-micrograph (magnification 4800X)

Another discrete structure is the well-known Kaoline-Water compound (soil) as shown above in Fig. 5. It reveals clearly the complexity of the spatial arrangement of the clay particles (~ 1 μm) and the existing voids between them, that are filled with water.

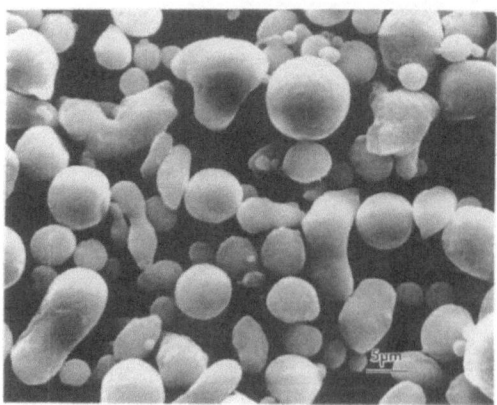

Fig. 6. Microstructure of Al-Glycol compound, SEM-micrograph (magn. 2000X)

A material, which is very useful in the study of the behaviour of "porous media" and the analytical simulation of this class of materials is shown in Fig. 6.

It indicates a SEM-micrograph of Aluminium particles ($\sim$ 3 μm dia.) which are dispersed in a diglicidyl-ether matrix. The flow through such a medium is discussed in Chapt. 4.

The classification of "structured fluids" is even more difficult. Such media have distinctive and unusual characteristics due to their large polyatomic structures. These complex fluids exhibit a variety of mechanical responses in contrast to "simple liquids" that have rather a smaller molecular structure. The latter consists of compact molecules which strongly interact only with their immediate neighbours. The "length scales" at which significant changes of the molecular structure during flow may occur are of the size of a molecule, i.e. a few Angstroms. But there is also a characteristic "time-scale" for the molecular motion, which is comparable to the time-interval between successive collision of a molecule with its neighbours. It is the "time and length scales" that are responsible for the similar flow behaviour of simple fluids. The stochastic mechanics approach concerning simple fluids and of more complex ones will be considered in Chapt. 4.

(iii) Idealized microstructures and fundamental concepts:

The micrographs of actual materials shown in Figs. (1–6) clearly indicate the complexity of their microstructure. Evidently, due to the geometrical and spatial arrangements of the latter, the analytical treatment will require some simplification or an "idealization". As already mentioned due to the limited scope of the present study, the characteristic groups of materials only, shown below can be considered. Thus the class of "polycrystalline solids" is shown schematically in Fig. 7(a).

It is seen from this figure that there exist a great number of "interfaces" (grain boundaries), which play a significant role in the description of the overall behaviour of the solid. In particular they affect the breakdown of the initial structure, when an external field is applied that induces certain critical internal stresses and interactions leading to the formation and propagation of "microcracks" and ultimately to the failure of the structure. Fig. 7(b) represents an idealization of a "binary structure" which often consists of a dispersion of "hard particles" in a softer or ductile matrix. Similar schematics can be used to characterize "multi-component systems" such as those mentioned before (high-temperature materials). Another idealization concerns "granular media" indicated by Fig. 8(a).

These materials are also complex in that their interactions include sliding and rolling of the elements with respect to each other and requires the introduction of a "generalized contact force" in the analytical treatment. Such materials are significant in geotechnical applications. The schematics of a "simple fluid" (idealized) is shown in Fig. 8(b), while Fig. 8(c) characterizes a "multi-phase fluid" in which the molecules may be of different sizes. A small region of fibrous

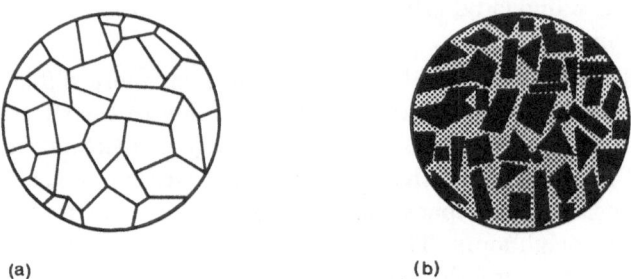

(a)   (b)

Fig. 7. Idealized microstructures of a polycrystalline and a binary structured solid.

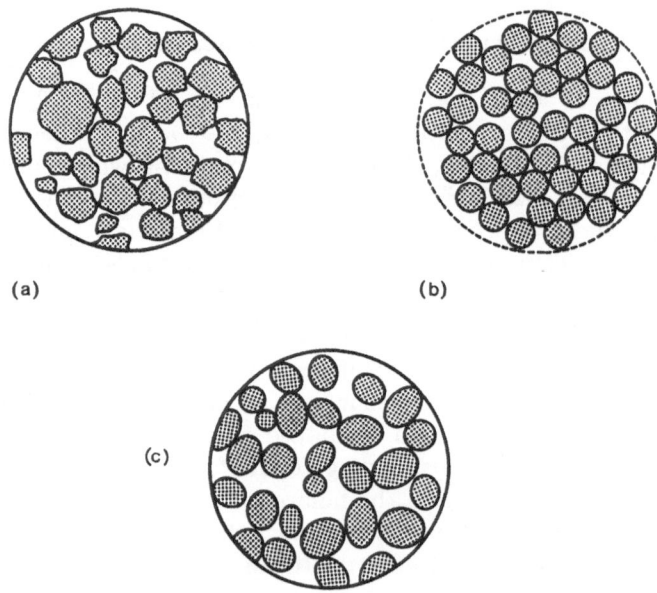

Fig. 8. Idealization of granular solids, simple fluids and multi-phase fluids.

and that of a polymeric material is indicated in Fig. 9(a),(b). The latter shows clearly the "entanglements" between longer chain molecules and the shorter ones of a second polymer which are mixed in a blend (see also Chapt. 4).

(a)   (b)

Fig. 9. Small regions of fibrous and polymeric structures.

Analogously to continuum mechanics the development of the stochastic mechanics of discrete media is based on certain principles that are briefly discussed below. The fundamental notions of probabilistic mechanics have been given previously in refs. [22, 23] in the form of postulates. They remain valid in the present analysis and hence are briefly re-stated here together with some additional definitions which are subsequently required. There are four postulates as follows:

P.1. : The stochastic mechanics of discrete media employs three measuring scales in which the smallest refers to an element of the structure. An intermediate scale is then used, called "meso-domain" that contains a statistical ensemble of structural elements. A finite number of meso-domains (nonintersecting) forms the macroscopic material body.

P.2. : All relevant field quantities pertaining to a structural element are regarded as "random variables" or functions of such variables.

P.3. : Internal stresses, strains, rates of strain, etc. are used in a "generalized form" to permit the inclusion of interaction forces, body forces, etc. in the analysis of the material response behaviour.

P.4. : Material functionals or "material operators" are employed in order to formulate "constitutive relations" for a specific material and given conditions.

In view of these postulates some remarks may be indicated.

(i) Although P.1. stresses the finiteness of a structural element, it is in certain cases convenient to adopt a somewhat larger unit as a primative base in the analysis. Such a unit will then be referred to as a "unit cell (U.C.)" or a "microdomain Γ". The usefulness of such structural units will be demonstrated when dealing with composite structures (see Chapt. 4).

(ii) Due to the recognition in P.2. of the field quantities as random variables or functions of such variables the material behaviour can be represented in terms of an "abstract dynamical system" $[X, \mathcal{F}, \mathcal{P}]$. In this representation, it is assumed that X is formed by a set of admissible "state-vectors" that describe the states of the structural elements or micro-domains as the case may be, during the deformation or flow of the material. Hence the "state-space" is a probabilistic function space or a subspace of the latter.

(iii) P.4. merely stipulates an "operator formalism" for the stochastic analysis instead of the conventional constitutive relations of continuum mechanics. This formalism permits the use of certain "observable parameters" in the constitutive relations which cannot be otherwise introduced in the former.

In general a set of "structural elements" is regarded as a "disordered set". Before discussing this type of sets it may be indicated to briefly review some basic definitions (see also [23]) that are required in the stochastic mechanics of discrete media.

Def. 1:

A "structural element" or "micro element" $^\alpha A \in \{^\alpha A\}$ (a set of mathematical manifolds). Each member of this set represents a microelement. Frequently microelements will be denoted simply by $\alpha, \beta \ldots$

As mentioned before, it is often necessary to use a somewhat larger unit than a microelement. Hence one can define a "microdomain" as follows:

Def. 2:

A "microdomain Γ" also referred to as a "unit cell (U.C.)" consists of a fixed number of structural elements $^\alpha A$. It has a larger volume than that of $^\alpha A$.

Since in continuum theory the notion of material points is employed, the latter can be defined in the present context by:

Def. 3:

A "material point $^\alpha a$" is defined as an element of $^\alpha A$, i.e.: $^\alpha a \in {}^\alpha A \in \{^\alpha A\}$. Thus $^\alpha a$ is a mathematical representation of an element or "unit" that is smaller than a microelement.

"It may refer to a molecule or atom depending on the requirements and scale used in the analysis. In considerations of the kinematics of discrete media the notion of "reference frames" is important. Hence

Def. 4:

$k \in \{K\}$: is a set of reference frames k, each of which is fixed within an individual microelement α or parts of it. The same definition applies to a microdomain Γ.

The Euclidean space $\mathbb{R}^3$ can be related to the k-frames by time-dependent orientation transformations.

Def. 5:

A "configuration" is the image of an element α at time t; $\mathbf{r}(^\alpha A, t) \in \mathbb{R}^3$ or briefly :$^\alpha \mathbf{r} \in \mathbb{R}^3$.

For the smaller elemental unit, i.e. the material point $^\alpha a$, one can speak of a "location" where:

Def. 6:

A "location" is the image of $^\alpha a$ at time t relative to the local frame $k \in \{K\}$;$^\alpha \mathbf{r}(^\alpha a, k, t) \subset \mathbb{R}^3$. Hence configurations generally refer to microelements and locations to material points within the elements.

It is to be noted that later in dealing with the molecular dynamics of simple fluids and those of a more complex structures, the above definitions (5,6) will be given in a somewhat different form.

Def. 7:

The "motion of an element" is defined by: $^\alpha A \triangleq \{\mathbf{r}(^\alpha a, k, t); -\infty < t < \infty\}$ in the discrete and time-continuous case. In stochastic mechanics these motions are considered as "stochastic processes $\{x_t\}$" defined for each microelement $^\alpha A$ or for a "material point $^\alpha a$ as the case may be.

Def. 8:

In accordance with postulate P.1 there are three measuring scales. Thus a "mesodomain M" is defined as a countable set of microelements: $M = \{^\alpha A\}$. Each member of the set of mathematical manifolds $\{M\}$ represents a meso domain. A "macro domain $\mathcal{M}$" is the union of disjoint meso domains and is the mathematical manifold representing the macroscopic material body. Two other definitions concerning the volume and mass density of these domains can be given.

Def. 9:

The "volume of a microelement $^\alpha v$" at time t is defined by:

$$(i) \quad ^\alpha v =^\alpha v(^\alpha A, t) = \int_{\mathbf{r}(^\alpha A, t)} d^3 \mathbf{r}; \quad ^\alpha v_0 \equiv v(^\alpha A, 0) = \int_{\mathbf{R}(^\alpha A, 0)} d^3 \mathbf{R}$$

in which $\mathbf{r}, \mathbf{R}$ refer to the current and initial positions of α in the Euclidean frame, respectively.

Similarly the "volume of a meso domain $^M v$" at time t:

(ii) $^M v \equiv v(M, t) = \int\limits_{\mathbf{r}(M,t)} d^3\mathbf{r}; \quad ^M v_0 \equiv v(M, 0) = \int\limits_{\mathbf{R}(M,0)} d^3\mathbf{R}$

and the volume of the macrodomain at time t:

(iii) $V \equiv^{\mathcal{M}} v = v(\mathcal{M}, t) = \int\limits_{\mathbf{r}(\mathcal{M},t)} d^3\mathbf{r}$ and by (i) and (ii):

$$V = \sum_{\{\mathcal{M}\}} v = \sum_{\{\mathcal{M}\}} \sum_{\{A\}} {}^\alpha v.$$

Def. 10:

The mass density of a "microelement $^\alpha A$" or α is defined by:

$^\alpha\mu \in M$: $^\alpha\mu$ is a scalar element of the set M representing the mass of $\alpha \in \{^\alpha A\}$; $^\alpha\rho \in \mathbb{R}$: $^\alpha\rho$ is a scalar element of the set $\mathbb{R}$ representing the mass density of $\alpha \in \{^\alpha A\}$ defined as $^\alpha\rho \, ^\alpha\mu/^\alpha v$.

Equivalently in terms of the Lebesgue-Stieltjes integral the mass density is defined by:

$$^\alpha\mu = \int\limits_{\mathbf{r}(\alpha,t)} {}^\alpha\rho d^3\mathbf{r} = {}^\alpha\rho \int\limits_{\mathbf{r}(\alpha,t)} d^3\mathbf{r}$$

Hence, accordingly the mass density of a meso domain is given by:

$$^M\mu = \int\limits_{\mathbf{r}(M,t)} {}^M\rho d^3\mathbf{r} = {}^M\rho\, ^M v = \sum_{\{^\alpha A\}} {}^\alpha\rho\, ^\alpha v$$

and thus the "mean density":

$$^M\rho = \int\limits_{X_\rho} \rho d^M\mathcal{P}(\rho); \int\limits_{X_\rho} d^M\mathcal{P}(\rho) = 1$$

in which the Lebesgue-Stieltjes integral extends over the subspace $X_\rho \in X$ on $\mathcal{M}$, which is embedded in the general state-space X and where $\mathcal{P}(\rho)$ is the Lebesgue measure on X_ρ. Similar definitions apply to the "macro domain $\mathcal{M}$". In the formalism of molecular dynamics to be discussed subsequently, the quantities defined by Def. (9,10) can be interpreted in a somewhat different manner (see Chap. 4). Most significant for the development of the stochastic mechanics of discrete media is the structure of the "state-space" and the definition of the measure $\mathcal{P}$. Thus the following definitions are given:

Def. 11:

    (i) the usual state-space denoted by Z is identified with a probabilistic function space X.

    (ii) The "state-vector $^{\alpha}\mathbf{z} \in Z$ or X" represents a set of r-parameters describing the mechanical states $i = 1 \ldots r$ of an element α of the material.

        Thus $^{\alpha}\mathbf{z}$ is an "outcome or elementary event" in the probability space X as a result of the "statistical experiment $^{\alpha}A$".

    (iii) The "event E" is a set of state vectors within an experimental range $\Delta\mathbf{z}$ of measurements such that:

$$E = \{^{\alpha}\mathbf{z}^i : \mathbf{z}^i <^{\alpha} \mathbf{z}^i < \mathbf{z}^i + \Delta\mathbf{z}^i; i = 1 \ldots r\}; E \in \mathcal{F}$$

        where $\mathcal{F}$ is the σ-algebra of "open spheres" or events of the set of state-vectors.

    (iv) $\mathcal{P}$ is the probability measure of the events E.

    (v) The triple $\mathcal{B} = \mathcal{B}[X, \mathcal{F}, \mathcal{P}]$ defines an "abstract dynamical system", where $\mathcal{B}$ usually is a topological vector space.

It is to be noted that the components of the state vector $^{\alpha}\mathbf{z}$ in general may be scalar or higher dimensional functions. For simplification of the analysis one may use one or more components of the state vector and thus attention must be given to the corresponding subspace of Z or X. In this case the "event structure" must be identified accordingly together with the appropriate probability measure. This will be further discussed in section 2.2 of this chapter dealing with the state-space representation of material systems.

2.2 Statistical Models of Discrete Media:

It is apparent from the preceding section that real materials exhibit a certain "disorder" and hence are characterized by the lack of a conventional geometrical order. The term disorder however has a broader meaning in that it includes also effects caused by the interactions between individual elements and certain internal mechanisms which contribute significantly to the disorder of the microstructure. Thus some remarks on this notion will be indicated.

(i) Disorder effects:

In order to distinguish between disorder effects, it becomes necessary to consider the "geometrical scales" of their occurrence and their respective influences on the overall microstructure of the medium. In polycrystalline solids for instance "microscopic disorder" refers to vacancies, inclusions, dislocations etc. occuring in individual crystals or elements of the structure and are dealt with at an atomic scale. Interaction effects at grain boundaries, i.e. bonding or debonding have to be considered also at this scale. For an ensemble of such elements in accordance with the given definitions a somewhat larger scale, i.e. the mesoscopic scale is appropriate. In the conventional analysis of "disorder effects" based on continuum theoretical considerations, the usual methods consist in the application of the "mean field" approach, the "local" approach and in the case of critical phenomena of "renormalization" methods. Generally, the mean field approach neglects the details of the spatial arrangement surrounding a "volume element" as defined by continuum theory. Hence accordingly one considers a "fictitious homogeneous" medium in which each volume element interacts equally with the others. In phenomena concerning solids or composite structures this notion leads to the necessity of introducing equivalent material characteristics or the concept of "effective material properties" of heterogeneous materials (see for instance Hashin [89], Beran [90], Kunin [91], and others). Consequently the presence of actual interactions between elements is neglected. This approach may be regarded as a sufficiently good approximation for the case of long-ranged or weak interactions or when the medium is not strongly sensitive to disorder effects. In the case of discrete fluids for instance, where the interparticle-pair interactions are in general considered in terms of a two part potential, i.e. a short-ranged "repulsive or core potential" and a long-ranged "attractive potential", the variance of the probability distribution does not decrease with a scaling correlation parameter and "fluctuation effects" cannot be neglected even for long ranged forces (see also Kac [92], Percus [93], Kac and Logan [94]).

(ii) The local approach:

From the point of view of stochastic mechanics the "local" approach is the more significant. In this type of analysis, one attempts in general to describe the behaviour of the elements of the microstructure at the molecular (atomic) level by approximating their real behaviour as closely as possible. By choosing appropriate interactions between the elements, one can then employ the methods of "molecular dynamics" to establish the sought macroscopic material characteristics. Several recent publications concerning disorder effects in discrete solids and the hydrodynamics of dispersed media should be mentioned here (see [95], [96], [97]). In accordance with the first postulate of stochastic mechanics, the "local" modelling is aimed at an approximation of the "local fields" within each microelement of the structure. In many cases a simplification is achieved by considering the element as a "continuum". The second step consists of establishing the distribution functions of the relevant field quantities over a mesoscopic domain. This then results in a set of distribution functions valid over the macroscopic material domain, which can be evaluated in accordance with the definitions (8, 9) of section (2.1).

It is to be noted, however, that in almost all practical applications, the required calculations include the use of "geometrical probabilities" to account for the given stochastic geometry of the medium. This will be further discussed later in the text. So far as the "critical behaviour" of discrete media is concerned, it exploits "self-similarity" (i.e. renormalization) and deals with the formulation of the invariance of some relevant physical quantities under a change of length scale. The continuum modelling of discrete media entails the description and history of the deformation and flow of such media at each mathematical point of the continuum, based on the assumption, that the involved functions are continuous and that only small spatial and temporal gradients occur. As mentioned before the representative volumes or RVE's, represent broadly speaking the smallest volumes of the discrete material, which responds as a continuum. The size of such "elementary volumes" depends on the choice of the level of difference between the statistical moments of a given field, when computed for the macro continuum and the RVE. In this context the "local representation" requires a "discretization" of the continuum field. This is often done in the mathematical analysis by introducing a "lattice structure" which replaces the otherwise "disordered" or only "partially ordered" given material geometry.

(iii) Molecular dynamics and correlation functions:

The molecular dynamics and lattice models of discrete media are based on the statistical mechanics of "particle systems" and their Hamiltonian representation. They are characteristic of the local approach and can be carried out either on the atomistic or the molecular level of the medium [93], [98]. This type of modeling is often used for discrete media such as simple fluids, i.e. for media that are primarily composed of "spherical molecules" which are chemically inert. Typical for this class of fluids are liquified rare gases and alkali metals. Their behaviour can be modelled on the assumption that the occuring interactions are due to interatomic potentials which are symmetric. However, the well-known Monte-Carlo method provides a larger range of applications particularly with regard to the "simulation" of the response of discrete media.

From a thermodynamics point of view, the properties of a molecular system can be expressed by the averages of certain functions of the coordinates and momenta of the molecules contained in that system. At equilibrium these averages are independent of time (see also Chapt. 1). Thus assuming for simplicity that all molecules are identical and have three degrees of freedom each, as for instance in a one-component monatomic fluid, the total energy in the absence of an external field is given by the Hamiltonian:

$$\mathcal{H} = \sum_{\alpha=1}^{N} \frac{1}{2\mu}|^{\alpha}\mathbf{p}(t)|^2 + U_N(^{\alpha}\mathbf{r}(t)\ldots {}^N\mathbf{r}(t)) \tag{2.1}$$

in which $^{\alpha}\mathbf{r}\ldots{}^N\mathbf{r}, {}^{\alpha}\mathbf{p}(t)\ldots{}^N\mathbf{p}(t)$ are the position and momentum vectors of the molecules or particles $\alpha = 1\ldots N, {}^{\alpha}\mu \equiv \mu$, respectively and U_N the potential energy due to their mutual interactions.

If for given initial conditions, i.e. the values of the 6N coordinates and momenta at time $t = 0$ can be specified, one can in principle determine the positions at a later time by means of 3N coupled differential equations of the form:

$$^{\alpha}\dot{\mathbf{p}} = \mu^{\alpha}\ddot{\mathbf{r}} = -\frac{\partial\mathcal{H}}{\partial^{\alpha}\mathbf{r}} = {}^{\alpha}\boldsymbol{\nabla} U_N(^N\mathbf{r}). \tag{2.2}$$

Hence, if $f(^N\mathbf{r}, {}^N\mathbf{p})$ is a function of the 6N phase-space coordinates and $\bar{f}$ the corresponding thermodymanic quantity, it can be represented by the statistical average or:

$$< f >_E = < f(^N\mathbf{r}, {}^N\mathbf{p}) > \tag{2.3}$$

Since from Boltzmann's theory (Chapt. 1) this average is taken as an average over time, then:

$$< f >_t = \lim_{\tau\to\infty} \frac{1}{\tau} \int_0^{\tau} f\{^N\mathbf{r}(t), {}^N\mathbf{p}(t)\}dt \tag{2.4}$$

In this manner one can determine for example the temperature of an "isolated system" from the time-averaged kinetic energy of the medium. For an isolated system the total energy is conserved, but the instantaneous kinetic energy will fluctuate with time (see also Kac and Logan [94]). The magnitude of these fluctuations is related to the specific heat. Similarly, one can determine the pressure in the fluid contained in a certain volume by the use of Clausius' virial function, i.e.:

$$\mathcal{V}(^N\mathbf{r}) = \sum_{\alpha=1}^{N} {}^{\alpha}\mathbf{r} \cdot {}^{\alpha}\mathbf{F}, \tag{2.5}$$

where $^N\mathbf{r}$ is the set of N-position vectors and $^{\alpha}\mathbf{F}$ the total force acting on the αth molecule or particle. This force originates from the interaction potential $\phi(\mathbf{r})$ in the absence of an external force field. It can be readily shown that

$< \mathcal{V}(\mathbf{r}) >_t = -3Nk_BT$, where k_B is Boltzmann's constant and T the temperature. Another way of carrying out the averaging can be done in terms of the Gibbs ensemble mentioned earlier. An ensemble in statistical mechanics is conventionally considered as a large number of replicas of the actual system, that may in general be infinite, which are all characterized by the same macroscopic parameters occupying different "microstates" and where the particles have different positions and momenta. This description is based on the distribution of representative "phase points" in a phase space Γ. Denoting the corresponding probabilty function by $P^{(N)}$, then:

$$P^{(N)}(^N\mathbf{r},{}^N\mathbf{p},t)d^N\Gamma \equiv P^{(N)}(^N\mathbf{r},{}^N\mathbf{p},t)d^N\mathbf{r}d^N\mathbf{p} \tag{2.6}$$

representing the probability of finding the system at time t in a microscopic state characterized by a phase point in Γ (6N-dimensional space). The infinitesimal volume in this space is:

$$d^N\Gamma = \prod_{\alpha=1}^{N} d^\alpha\mathbf{r}d^\alpha\mathbf{p} \tag{2.7}$$

Under equilibrium conditions this probability function is time independent and is designated by $P_0^{(N)}(^N\mathbf{r},{}^N\mathbf{p})$. Hence the statistical average of any function $f(\mathbf{r},\mathbf{p})$ in terms of the ensemble average is given by:

$$< f >_E = \int \int f(^N\mathbf{r},{}^N\mathbf{p})P_0^{(N)}(^N\mathbf{r},{}^N\mathbf{p})d^N\mathbf{r}d^N\mathbf{p} \tag{2.8}$$

with $P_0^{(N)}$ normalized, or:

$$\int \int P_0^{(N)}(^N\mathbf{r},{}^N\mathbf{p})d^N\mathbf{r}d^N\mathbf{p} = 1.$$

Hence the internal energy U can be expressed in terms of the average Hamiltonian of the system by:

$$U = < \mathcal{H} >_E = \int \int \mathcal{H}(^N\mathbf{r},{}^N\mathbf{p})P^{(N)}(^N\mathbf{r},{}^N\mathbf{p})d^N\mathbf{r}d^N\mathbf{p}. \tag{2.9}$$

The explicit form of the probability function depends on the type of ensemble considered in the averaging procedure (see also Zubarev [99], Balescu [100], Ishi [101], Prigogine [102] among others).

It is well-known that there are four types of ensembles in statistical mechanics, i.e. the "canonical" in which the corresponding macroscopic parameters are (N,V,T) or the number of particles in the volume V, the isothermal-isobaric ensemble with (N,P,T), (P being the pressure), the grand-canonical and micro canonical ensembles. Of special interest in the present analysis is the canonical ensemble representing a collection of systems or subsystems that contain a fixed number of particles or molecules ($\alpha = 1 \ldots N$) under the same conditions as the macroscopic volume and temperature. Assuming identical particles, the equilibrium probability function for such an ensemble is given by:

$$P_0^{(N)}(^N\mathbf{r}, {}^N\mathbf{p}) = \frac{1}{N!}\frac{1}{h^{3N}}\frac{1}{Q_N(V,T)} \exp[-\beta\mathcal{H}(^N\mathbf{r}, {}^N\mathbf{p})] \tag{2.10}$$

where h is the Planck constant and the factor $N!$ is due to the assumption that all particles are indistinguishable. The quantity $Q_N(V,T)$ is a normalizing factor known as the "partition function", which is defined by:

$$Q_N(V,T) = \frac{1}{N!}\frac{1}{h^{3N}} \int \int \exp[-\beta\mathcal{H}(^N\mathbf{r}, {}^N\mathbf{p})]d^N\mathbf{r}\,d^N\mathbf{p} \tag{2.11}$$

The multiplication constant $(h^{3N})^{-1}$ ensures in relations (2.10, 2.11) that Q_N and $P_0^{(N)}$ are dimensionless and assume the corresponding expressions of quantum statistical mechanics. The above equations and assumption of the system to consist of indistinguishable particles is usually employed in the considerations of simple fluids. Another quantity of interest here is the "configuration integral Z_N" or:

$$Z_N(V,T) = \int_V \exp[-\beta U_N(^N\mathbf{r})]d^N\mathbf{r} \tag{2.12}$$

where the integration extends over each position vector $^\alpha\mathbf{r}$; ($\alpha = 1 \ldots N$) within the volume V containing the ensemble. From thermodynamics both these quantities are related by:

$$Q_N(V,T) = \frac{Z_N}{N!\lambda^{2N}} \tag{2.13}$$

in which λ is the Broglie thermal wave length, i.e. $[\frac{2\pi h^2}{\alpha_\mu}]^{1/2}$ and β from before, is $\beta = [k_B T]^{-1}$, (see also Hirschfelder, Curtiss and Bird [103], Boon and Yip [104] and others).

Although the equilibrium probability functions permit a rigorous analysis of the molecular dynamics of discrete fluids, it is often sufficient in the calculation of equilibrium properties to employ lower order distributions or densities. This is particularly so in the case of one-component monatomic fluids. Thus, if such a fluid consists of $\alpha = 1 \ldots N$ identical molecules, relation (2.10) can also be written as:

$$P_0^{(N)}(^N\mathbf{r}, {}^N\mathbf{p}) = (\frac{\lambda}{h})^{3/2} \exp[-\beta K_N(^N\mathbf{p})]Z_N^{-1} \exp[-\beta U_N(^N\mathbf{r})] \tag{2.14}$$

in which $K_N(^N\mathbf{p})$ denotes the kinetic energy term of the Hamiltonian in (2.10). The probability function can be factorized into 3N-independent Maxwellian distributions for the components of the momenta and into a probability function for the coordinates. However the latter does not in general separate due to the correlation between particle positions. Hence using the probability of simultaneously finding molecule 1 in the volume $d^1\mathbf{r}$ around $^1\mathbf{r}$ and molecule 2 in $d^2\mathbf{r}$ around $^2\mathbf{r}$, one has:

$$P_N^{(N)}(^1\mathbf{r}\dots{}^N\mathbf{r})d^1\mathbf{r}\dots d^N\mathbf{r} = Z_N^{-1}\exp[-\beta U_N(^1\mathbf{r}\dots{}^N\mathbf{r})]d^1\mathbf{r}\dots d^N\mathbf{r} \qquad (2.15)$$

This leads to the notion of a multi or n-body probability density $P_N^{(n)}$ obtained from $P_N^{(N)}$ by integration over the coordinates of $(N-n)$ particles so that:

$$P_N^{(n)}(^1\mathbf{r}\dots{}^N\mathbf{r}) = \int\dots\int P_N^{(N)}(^1\mathbf{r}\dots{}^N\mathbf{r})d^{n+1}\mathbf{r}\dots d^N\mathbf{r}. \qquad (2.16)$$

Thus the mean value of a function f of the n-particle coordinates becomes:

$$< f(^n\mathbf{r}) > = Z_N^{-1}\int\exp[-\beta U_N(^N\mathbf{r})]f(^n\mathbf{r})d^N\mathbf{r}$$

$$= \int P_N^{(n)}(^n\mathbf{r}')f(^n\mathbf{r}')d^n\mathbf{r}' \qquad (2.17)$$

Introducing the mutual distances between n-particles, (2.16) can formally be written as:

$$P_N^{(n)}(^1\mathbf{r}\dots{}^N\mathbf{r}) = < \delta(^1\mathbf{r} - {}^1\mathbf{r}')\dots\delta(^n\mathbf{r} - {}^n\mathbf{r}') > . \qquad (2.18)$$

In the limit when the distances between the particles $|^\alpha\mathbf{r} - {}^\beta\mathbf{r}|$ or $|\mathbf{r} - \mathbf{r}'| \to \infty$ for all $1 \le \alpha, \beta \le n$, $P_N^{(n)}$ can be approximated by the product of the n-single particle probability density, i.e.:

$$P_N^{(n)}(^1\mathbf{r}\dots,{}^N\mathbf{r}) \simeq P_N^{(1)}(^1\mathbf{r})\dots P_N^{(1)}(^N\mathbf{r}) \qquad (2.19)$$

where in this limit the position of each particle $\alpha,(\alpha = 1\dots N)$ is independent of the positions of the $(n - 1)$ particles.

Hence one can define an "n-particle" distribution function as follows:

$$G_N^{(n)}(^1\mathbf{r}\dots,{}^n\mathbf{r}) = \frac{P_N^{(n)}(^1\mathbf{r}\dots,{}^n\mathbf{r})}{\prod\limits_{\alpha=1}^{n} P_N^{(1)}(^\alpha\mathbf{r})}. \qquad (2.20)$$

This indicates that $G_N^{(n)}(^n\mathbf{r}) \to 1$ for all n, when the mutual distances between the n-particles increase to infinity (thermodynamic limit). Of special interest here are the "single-particle" and the "pair-distribution" function. The latter can be deduced by noting that the potential energy U_N arises from the mutual interactions so that eqn. (2.20) can also be written as:

$$G_N^{(n)}(^n\mathbf{r}) = U_N^{(n)} P_N^{(n)}(^n\mathbf{r}) \tag{2.21}$$

and hence the pair-distribution can be expressed by:

$$G_N^{(2)}(^\alpha\mathbf{r},{}^\beta\mathbf{r}) = U^{(2)} P_N^{(2)}(^\alpha\mathbf{r},{}^\beta\mathbf{r}); \quad (\alpha = 1\ldots N) \tag{2.22}$$

Equivalently the above function is also given by:

$$G_N^{(2)}(^\alpha\mathbf{r},{}^\beta\mathbf{r}) = Z_N^{-1} U^{(2)} \int \ldots \int \exp[-\beta U_N(^N\mathbf{r})] d^1\mathbf{r} \ldots d^{(\alpha-1)}\mathbf{r} d^{(\alpha+1)}\mathbf{r} \ldots d^N\mathbf{r};$$
$$(\alpha = 1\ldots N) \tag{2.23}$$

Evidently for an isotropic system, one has:

$$G_N^{(2)}(^\alpha\mathbf{r},{}^\beta\mathbf{r}) = G(|^\alpha\mathbf{r} - {}^\beta\mathbf{r}|). \tag{2.24}$$

The function $G(\mathbf{r})$ is usually referred to as the "radial distribution function" and is used in the molecular dynamics of particle systems that are subjected to central force pair interactions. In this case the potential energy for the ($\alpha = 1\ldots N$) particles and the system Hamiltonian become:

$$\left.\begin{array}{l} U_N(^N\mathbf{r}) = \sum_{\alpha<\beta}^N \phi(^{\alpha\beta}\mathbf{r}) = \dfrac{1}{2}\sum_{\alpha\neq\beta}^N \phi(^{\alpha\beta}\mathbf{r}) \\[3mm] \mathcal{H} = \displaystyle\sum_{\alpha=1}^N \dfrac{{}^\alpha\mathbf{p}^2}{2\mu} + \dfrac{1}{2}\sum_{\alpha,\beta}{}'\phi(^{\alpha\beta}\mathbf{r}); \\[3mm] \text{(where } \Sigma'\text{: no summation over indices)} \end{array}\right\} \tag{2.25}$$

The molecular dynamics method can be regarded as a direct method for the simulation of the material characteristics (see also Hansen and McDonald [105], Boon and Yip [104] among others). In particular in the case of solids, if it is adapted to this level of description (see for instance MacPherson [106]), requires however for "covalent bonding" between the structural elements an appropriate choice of the interaction potential. The latter becomes rather less important in considerations of the behaviour of metallic structures. Closely related to the molecular dynamics method is the "Monte-Carlo" technique frequently used in the simulation of the material behaviour. It applies essentially to stationary processes of a system of N-interacting particles or microelements with a specified interaction potential. The system is then given a set of arbitrarily chosen intitial coordinates and a sequence of configurations of the elements is generated by admitting random displacements to occur. However, a decision must be made whether a particular configuration is acceptable or must be rejected, since the obtained results must comply with the configurations corresponding to an as yet unknown probability density function of the ensemble. The particular method in the Monte-Carlo simulation usually employed in liquid state physics goes back to Metropolis *et al* [107]. It consists of generating a set of "molecular configurations $\mathcal{A} \subset \mathcal{C}$" ($\mathcal{C}$ being the configuration space $\mathcal{C} \subset X$) for a specific "state i" in correspondence to the random displacements of the N-particle ensemble (eqn. 2.17). If the configurations are discrete and finite in number, the formal expression of the mean value of a dynamical variable or function f can be approximated as follows:

$$< f > \; \cong \frac{\sum\limits_{i=1}^{m} \exp[-\beta U_N(i)] f(i)}{\sum\limits_{i=1}^{m} \exp[-\beta U_N(i)]} \tag{2.26}$$

where $U_N(i)$ denotes the potential energy of the particle configuration in the "state i" symbolically written as (i), m is the total number of configurations in the set $\mathcal{A}$ and β the Boltzmann factor from before. Since m is usually rather large, it is convenient for computational purposes to consider a smaller number of configurations, i.e. $r \ll m$.

Hence one can use the approximation to the probability distribution function $P(j)$ for the state j, as follows:

$$P(j) \cong \frac{\exp[-\beta U_N(j)]}{\sum\limits_{i=1}^{r} \exp[-\beta U_N(i)]} \; . \tag{2.27}$$

A selection of possible configurations in the set $\mathcal{A} \subset \mathcal{C}$ can then be made so that the average of the function f is given by:

$$< f > \; \cong \frac{1}{r} \sum\limits_{i=1}^{r} f(i). \tag{2.28}$$

From a stochastic mechanics point of view the above approximation necessitates the use of a Markov chain so that the unweighted average of f over all states will converge, for a sufficiently large number ($r \leq m$), to the theoretical value of f for a canonical ensemble. This may be achieved by employing a "recurrent chain" (see also [42], [83]). Recalling from Chapt. 1, the Chapman-Kolmogorov equations given there, one can express the transition probabilities by:

$$P_{ij}(t_r + t_s) = \sum\limits_{k} P_{ik}(t_r) P_{kj}(t_s)$$

corresponding to the one-step transition from a state i to j. One can also use an n-step transition (eqn. 1.183). Thus, if the homogeneous Markov chain is considered at discrete time instants as required by the Monte-Carlo method, then $t = n.h; (n = 0, 1, 2 \ldots; h > 0)$ and the probabilities $P_{ij}(nh)$ for n-steps are uniquely defined from the one-step probabilities ($P_{ij} = P_{ij}(h)$) so that:

$$P_{ij}(nh) = \sum_k P_{ik}P_{kj}[(n-1)h] = \sum_k P_{ik}[(n-1)h]P_{kj},$$

$$(i,j = 1,2\ldots) \text{ for all } n = 1,2\ldots$$

Moreover the transition probabilities satisfy the continuity property, and are continuously differentiable. They have the limits given by eqn. (1.185) i.e.:

$$\lim_{h\to 0} \frac{P_{ij}(h) - P_{ij}(0)}{h} = q_{ij}; (i,j = 1,2\ldots) \tag{2.29}$$

where q_{ij} for $i \neq j$ is finite.

Thus assuming that the transition matrix $\underline{P}$ is time-independent and that its elements

$$P_{ij} \geq 0; \sum_{j=1}^r P_{ij} = 1$$

a recurrence relation is obtained for the n-step transition probabilities in the form of:

$$P_{ij}^{(n)} = \sum_{k=1}^r P_{ik}^{(n-1)} P_{kj}. \tag{2.30}$$

If all states in (2.29, 2.30) belong to the same ergodic class of sets (see also Yosida and Kakutani [108], [109], [110]), then for all pairs i,j a value of n can be found such that $P_{ij}^{(n)} > 0$ with the limit:

$$P_j^* = \lim_{n\to\infty} P_{ij}^{(n)} \text{ for all } j \tag{2.31}$$

that are independent of i. Thus the following conditions will be satisfied:

$$P_j^* > 0; \sum_{j=1}^r P_j^* = 1; \ P_j^* = \sum_{i=1}^r P_i^* P_j \tag{2.32}$$

Since in the simulation method the limit is perscribed, it follows from (2.30–2.32) that:

$$P_j^* \equiv P(j) \text{ of eqn. (2.27).}$$

Evidently, the stochastic matrix required for the simulation technique to be carried out must be chosen so as to satisfy the conditions of (2.32). A more detailed discussion on the computer simulation concerning molecular fluids are given by Hansen and MacDonald [105], Hess [111], Binder [112] and others.

A frequently used approach in the construction of continuum models of discrete media consists of employing so-called "lattice models". In such models the notion of correlation functions is of utmost significance. Generally the analysis of microdynamical processes occuring in solids and liquids, which are near the thermodynamic equilibrium, needs the consideration of time-correlation functions. In fluids time-correlation functions are equally important for the single particle motion as well as the collective mode of motion of a large number of particles or molecules. Such functions form also a link between the microscopic and the macroscopic or phenomenological description of conventional hydrodynamics. Hence it is necessary first to consider some definitions and the general formulation of these functions, which lead further to a linear response theory and the "fluctuation theory". They also form a natural transition to the probabilistic approach in molecular dynamics.

In a system consisting of $\alpha = 1 \dots N$ particles or molecules any dynamic variable Y associated with the system will be a function of the generalized coordinates and momenta in the phase-space. It can be a scalar, vector or tensor valued quantity the time evolution of which can be expressed in terms of the well-known Liouville operator. It is often convenient to use "local" dynamical variables that can in general be expressed by:

$$Y(^{\alpha}\mathbf{r}, t) = \sum_{\alpha=1}^{N} {}^{\alpha}y(t)\delta(\mathbf{r} - {}^{\alpha}\mathbf{r}(t)) \tag{2.33}$$

where $^{\alpha}y$ may be a physical quantity such as the mean $^{\alpha}\mu$ of the molecule α, its linear or angular momentum etc., and $^{\alpha}\mathbf{r}(t)$ the position vector to its centre of mass (C.M.).

In a quantum mechanical system $^{\alpha}y$ and $^{\alpha}\mathbf{r}$ in general do not commute and hence (2.33) will be subject to an appropriate symmetrization (see for instance Gudder [17]). Frequently the spatial Fourier components of the variable $Y(^{\alpha}\mathbf{r}, t)$ are required, i.e.:

$$Y(\mathbf{k}, t) = \int Y(\mathbf{r}, t)e^{-i\mathbf{k}\mathbf{r}} - \sum_{\alpha=1}^{N} {}^{\alpha}y(t)e^{-i\mathbf{k}\cdot{}^{\alpha}\mathbf{r}(t)} \tag{2.34}$$

in which $\mathbf{k}$ is the wave vector. If the dynamical variable is conserved, it will satisfy the continuity condition so that:

$$\frac{\partial Y(\mathbf{r}, t)}{\partial t} + \boldsymbol{\nabla} \cdot \mathbf{j}(\mathbf{r}, t) = 0 \tag{2.35}$$

where $\mathbf{j}$ designates the "current" related to the density of the variable Y. This relation also expresses the constancy of Y_{total}, which is equal to $\sum_{\alpha=1}^{N} {}^{\alpha}y$ in a continuous medium.

An example in this context is the "number density ρ", i.e. ($^\alpha y = 1$), at any point in the discrete medium, which can be written as:

$$\rho(\mathbf{r}, t) = \sum_{\alpha=1}^{N} \delta(\mathbf{r} - {}^\alpha\mathbf{r}(t)) \tag{2.36}$$

and by taking the Fourier transform of ρ, one obtains:

$$\rho(\mathbf{r}) = \sum_{\alpha=1}^{N} \exp[-i\mathbf{k} \cdot {}^\alpha\mathbf{r}] \tag{2.37}$$

Hence the average density at an arbitrary point $\mathbf{r}$ of the medium given by:

$$\left. \begin{array}{l} < \rho(\mathbf{r}) >= \dfrac{N}{Z} \displaystyle\int \ldots \int \exp[-\beta U_N] d^3\mathbf{r} \ldots d^N\mathbf{r} \\[2mm] \text{is approximated by } \rho^{(1)}(\mathbf{r}) \equiv \rho(\mathbf{r}), \end{array} \right\} \tag{2.38}$$

which is also the mean number density at any point of the continuous medium. The associated current is then:

$$j_\rho(\mathbf{r}, t) = \sum_{\alpha=1}^{N} \frac{{}^\alpha\mathbf{p}}{{}^\alpha\mu} \delta(\mathbf{r} - {}^\alpha\mathbf{r}(t)) \tag{2.39}$$

where $^\alpha\mu$ is the mass of the αth molecule and $^\alpha\mathbf{p}$ the linear momentum at its C.M. Taking the Fourier transform of a conserved quantity, the continuity relation from before (2.35) becomes:

$$\frac{\partial Y(\mathbf{k}, t)}{\partial t} + i \cdot \mathbf{k}j(\mathbf{k}, t) = 0. \tag{2.40}$$

In the molecular dynamics of simple fluids the pair-correlation function is of further significance since it permits the establishment of relations for the physical properties of such fluids. In order to establish this function in the time-independent and time-dependent case, it is necessary to introduce first the equilibrium density-density correlation function. The latter can be defined by the "relative distance" or the separation between the molecules, i.e. $|^{\alpha\beta}\mathbf{d}| = |^\alpha\mathbf{r} -^\beta \mathbf{r}|$. The correlation function is then defined by:

$$R_0(\mathbf{r}, \mathbf{r}') = \frac{1}{N} < [\rho(\mathbf{r}') - < \rho r' >][\rho(\mathbf{r} + \mathbf{r}') - < \rho(\mathbf{r} + \mathbf{r}') >] > \tag{2.41}$$

showing the fluctuations of the number density $\rho(\mathbf{r})$ around the position vector $\mathbf{r} \equiv {}^\alpha\mathbf{r}$ and $\mathbf{r}' \equiv {}^\beta\mathbf{r}$. Since in a "uniform system" $R_0(\mathbf{r}, \mathbf{r}')$ depends only on the relative distance $|^{\alpha\beta}\mathbf{d}|$, $R_0(\mathbf{r})$ will be a function that is independent of the choice of the origin so that:

$$R_0(\mathbf{r}) = \frac{1}{N} \int < \rho(\mathbf{r}')\rho(\mathbf{r} + \mathbf{r}') > d\mathbf{r}' - \rho \tag{2.42}$$

or equivalently

$$R_0(\mathbf{r}) = \frac{1}{N} \sum_{\alpha \neq \beta}^{N} < \delta(\mathbf{r} + {}^\alpha\mathbf{r} - {}^\beta\mathbf{r}) > + \delta(\mathbf{r}) - \rho. \tag{2.43}$$

By using the pair distribution function from before (2.24), one can also write:

$$R_0(\mathbf{r}) = \rho G^{(2)}(\mathbf{r}) + \delta(\mathbf{r}) - \rho = \rho R_0^{(2)}(\mathbf{r}) + \delta(\mathbf{r}) \tag{2.44}$$

in which $R^{(2)}(\mathbf{r})$ denotes the equilibrium pair-correlation function of ${}^{\alpha\beta}R$ in terms of the distance $|{}^{\alpha\beta}\mathbf{d}|$. Hence ${}^{\alpha\beta}R$ is defined by:

$$^{\alpha\beta}R \equiv R^{(2)}({}^{\alpha\beta}\mathbf{d}) = G^{(2)}({}^{\alpha\beta}\mathbf{d}) - 1 \tag{2.45}$$

indicating that in a uniform system and where $G^{(2)}(\mathbf{r})$ has an asymptotic behaviour, $R^{(2)}(\mathbf{r}) \to 0$, when $|\mathbf{r}| \to \infty$. In the stochastic mechanics of discrete fluids $|{}^{\alpha\beta}\mathbf{d}|$ can also be regarded as a metric in the configuration space $\mathcal{C} \subset X$. In the classical theory, an equilibrium time-correlation function pertaining to two dynamical variables Y, Z is often considered. It is defined by:

$$R_{YZ}(t', t'') = < Y(t')Z(t'') > \tag{2.46}$$

where the angular bracket either means an ensemble average over the initial conditions or an average over time. This depends whether a "generalized Langevin" equation approach or a "kinetic theory" formalism is used in modelling of the fluid behaviour. It is assumed however that the system under consideration is ergodic in the sense of Khintchine [20]. By considering the time-average of the above variables one obtains:

$$< Y(t')Z(t'') > = \lim_{T \to \infty} \frac{1}{T} \int_0^T Y(t' + s)Z(t'' + s)ds \tag{2.47}$$

It is apparent that if these variables are also subject to a spatial variation, the time-correlation function must be formulated in both time and space so that the "non-local" form of this function is then:

$$R_{YZ}(\mathbf{r}', \mathbf{r}''; t', t'') = < Y(\mathbf{r}', t')Z(\mathbf{r}'', t'') > \tag{2.48}$$

In terms of the Fourier components of the variables, which in general may be complex quantities, the time-correlation function is then expressed by:

$$\left. \begin{aligned} R_{YZ}(\mathbf{k}', \mathbf{k}''; t', t'') &= < Y(\mathbf{k}', t')Z^*(\mathbf{k}'', t'') > \\ &= < Y(\mathbf{k}', t')Z(-\mathbf{k}'', t'') > \end{aligned} \right\} \tag{2.49}$$

since the variables ${}^\alpha y$ in (2.33) are real. As already mentioned in chapt. 1, if a dynamic variable is correlated with itself, the time-correlation function is the auto-correlation function and hence:

$$R_{YY}(t', t'') = < Y(t')Y(t'') > . \tag{2.50}$$

In particular by considering the auto-correlation function of the Fourier components of the particle density which defines a so-called "structure factor $S(\mathbf{k})$" of the fluid, one has:

$$S(\mathbf{k}) = \frac{1}{N} < \rho_{\mathbf{k}}\rho_{-\mathbf{k}} > \tag{2.51}$$

which can also be obtained from the Fourier transform of the pair-distribution function (2.51, 2.31) i.e.:

$$S(\mathbf{k}) = \frac{1}{N} \sum_{\alpha=1}^{N} \sum_{\beta=1}^{N} \exp[-i\mathbf{k} \cdot {}^{\alpha}\mathbf{r}] \exp[i\mathbf{k}^{\beta}\mathbf{r}]. \tag{2.52}$$

It can be readily shown that this leads to the form:

$$S(\mathbf{k}) = 1 + \rho \int \exp[-i\mathbf{k} \cdot \mathbf{r}] G(\mathbf{r}) d\mathbf{r} \tag{2.53}$$

where $G(\mathbf{r})$ is the radial distribution function form before. Using the Fourier transform of $S(\mathbf{k})$, this distribution function is then:

$$\rho G(\mathbf{r}) = \frac{1}{(2\pi)^3} \int [S(\mathbf{k}) - 1] \exp(i\mathbf{k} \cdot \mathbf{r}) d\mathbf{k}. \tag{2.54}$$

The above factor $S(\mathbf{k})$ is referred to as the "static structure factor". It can be established directly from scattering experiments (see for instance Lovesey [113], Dore et al [114], [115], [116] and others). In an isotropic fluid where $G(\mathbf{r}) \equiv G(r)$; $S(\mathbf{k}) \equiv S(k), G(\mathbf{r})$ describes the average distribution of the molecular separation in the fluid. Of greater interest however is the "dynamic structure factor $S(\mathbf{k}, \omega)$", which is a function of both the wave factor $\mathbf{k}$ and the frequency ω of the wave motion. In this case one has to consider the correlation in space and time as well as a scattering function. Thus the dynamic structure factor can be defined as follows:

$$\left. \begin{aligned} S(\mathbf{k}, \omega) &= \frac{1}{2\pi N} \int_{-\infty}^{\infty} \exp(i\omega t) < \rho_{\mathbf{k}}(t)\rho_{-\mathbf{k}}(t) > dt \\[2mm] &= \frac{1}{2\pi} \int_{-\infty}^{\infty} \exp(i\omega t) F(\mathbf{k}, t) dt \end{aligned} \right\} \tag{2.55}$$

in which

$$F(\mathbf{k}, t) = \frac{1}{N} < \rho_{\mathbf{k}}(t)\rho_{-\mathbf{k}}(t) > \tag{2.56}$$

is known as an intermediate scattering function. For more details the reader is referred to the above references. Relation (2.56) shows by comparing it with (2.52) that for $t = 0$: $F(\mathbf{k}, 0) = S(\mathbf{k})$ or the static structure function. Although no calculation of $S(\mathbf{k})$ is intended here, some basic properties of it should be mentioned.

There are three functions that are conventionally used to describe spatial correlations in fluids. Thus apart from $S(\mathbf{k})$ one has the pair-correlation function, which is related to the structure function by:

$$S(\mathbf{k}) = 1 + \rho R^{(2)}(\mathbf{k}). \tag{2.57}$$

Since $F(\mathbf{k}, 0) = 1 + \rho \int d^2 \mathbf{r} \cdot e^{i\mathbf{k} \cdot \mathbf{r}} [R_0(r) - 1] = S(k)$, it is seen that $R^{(2)}(r)$ is the Fourier transform of $[R_0(r) - 1]$. A direct correlation function may also be considered of the form:

$$\rho R(k) = \frac{S(k) - 1}{S(k)}$$

or

$$R^{(2)}(k) = R(k) + \rho R(k) R^{(2)}(k) \tag{2.58}$$

which is known as the Ornstein-Zernicke relation and which is a fundamental relation in the theory of pair distribution functions (see for instance Hansen and McDonald [105], Yvon [117]).

The behaviour of the structure factor $S(k)$ is generally analogous to that of the pair distribution function. Thus, if k is large, the asymptotic value of $S(k)$ approaches unity indicating the vanishing pair-correlation at short wave length. In the other limit for small k, one obtains the known compressibility relation linking the structure factor to the isothermal compressibility of a dilute fluid (see for instance Egelstaff [116]).

The above brief remarks show that the particle density or the number density is a characteristic of general interest in molecular dynamics of fluids. Experimental work concerned with the determination of various frequency domains (k, ω) of discrete fluids can be grouped according to these two parameters, i.e. the characteristic frequency ω_c and the wave number k_c, respectively. The first corresponds to the reciprocal collision time $(\tau_c \sim 10^{12} - 10^{13} \ sec^{-1})$ and the second to the reciprocal of the molecular distance $(^{\alpha\beta}d_c \sim 10^8 \ cm^{-1})$. Hence the domain of application of the molecular dynamics is bounded by the values $\omega_c \sim \tau_c^{-1}$ and $k_c \sim d_c^{-1}$ or the region of short wave length and high frequencies. It should be pointed out that often, instead of actual experiments, only simulation methods can be employed to assess the behaviour of discrete fluids.

(iv) Lattice Models:

It has been mentioned earlier that "continuum models of discrete media" are largely based on lattice constructs in which the continuous medium is replaced by a set of points forming a grid or lattice. The "local" continuum laws are then implemented at each lattice point or "node" and a few surrounding points. It is apparent, that the scale at which this description occurs is somewhat larger than that used in molecular dynamics, but it is convenient so long as the phenomena involved can be represented by continuous vector fields. It implies however, that the continuum has to be "discretized" so that all spatial sites are equivalent and have the same number of neighbours conforming to the topological structure of a "regular lattice". The foremost and best known lattice model was introduced in 1925 by Ising [118]. It is rooted in statistical mechanics and concerned with originally ferro-magnetic materials. Since then this model has been used in the analysis of a great variety of materials particularly with reference to their "critical behaviour". The wide applicability of Ising's model is due to the fact that continuum models of various materials are only variants of this model, if an appropriate lattice for the phenomena can be chosen. The latter are however equally representable by the use of a "probabilistic formalism" as shown subsequently.

Ising considered first a sequence of points $x = \{0, 1, 2 \ldots n\}$ or sites (1-dimensional model), at which at any given time a small "dipole or spin s" is in one of two positions at the configuration $x \in X$. These configurations in general form a set $\{x\}$ in the configuration space $\mathcal{C} \subset X$ or a probabilistic function space and can be identified with the set of elementary outcomes $\omega \in \Omega$ (sample space) contained in X. The non-observable spins s can be expressed by:

$$s_j(x) = \begin{cases} +1, & \text{if } x \text{ is positive on } \mathbb{R}^1 \\ -1, & \text{if } x \text{ is negative on } \mathbb{R}^1 \end{cases}$$

Ising then assigned an energy $U(x)$ to each site, i.e.:

$$U(x) = -J \sum_{i,j} s_i(x) s_j(x) - mH \sum_i s_i(x) \tag{2.59}$$

in which the first sum over all pairs i, j (one unit apart) represents the interaction energy of the spin s and where $m > 0$, J are material characteristics. The second term in (2.59) refers to the effect of the magnetic field with the intensity H. Since the probability of a configuration $x \in \{x\} \subset \mathcal{C}$ is proportional to: $e^{-\beta U(x)}$, the probability measure on the configuration space $\mathcal{C}$ is given by:

$$\left. \mathcal{P}(x) = Z^{-1} \exp[-\beta U(x)]; \quad Z = \sum_{x \in \mathcal{C} \subset X} e^{-\beta U(x)} , \right\} \tag{2.60}$$

where Z is the normalizing constant or partition function.

Of greater interest here is the case for which the spins s are located on a set of lattice points in two or more dimensions. Thus, by considering a regular 2-dimensional or "square lattice" and a point "i" that replaces the coordinates x_1, x_2 or i, j (integers) with the above probability measure, then the "local energy $U_i(x)$" can be written as:

$$U_i(x) = -\frac{J}{2} \sum_{|j-i|=1} s_i(x)s_j(x) - mHs_i(x) \qquad (2.61)$$

and the probability measure as:

$$\mathcal{P}(x) = Z^{-1} \prod_i \exp[-\beta U_i(x)]. \qquad (2.62)$$

Hence the relative probability of a configuration $x \in \{x\} \subset \mathcal{C}$ is obtained by taking the product over all points of the lattice and using the local energy at each point to establish the weight at that point.

Although the above relations concern magnetic fields, they apply analogously to various other physical systems such as gases, binary structures, "cell structures" etc. The probability measure in (2.60) is also called a "Gibbs measure" and will be further discussed later in the text (see Chapt. 3). The relative probabilities are of significance in stochastic mechanics in conjunction with "random walks" on square lattices or in the more general case on infinite lattices in the d-dimensional Euclidean space. Thus, in this type of lattice a "walker" jumps at regular time intervals separated by $\Delta t = 1$ from one site to a neighbouring one in a random fashion. If in the lattice each site x has q neighbours (q is also called the coordination number), then the probability of attaining any particular neighbouring site is $1/q$ in which successive jumps are regarded as statistically independent events.

By considering the frequently used "cubic-lattice", which in general is generated by d-orthonormal vectors $e_1 \ldots e_d$ (corresponding to a hyper lattice), where $e_{(k)}e_{(\ell)} = \delta_{k\ell}$ and the sites are located at $\mathbf{x} = x^k e_{(k)}$ (the x^k coordinates are only integers), the conditional probability for the walker to be at the site $\mathbf{x}_1$ at time t_1 when the initial position is $\mathbf{x}_0$ at time t_0, can be expressed by:

$$P\{\mathbf{x}_1, t_0; \mathbf{x}_0, t_0\} = \delta_{\mathbf{x}_1 \mathbf{x}_0} \qquad (2.63)$$

for $t_1 = t_0$ and where $\delta_{\mathbf{x}_1 x_0}$ is an abbreviation of $\prod_{(k)=1}^d = \delta_{x_1^{(k)} x_2^{(k)}}$. Assuming translational invariance in space and time, then the probability P defined for $t_1 \geq t_0$, depends only on the difference $\mathbf{x}_1 - \mathbf{x}_0, t_1 - t_0$, and where for a fixed time t_1 the normalization condition is given by:

$$\sum_{\mathbf{x}_1} P\{\mathbf{x}_1, t_1; \mathbf{x}_0, t_0\} = 1; P \geq 0. \qquad (2.64)$$

It has been shown by C. Itzykson and J.M. Drouffe [119] in their rigorous treatment of statistical field theory that includes interacting fields and random walks, that the random walk considered above can be used in its discrete formulation to arrive at a "recurrence relation" between the probabilities in the following form:

$$P\{\mathbf{x}, t+1; \mathbf{x}_0, t_0\} = \frac{1}{2d} \sum_{\mathbf{x}' \in C \subset X} P\{\mathbf{x}', t; \mathbf{x}_0, t_0\} \tag{2.65}$$

when the walker reaches the site $\mathbf{x}$ only, if one unit earlier he was on the neighbouring site $\mathbf{x}' = \mathbf{x} \pm \mathbf{e}_{(k)}$. The probability P is completely determined by relations (2.63–2.65). A more detailed discussion on random walks can be found in Montroll and West [120] and in [119].

(v) Percolation Models:

In the analysis of the behaviour of various discrete media, one frequently employs other than the above described lattice models. Such models are known as "percolation models" and are of particular interest in the study of critical phenomena. A continuum percolation theory has been developed since 1957 (see also Broadbent and Hammersley [121], Domb and Green [122], Stauffer [123] and others). In most formulations "linear graph theory" was used and hence it may be necessary to digress briefly to give a few definitions and notions of graph theory. A linear graph $\mathbf{G} = (\mathbf{V}, \mathbf{E})$ in general consists of a set of vertices $\mathbf{V}$ and a set of edges or arcs $\mathbf{E}$, where $\mathbf{V} = \{x_1, x_2 \cdots x_n\}$ and $\mathbf{E} = \{e_1, e_2, \cdots e_n\}$, respectively. A "directed graph" or digraph D is thus defined by a set of points $\mathbf{V}(\mathbf{D})$ or the vertices of $\mathbf{D}$ together with the set $\mathbf{E}(\mathbf{D})$ or "ordered pairs" and a set of edges of $\mathbf{D}$. An edge (e_1, e_2) is conventionally represented as a line or path directed from $e_1 \rightarrow e_2$. From a topological point of view there is a unique digraph $\mathbf{D}(T)$ for any topology T. This means that all the information about the topological space is implicit in $\mathbf{D}(T)$ in the sense, if $\mathbf{D}(T)$ is given then a basis $\mathcal{B}_e$ and hence the entire topology can be obtained from it. A digraph $\mathbf{D}(T)$ with the characteristic that $(e_1, e_2) \in \mathbf{E}(\mathbf{D})$ and $(e_2, e_3) \in \mathbf{E}(\mathbf{D})$ implies that $(e_1, e_3) \in \mathbf{E}(\mathbf{D})$, where the latter is called a "transition graph". Although the structure of a topological space X has been discussed in terms of open sets (Chapt. 1), for the consideration of graph theory an alternative way of defining the structure of X is given here. Thus by considering the pair-wise relations of any open sets of the space X, they can be related in three different ways, i.e.:

(i) they may be disjoint,

(ii) they may intersect, but neither is a subset of the other,

(iii) and finally, one is the subset of the other.

If the last relation holds one says that the two sets are comparable otherwise they are incomparable. Thus one can introduce a "partial order" relation for the sets, namely:

(i) $x \subset x$ (reflexive); $x \in X$

(ii) $x \subset y$ and $y \subset x$ implies: $x = y$ (antisymmetric);

$x, y \in X$

(iii) $x \subset y$ and $y \subset z$ implies: $x \subset z$ (transitive)

$x, y, z \in X$.

$$\left. \right\} \quad (2.66)$$

A partial order set or "poset" is thus defined by a pair of elements that may be comparable or incomparable. There are other ways of recognizing partial ordering besides set inclusion (see for instance Kuratowski and Mostowski [25]).

Open sets of any topology form a "lattice" when partially ordered by set inclusion. By the definition of a topology (Chapt. 1), if the sets A, B are both members of the topology T, then $A \cup B$ and $A \cap B$ are also members. Since $A \cup B$ is the smallest member of T that includes A and B, one has also by mathematical logic: $A \vee B = A \cup B$ and $A \wedge B = A \cap B$. Hence the open sets of a finite topology form a "lattice $\mathcal{L}$" in which the "join and meet" operations are identical to the union and intersection, respectively. However not all lattices $\mathcal{L}$ represent topologies unless they satisfy the "distributive property", i.e.:

(i) $x \wedge (y \wedge z) = (x \wedge y) \vee (x \wedge z)$;

(ii) $x \vee (y \wedge z) = (x \vee y) \wedge (x \vee z)$ for any element $(x, y, z)y \in \mathcal{L}$

$$\left. \right\} \quad (2.67)$$

The lattice of the discrete topology on "n-points" is the direct product of n-two element lattices and is called a Boolean algebra. The latter will subsequently be further discussed.

It is to be noted however, that the space resulting from the product of two topological lattices is not the same as the topological product space $X = X_1 \otimes X_2$ as discussed in Chapt. I (section 1.2). The point set on which $X_1 \otimes X_2$ is defined is a "product set", whilst the "$\mathcal{L}$ set" is a union of sets.

A graph $\mathbf{G}$ is said to be "strongly connected" or strong, if for any two distinct vertices x_i, x_j, there exists at least one path going from $x_i \to x_j$. A "maximal subgraph of $\mathbf{G}$" is a strong subgraph that is not contained in any other strong subgraph of $\mathbf{G}$. Such a subgraph is also called a strong component of $\mathbf{G}$. Hence, if $\mathbf{y} = \{\mathbf{y}_1, \mathbf{y}_2, \dots \mathbf{y}_k\}$ are the strong components of $\mathbf{G}$ in which each subgraph $\mathbf{y}_i$ has the vertices $\{x_{i_1}, x_{i_2} \dots x_{i_k}\}; (i = 1, 2 \dots k)$ and $\mathbf{g}$ is the set of edges between these vertices, then the graph $\mathbf{G} = (\mathbf{y}, \mathbf{g})$ is the "condensed graph of $\mathbf{G}$". Similarly the digraph $\mathbf{D}(T)$ resulting from contracting each component to a point is called the "condensation of the digraph".

For a given graph $\mathbf{G}$ one can define an "adjacency matrix $\mathbf{A}$" that has the elements:

$$a_{ij} = \left\{ \begin{array}{ll} 1, & \text{if the edges } (e_i, e_j) \text{ exist in } \mathbf{G} \\ 0, & \text{otherwise} \end{array} \right\} \quad (2.68)$$

indicating the possibility of moving between any two vertices along an edge in "one step". If more than one step is involved one can define a "reachability matrix $\mathbf{R}$" the elements of which are then given by:

$$
r_{ij} = \begin{cases} 1 & \text{if } x_j \text{ is reached from } x_i \text{ without restriction} \\ & \text{in the number of steps} \\ 0, & \text{otherwise.} \end{cases} \tag{2.69}
$$

The above forms can be obtained in a more general way by considering topological combinatorics (see also Christofides [124], Harary [125] and others). Thus, by recognition of the fundamental theorem concerning the "cardinality" of a finite topology, which is equal to the number of independent vertices of the graph of that topology, one has:

$$
\mathbf{G} = \sum_{\mathbf{r}} (\prod_{ij} (1 - a_{ij} r_i r_j) \tag{2.70}
$$

in which $a_{ij} \in \mathbf{A}$ or the adjacency matrix of $\mathbf{G}$, $x_i \in \mathbf{V}(\mathbf{G})$ and $r_i \in \{0,1\}$.

It is evident from the above sketched fundamentals of graph theory, that even in the case of a single crystal or microelement of a polycrystalline solid for instance, because of existing defects, some of the "sites" (vertices) may be occupied and others not. The occupied sites will fall into a number of sets or "clusters". Two such sites belong to the same cluster, if there is a path of occupied sites connecting them. The size of such clusters will depend on the fraction p of sites that are occupied. Hence one may expect that the mean size of the clusters to increase with p and for a "finite crystal" to attain a maximum when $p = 1$.

One can also expect a "critical concentration p_c", which is the largest value of p for which a given occupied site with probability one belongs to a cluster of finite size. For $p > p_c$ there is a non-zero probability that a given occupied site belongs to an "unbounded cluster". The latter is called the "percolation probability $P(p)$", it is zero if $p \leq p_c$. This situation arises in many physical systems. A related problem is one in which the "bonds" rather than the "sites" may be in one of two possible states. These two models are then called the site and the bond problem, respectively. Considering the site problem first, the vertices of the graph are the possible locations or configurations of a particle. An edge designated by $(i,j) \in \mathbf{E}(\mathbf{G})$ is occupied, if both its vertices are occupied by a particle. As stated above two particles belong to the same cluster, if there is a connecting path between them. Although according to statistical mechanics, a crystal lattice would be represented by an infinite graph $\mathbf{G}$, there being a vertex for every lattice site, one can approach the infinite case by a sequence of finite lattices. The edges are usually considered as the nearest neighbour pairs of sites based on a central force interaction, although in certain cases it may be necessary to include second and higher order neighbour conditions.

A subset $V^* \subset V$ consisting of all occupied vertices defines a "section graph $G^* = (V^*, E^*)$, where E^* consists of all edges with both vertices in V^*. Thus each component of G^* corresponds to a cluster of particles in this particular configuration. In general one is often interested in the random function $f(\mathbf{V}^*, \mathbf{G}^*)$ with the positive integer values on $\mathbb{Z}_+$, the mean value of which may be expressed as follows:

$$< f(\mathbf{G}^*) >= \sum_{\mathbf{V}^* \subset \mathbf{V}} P(\mathbf{V}^*) f(\mathbf{V}^*, \mathbf{G}^*) \tag{2.71}$$

where $P(\mathbf{V}^*)$ is the probability that the vertices in $\mathbf{V}^*$ are occupied and the vertices $\mathbf{V} - \mathbf{V}^*$ are vacant. Evidently there are $2^{|\mathbf{V}|}$ terms contained in the above sum and $P(\mathbf{V}^*)$ is normalized to unity. The probability that a vertex i is occupied can be stated as:

$$p_i =< r_i(\mathbf{G}^*) > \tag{2.72}$$

in which

$$r_i(\mathbf{V}^*) = \begin{cases} 1 & \text{for } i \in \mathbf{V}^* \\ 0 & \text{otherwise .} \end{cases} \tag{2.73}$$

If the vertices are occupied independently, one has:

$$< r_i(\mathbf{G}^*), r_j(\mathbf{G}^*) > \tag{2.74}$$

indicating that the deviation from this value means a correlation. In general for the random case all vertices will be occupied independently with probability p.

In the second or "bond" problem the edges of $\mathbf{G}$ will be in one of the two states, namely either open or closed. Thus the open edges $\mathbf{E}^* \subseteq \mathbf{E}$ define then a partial graph $\mathbf{G}^* = (\mathbf{V}^*, \mathbf{E}^*)$ in which the vertex set is identical with that of $\mathbf{G}$. The definition in (2.71) still holds but $\mathbf{V}$ has to be replaced by $\mathbf{E}$. The probability that an edge is open can be stated as before, i.e.:

$$\left. \begin{aligned} p_{ij} &=< r_{ij}(\mathbf{G}^*) > \\ \text{where } r_{ij} &= \begin{cases} 1 & \text{for } (i,j) \in \mathbf{E}^* \\ 0 & \text{otherwise.} \end{cases} \end{aligned} \right\} \tag{2.75}$$

For a general graph an important function is the "pair connectedness" of the vertices i and j, denoted by $P_{ij}(p, \mathbf{G})$ which is defined by:

$$\left. \begin{aligned} P_{ij}(p; \mathbf{G}) &=< t_{ij}(\mathbf{G}) > \\ \text{where } t_{ij} &= \begin{cases} 1, & \text{if } (i,j) \in \mathbf{E} \\ 0 & \text{otherwise.} \end{cases} \end{aligned} \right\} \tag{2.76}$$

This function is often used in the "network theory" representation of various media. The connectedness of a network is usually defined for the bond problem only, and is the probability that the partial graph specified by the open edges

E^* is connected. It has been shown by Broadbent and Hammsersley [121] that for a restricted class of graphs a "random medium" has a critical probability given by:

$$p_c(i) = \sup_{P(i,p)=0} p$$

with the characteristic that $p_c(i) = p_c(j)$, if there is a finite chain connecting the vertices i and j. Hence a "random medium" is a medium in which the vertices (edges) are occupied (open) independently with probability p. The percolation probability can also be established by the use of random walks. Thus if $S_n(i)$ denotes the set of "n-stepped" self avoiding walks that start at a vertex i of the random medium, then in the bond problem a walk is open, if all edges are open. Thus the "percolation probability" can be expressed by:

$$P^{(b)}(i,p) = \lim_{n \to \infty} P_n^{(b)}(i,p) \tag{2.77}$$

in which $P^{(b)}(i,p)$ is the probability that at least one of the set $S_n(i)$ is open. A similar form expresses the case of the "site" problem when (b) is replaced by (s) and where the open walk is such that all vertices of the walk are occupied. As mentioned earlier since for a connected and "undirected graph" a walk in both directions between any two vertices i, j is possible:

$$P(i,p) \geq p^n P(j,p) \tag{2.78}$$

where for both the bond and site problem, n is the number of steps in the "shortest walk" between i and j.

Another basic quantity of interest in percolation problems is the distribution "n_s" of clusters of a size s per bond. Since from before "p" is the probability when each bond is randomly occupied in the bond percolation and $q = 1 - p$, when it is absent, one has by normalization that:

$$q + \sum_s sn_s + P_\infty(p) = 1 \tag{2.79}$$

in which $P_\infty(p)$ is the probability that a given site belongs to an "infinite cluster". This distribution as a function of the concentration p for an infinite large regular lattice is indicated in Fig. 10(a). This figure also shows that above the critical value p_c, the function $P_\infty(p)$ is non-zero, whilst for $p \leq p_c$ it is identically zero. Thus this function may be regarded as an "order parameter" and p as a "control parameter".

The notion of the latter will be further considered in the "state-space representation" of discrete media in the following section (2.3). Another aspect of graph theory should be remarked upon that concerns its application to problems of "clusters" of micro elements in a discrete medium. Thus be denoting the "cluster size $s(x_i, x_j)$", where x_i, x_j are the coordinates of a "bond (i, j)"

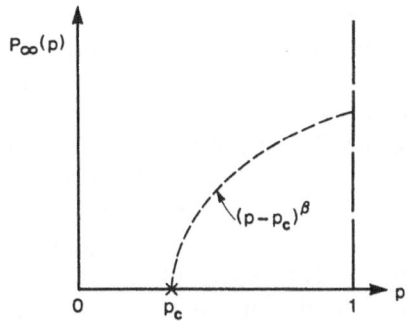

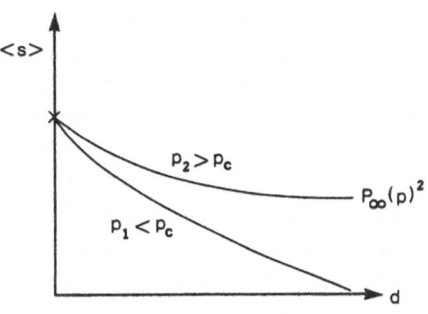

(a) density of "infinite cluster"
      P_∞ as a function of p

(b) average cluster size $<s>$ for two
      values of the percolation concentration p

Fig. 10.

and the distance: $d = |x_i - x_j|$; $s(\cdot)$ will be unity, if the sites x_i, x_j belong
to the same cluster of occupied bonds and zero otherwise. The average of the
random configurations $x_i, x_j \in \mathcal{C} \subset X$ is then given by:

$$< s(0, d) > \equiv < s > \tag{2.80}$$

and decays to its infinite value over a certain length scale ξ such that:

$$\xi \simeq \Delta p^{-s} \tag{2.81}$$

This is indicated in Fig. (10b). The probability $P_\infty(p)$, where p is referred
to as percolation concentration, is also related to the "renormalization group
analysis" for the critical behaviour of various media (see also Domb [122]).

2.3 Probabilistic models of discrete media.

(i) Introduction:

In the preceding section some statistical models based on regular lattices of the
$\mathbb{R}^d$ space have been briefly considered. It is the principal aim of this and the
following sections to introduce probabilititistic models of discrete media that are
based on the theory of Markov processes. Such models lead to a more detailed
description of the material behaviour and permit the inclusion of "interaction
effects" between the elements of the microstructure. These interaction effects
are caused by changes in forces and fluxes acting on individual elements of the

microstructure and by the occurrence of a multitude of internal surfaces (inter-faces) at which relevant quantities become discontinuous. Although the previ-ously discussed models are based on the principles of statistical mechanics, they represent essentially "networks of particles" (mass points) and do not reflect the discreteness and random nature of the geometrical and physical characteris-tics of a large class of discrete media. Hence the stochastic mechanics approach employs models in which, from the onset, the relevant field quantities are con-sidered as random variables or functions of such variables. Furthermore the formulation uses a representation in terms of an "abstract dynamical system" $[X, \mathcal{F}, \mathcal{P}]$ as given by the definitions of section 2.1. This representation will be maintained throughout the analysis due to its suitability in the representation of the deformation and flow of discrete media. A "dynamical theory" in general is based on an equation of motion, which from a statistical mechanics point of view, is either an equation for the time evolution of the dynamic variables or the evolution of their "probability distribution". If both these descriptions can be given they must be consistent. As known from the deterministic formula-tion of classical mechanics these two representations correspond to Newton's or Hamilton's equations and the second description to Liouville's equation. In the stochastic analysis the first type of representation is associatd with the "Brownian motion" as dealt with by the theory of Markov processes or equiv-alently with the approach by means of the "Langevin equation". The second type of representation is associated with the Kolmogorov differential equation (forward) or the "Fokker-Planck" equation. In view of the above statement it will be necessary to consider these representations in some detail. First however, it may be instructive to consider the notions of "observables" and "states" of the elements of the discrete media, which play a significant role in the analysis.

(ii) Observables and states:

(a) Observables:

In analogy to stochastic quantum mechanics two concepts of classical statis-tical mechanics are fundamental in the formulation of stochastic mechanics of discrete media. These notions concern "observables" and "states" of a medium. Thus considering a given medium to be in a specific state or condition, it is usually subjected to some external field and an appropriate experimental pro-cedure to obtain measurements of "observable quantities" pertaining to the elements of the structure.

For a given medium in the "state z" of the state-space $Z = \{z, z_1, z_2 \ldots\}$, one may measure a single observable "A" or a set of observables $\mathcal{A} = \{A, A_1, A_2 \ldots\}$. Although the result of measuring $A \in \mathcal{A}$ may give a certain number, in general, it will be necessary to perform a number of experiments so that a statistical distribution of A is obtained. Thus given $A \in \mathcal{A}$ and the corresponding state $z \in Z$ one obtains the distribution $P(A, z, E)$, which can be interpreted as the probability of the observable A having a value in the Borel set E (even set), when the system is in the state z. This distribution is also a probability measure $\mathcal{P}$ on the space $\mathbb{R}^1$. In dealing with stochastic quantum mechanics (see for instance Gudder [17]), the notion of an event structure can be used in which the events of the physical system are taken as the primitive axiomatic elements. The events correspond to physical phenomena which may or may not occur. In particular for the type of observables which attain at most the two possible values 0 and 1, one can speak of a generalized event of the observable A having the distribution:

$$P\{A\} = 1 \text{ for every} z \in Z. \tag{2.82}$$

In the stochastic mechanics formulation which is aimed at the representation of the behaviour of discrete media, it may be useful to mention here the "propositional logic" employed in quantum mechanics. This logic can be defined (see also Kossakowski [126], MacKey [127], Yosida [24] and others) as a "partially ordered set $\mathcal{E}$" with the smallest and greatest element 0 and 1, respectively and which has the following characteristics:

(i) The set $\mathcal{E}$ is orthocomplemented, i.e. there is a map $a \mapsto a'$ of $\mathcal{E}$ into itself such that $a \leq b$ implies $b' \geq a', (a')' = 1$ and $a \wedge a = 0$ for all $a \in \mathcal{E}$. The symbol $\wedge$ designates the greatest lower bound of a, a' on $\mathcal{E}$.

(ii) For each sequence $a_1, a_2 \ldots$ of pair-wise orthogonal "propositions", there exists in $\mathcal{E}$ its least upper bound $\vee_{i=1}^{\infty} a_i$; two propositions $a, b \in \mathcal{E}$ are orthogonal or symbolically $a \perp b$, if $a \leq b'$ or equivalently $b \leq a'$.

(iii) $\mathcal{E}$ is orthomodular, i.e. $a \leq b$ implies $b = a \vee c$, (sup of a and c) for some $c \in \mathcal{E}, c \perp a$.

Hence a "statistical state" can be defined as a non-negative real function z on $\mathcal{E}$ such that $z(1) = 1$ and

$$z\left(\vee_{i=1}^{\infty} a_i\right) = \sum_{i=1}^{\infty} z(a_i) \tag{2.83}$$

for each sequence of mutually orthogonal propositions. It follows from (iii) that $a \leq b$ always leads to $z(a) \leq z(b)$ and thus maps $\mathcal{E}$ into the interval $[0, 1]$. An observable A can be considered as a map from the Borel sets of the real line $\mathbb{R}^1$ into $\mathcal{E}$, i.e. $\mathcal{A}: \mathcal{B}(\mathbb{R}^1) \to \mathcal{E}$ such that:

(i) $A(\mathbb{R}^1) = 1$

(ii) If $E \cap F = \emptyset$, then $A(E) \perp A(F)$

(iii) If $E_i \in \mathcal{B}(\mathbb{R}^1)$ is a sequence of mutual disjoint sets, then:

$$A\left(\cup_{i=1}^{\infty} E_i\right) = \vee_{i=1}^{\infty} A(E_i). \tag{2.84}$$

The number $z(A(E))$ can be regarded as the probability that a measurement of an observable A of the system in the state z will lead to a value in the Borel set E. One can thus write the expected value of A in the state z as:

$$< A, z > = \int_{\mathbb{R}^1} tz(A(dt)) \qquad (2.85)$$

if this integral exists. The observable A will be bounded, if its spectrum or $sp\, A$, which is defined as the smallest closed set $E \subset \mathbb{R}^1$ such that $A(E) = 1$, is bounded. One can thus define a norm for A by:

$$\| A \| = \sup\{|t| : t \in spA\}. \qquad (2.86)$$

In the present formulation only "bounded observables" will be considered. It is to be noted that generally in order to distinguish between quantum and classical mechanics a specific logic in each of these theories is required. Thus in the terminology of quantum mechanics "events" can be thought of as propositions that correspond to true or false experiments. However in the present analysis stressing the stochastic properties of a medium the term event will be used throughout. Hence in the terminology of classical mechanics propositions are Borel sets of the "phase-space" of the system. Since Z in stochastic mechanics is considered as a set of all "thermo-mechanical states" of the medium, the preferred term is "state-space". In most cases of the subsequent formulation the topological structure of Z will be assumed to be of a simple form, i.e. excluding more complicated structures. In general the set Z of all states of the medium may be considered as a norm-closed convex subset of a suitably ordered (partially) Banach space. The evolution of states with time from the point of view of Markovian or non-Markovian processes will be considered in the following. However, it is necessary first to remark on the fundamental structure of the state-space Z or correspondingly on the probabilistic function space X.

(b) The "state-space" representation:

In order to establish probabilistic models of the behaviour of discrete media, the concept of the state-space Z as a probabilistic function space should be clarified further. This notion is well founded in quantum and classical statistical mechanics (see for instance Khintchine [20], MacKey [127] and others). The mechanical states of microelement "α" of the medium can be represented by an "r-dimensional state vector", the components of which are real valued functions of the geometrical and thermomechanical parameters characterizing a specific medium (see also [22, 23]). In the probabilistic formulation, stochastic thermodynamics is characterized by the probability measure space $[X, \mathcal{F}, \mathcal{P}]$, where $\mathcal{F}$ is the Kolmogorov algebra as defined earlier (see also Muschik [128]). More specifically the triple $[X, \mathcal{F}^z, \mathcal{P}^z]$ defines the state space of a microelement

$^\alpha A$ or α of the medium. For the corresponding representation of the larger domain of the material, i.e., the "meso-domain" (Def. 8) a set of state vectors is required, or $\{^\alpha z \colon {}^\alpha z_i, \alpha = 1 \ldots N; i = 1 \ldots r\}$. The latter form the state-space $Z \equiv X$, which is assumed to be locally compact.

An analogous structure of Z can be given to that of the more general state-space of classical mechanics by noting however, that the vector $^\alpha z$ can only be specified within certain limits. This is due to the experimental constraints and the accuracy with which relevant obervations can be carried out. Thus the state of a microelement α will be represented by:

$$z < {}^\alpha z < z + \Delta z$$

where $^\alpha z$ is a specific value for the element α and "Δz" the range of experimental observations. This indicates that by pre-setting this range for a particular experiment, one can only state the number of microelements that will have their mechanical states within that range. One obtains therefore a subset $E_n \subset Z = X$, which includes the states within this range only. The subset of events E_n can be regarded as an open sphere so that:

$$\left. \begin{aligned} &E_n = \{z_n <^\alpha z \le z_n + \Delta z_n\}; \ \cup_n E_n = X, \\ &E_n \cap E_k = \emptyset, n \ne k \end{aligned} \right\} \quad (2.87)$$

If X is locally compact, the above subsets are also compact and bounded under closure. With this interpretation one has a class of sets $\mathcal{F}$ with the following properties:

$$\left. \begin{aligned} &\text{(i)} \ E_n \in \mathcal{F} \Rightarrow E_n' \in \mathcal{F}; E' = \text{complement of } E \\ &\text{(ii)} \ E_n \in \mathcal{F} \Rightarrow \cup_1^\infty E_n \in \mathcal{F} \\ &\text{(iii)} \ X \in \mathcal{F} \end{aligned} \right\} \quad (2.88)$$

showing that the class $\mathcal{F}$ forms a σ-algebra, as discussed in Chapt. 1. The elements E_n of $\mathcal{F}$ are Borel sets and the space X together with $\mathcal{F}$ so defined forms a measurable space $[X, \mathcal{F}]$. Since the main interest in stochastic mechanics lies in the random motion of the microelements during deformation and flow of discrete media, an appropriate measure on the subsets of X must be chosen such that

$$0 \le \mathcal{P}\{E_n\} \le 1; \ \mathcal{P}\{E_n\} = 0, \text{ if } E_n = \emptyset \text{ and } \mathcal{P}\{X\} = 1. \quad (2.89)$$

This indicates in accordance with probability theory that $\mathcal{P}$ is the distribution of the relevant field quantities. Thus by using this measure one can designate a measure space by the triple $[X, \mathcal{F}, \mathcal{P}]$ which can be interpreted as a probability space.

As mentioned earlier, since the state vector $^\alpha z$ contains several components each of which belongs to a subspace of X, it is often convenient to use these subspaces in the analysis. Thus, for instance, one may choose in the study of the deformations of a discrete solid a "deformation space $\mathcal{U}$" and a corresponding "stress space $\sum$" as subspaces of X. Similarly in the flow dynamics of discrete fluids one can use a "configuration space $\mathcal{C} \subset$ X" or the "velocity space $\mathcal{V} \subset$ X" and a corresponding "force or stress space $\sum$". Consequently the state vectors in either case may be expressed in an approximate form as follows:

$$
{}^{\alpha}\mathbf{z} = \begin{pmatrix} {}^{i}\boldsymbol{\sigma}(t) \\ {}^{s}\boldsymbol{\sigma}(t) \\ {}^{i}\mathbf{u}(t) \\ {}^{s}\mathbf{u}(t) \\ \vdots \end{pmatrix} \quad ; \quad {}^{\alpha}\mathbf{z} = \begin{pmatrix} {}^{\alpha}\mathbf{r}(t) \\ {}^{\alpha}\mathbf{v}(t) \\ {}^{\alpha}\boldsymbol{\sigma}(t) \\ \vdots \end{pmatrix} \left. \begin{matrix} \\ \\ \\ \\ \\ \end{matrix} \right\} \qquad (2.90)
$$

(a) solids   (b) fluids

where in (a) the symbols ${}^{i}\boldsymbol{\sigma}(t)$, ${}^{s}\boldsymbol{\sigma}(t)$ refer to the microstresses within a structural element α and on its boundary, respectively. The quantities ${}^{i}\mathbf{u}, {}^{s}\mathbf{u}$ represent the corresponding local deformations. The column vector in (b) pertaining to fluids contains the configuration vector ${}^{\alpha}\mathbf{r}(t)$, the velocity vector ${}^{\alpha}\mathbf{v}(t)$ of an element (molecule) of the discrete fluid and a microstress ${}^{\alpha}\boldsymbol{\sigma}(t)$ that may include in general the contributions of interaction effects, as well as other significant dynamic variables.

Hence the following subspaces of the probabilistic function space X may be chosen:

$$
\left. \begin{aligned}
&{}^{i}\boldsymbol{\sigma}(t) \in {}^{i}\Sigma, \; {}^{s}\boldsymbol{\sigma}(t) \in {}^{s}\Sigma; \; {}^{i}\Sigma \oplus {}^{s}\Sigma = \Sigma \subset Z = \text{X}, \\
&{}^{\alpha}\boldsymbol{\sigma}(t) \in \Sigma \subset \text{X}, \\
&{}^{i}\mathbf{u}(t) \in {}^{i}U, {}^{s}\mathbf{u}(t) \in {}^{s}U; \; {}^{i}U \oplus {}^{s}U = \mathcal{U} \subset Z = \text{X} \\
&{}^{\alpha}\mathbf{r}(t) \in \mathcal{C} \subset Z; \; {}^{\alpha}\mathbf{v}(t) \in \mathcal{V}, \subset Z.
\end{aligned} \right\} \qquad (2.91)
$$

Evidently, by representing the behaviour of a discrete medium in terms of an abstract dynamical system, the probability space X and any of its subspaces must be given a suitable structure and an appropriate measure. Since in stochastic mechanics all field variables are considered as random variables, it is readily recognized that their distribution function in terms of the event sets will have the following properties:

$$
\left. \begin{aligned}
&\text{(i) } 0 \leq P\{E_n\} \leq 1 \text{ for all } E_n \in \mathcal{F} \\
&\text{(ii) } P\{E_1 \cup E_2\} = P\{E_1\} + P\{E_2\} - P\{E_1 \cap E_2\}, \\
&\qquad \text{if } E_1, E_2 \in \mathcal{F} \text{ and } E_1 \cap E_2 \neq \emptyset \\
&\text{(iii) } P\{\cup_r E_r\} = \sum_r P\{E_r\}, \text{ if } E_r \cap E_s = \emptyset, r \neq s \\
&\text{(iv) } P\{\text{X}\} = 1.
\end{aligned} \right\} \qquad (2.92)
$$

Showing that the distribution function $P\{E_n\}$ indeed satisfies all the properties of a measure, i.e. that it is an extended real-valued, non-negative and countably additive set function. Hence introducing P as a probability measure, one can construct the probability space $[X, \mathcal{F}, \mathcal{P}]$. The state of a microelement within the range given by (2.87) becomes an event $E(\mathbf{z})$ with the probability measure:

$$\mathcal{P}\{E(\mathbf{z}_n)\} = \mathcal{P}\{\mathbf{z}_n <^\alpha \mathbf{z} < \mathbf{z}_n + \Delta\mathbf{z}_n\}; \ \mathcal{P}\{X\} = 1. \tag{2.93}$$

Hence each of the subspaces of X will be measurable, if and only if E is the union of disjoint open sets E_n. Thus one may consider for instance that part of E which relates to the deformations $^\alpha\mathbf{u}$ of an element in a polycrystalline solid. In this case the event will be specified by:

$$E(\mathbf{u}_n) = \{\mathbf{u}_n <^\alpha \mathbf{u} < \mathbf{u}_n + \Delta\mathbf{u}_n\}; \ \cup_{n=1}^\infty E_n = X, \tag{2.94}$$

and the corresponding probability measure is then:

$$\mathcal{P}\{E(\mathbf{u}_n)\} = \mathcal{P}\{\mathbf{u}_n < \mathbf{u}_n + \Delta\mathbf{u}_n\}; \ P\{\mathcal{U}\} = 1. \tag{2.95}$$

In stochastic mechanics this measure is called the probability density of the microdeformations. In most cases, however, the distribution functions are only experimentally accessible (see also [22] and Chapt. 4). It is necessary therefore to consider cumulative measures that can be defined in terms of the distribution function of the microdeformations, i.e.:

$$P^{\mathbf{u}} = \mathcal{P}\{E(\mathbf{u}_n)\} = \mathcal{P}^{\mathbf{u}}\{^\alpha\mathbf{u} < \mathbf{u}\}; \ \mathcal{P}^{\mathbf{u}}\{\mathcal{U}\} = 1 \tag{2.96}$$

and analogously, a cumulative measure can be given for any other field variable. Generally by using the state vector $\mathbf{z}$, one has:

$$\mathcal{P}^{\mathbf{z}} = P\{E(\mathbf{z})\} = \mathcal{P}\{^\alpha\mathbf{z} < \mathbf{z}\}; P\{X\} = 1. \tag{2.97}$$

Since the above measure is closely related to the distribution function of probability theory, one can employ the probability space of "microdeformations" in a structured solid or the "velocity space" in a discrete fluid by the triple $[\mathcal{U}, \mathcal{F}^{\mathbf{u}}, \mathcal{P}^{\mathbf{u}}]$ and $[\mathcal{V}, \mathcal{F}^{\mathbf{v}}, \mathcal{P}^{\mathbf{v}}]$, respectively. The measures $\mathcal{P}^{\mathbf{u}}, \mathcal{P}^{\mathbf{v}}$ etc. must satisfy however the regularity conditions (Chapt. 1). To show that the measure $\mathcal{P}^{\mathbf{u}}$, for instance, is regular, consider the σ-algebra $\mathcal{F}^{\mathbf{u}}$ of the deformation space $\mathcal{U}$. Since $\mathcal{F}^{\mathbf{u}}$ contains the open sets $E(\mathbf{u}_n)$, it can be shown that the intersection of a countably finite number of open sets $E(\mathbf{u}_n)$ forms a closed set $\overline{E}(\mathbf{u}_n) \in \mathcal{F}^{\mathbf{u}}$. In accordance with the definition of regularity, $\mathcal{P}^{\mathbf{u}}$ is also a measure on $\overline{E}(\mathbf{u}_n) \in \mathcal{F}^{\mathbf{u}}$, so that $\mathcal{P}^{\mathbf{u}}$ satisfies the condition of regularity. Thus the random deformation vector $^\alpha\mathbf{u} \in \mathcal{U}$ can be defined as a $\mathcal{P}^{\mathbf{u}}$-regular measurable function in $\mathcal{U}$ (see also Doob [50] and Billingsley [59]). The identification of $[\mathcal{U}, \mathcal{F}^{\mathbf{u}}, \mathcal{P}^{\mathbf{u}}]$ with the space of all $\mathcal{P}^{\mathbf{u}}$-regular measurable functions of $\mathbf{u}$ and similarly of $[\mathcal{V}, \mathcal{F}^{\mathbf{v}}, \mathcal{P}^{\mathbf{v}}]$ as the space of all regular measurable functions of $\mathbf{v}$ with a probabilistic function space forms the basis for the mathematical structure of the stochastic mechanics of the deformation and flow of discrete media.

In the case of solids, for instance, this leads directly by recognizing the duality with the corresponding stress-space to a "constitutive mapping" and the representation of the response relations in an operational form. For a given structured solid a mapping between the spaces $\mathcal{U}$ and Σ will always exist, whereby the topological structure of one of these spaces, say Σ, may be given in terms of a non-degenerate bilinear form with respect to this mapping. The above function space approach and the mapping between subspaces of the former is considered below.

(iii) The state-space and constitutive maps:

It has been mentioned earlier that the formulation of the stochastic mechanics of discrete media employs an operational formalism. The latter is well-known in mathematical physics and is frequently used when dealing with random equations. This type of equations have become the subject of intense research (see for instance Bharucha-Reid [129], Itô [130], Arnold [131], Hanš [132], Spaček [133], McKean [134] and others). In the application of the operational formalism to the stochastic mechanics of discrete solids, for instance, it is convenient to use a "constitutive mapping" in form of a "material functional" or "material operator" as indicated by the postulate P.4 of section 2.1. Such an operator reflects on the one hand the material properties of a specific solid and on the other, allows for the use of the "duality" of the stress and deformation space. It is evident, that for the use of these subspaces of Z and the corresponding constitutive maps, the duality of linear vector spaces is significant. The mathematical structure of a pair of linear spaces placed in duality by a bilinear form is analogous to the principle of virtual work in classical mechanics. To illustrate the state-space representation concerning a polycrystalline solid and its "elastic response" consider the diagramatic sketch of Fig. 11. Here the subspaces $\Sigma, \mathcal{U}$, respectively are linked by the mapping or operator "M" such that:

$$\left. \begin{aligned} M &: \Sigma \to \mathcal{U}; \ \Sigma = {}^i\Sigma \cup {}^s\Sigma \subset Z \ , \\ M^{-1} &: \mathcal{U} \to \Sigma; \ \mathcal{U} = {}^iU \cup {}^sU \subset Z \ . \end{aligned} \right\} \quad (2.98)$$

The mapping shown in Fig. 12 refers to the inelastic behaviour of a multi-component system to be discussed further in Chapt. 3, 4. The subspaces of interest here are the induced strain field $\mathcal{E}^E$ (elastic) and the inelastic field $\mathcal{E}^{In}$ as well as the subspace $\mathbb{R}^d$ required to describe the transient response of such materials.

Another example of the application of the stochastic state-space where the velocity subspace of Z, i.e. $\mathcal{V} \subset Z$ is important will be used in Chapt. 4 by considering a stochastic transport theory of porous media. In view of the postulate P.1. of the stochastic mechanics theory, constitutive maps concerning the

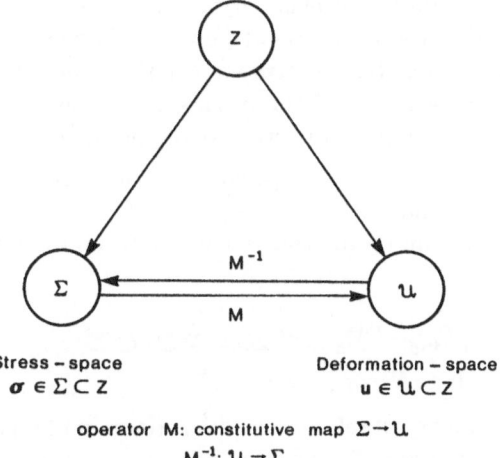

Stress – space
$\sigma \in \Sigma \subset Z$

Deformation – space
$u \in \mathcal{U} \subset Z$

operator M: constitutive map $\Sigma \rightarrow \mathcal{U}$
$M^{-1}: \mathcal{U} \rightarrow \Sigma$

Fig. 11. Stochastic state-space Z with stress space Σ, deformation space $\mathcal{U}$ and the constitutive map M, M^{-1}.

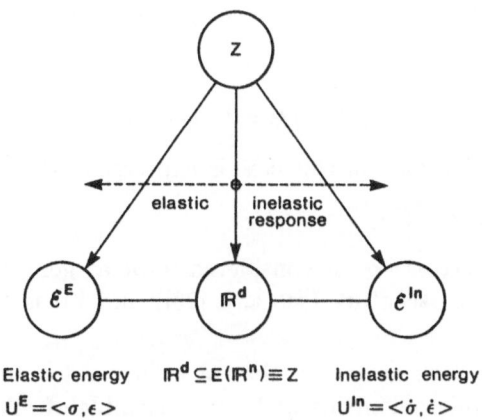

Elastic energy $\mathbb{R}^d \subseteq E(\mathbb{R}^n) \equiv Z$ Inelastic energy
$U^E = <\sigma, \epsilon>$ $U^{In} = <\dot{\sigma}, \dot{\epsilon}>$

Fig. 12. Stochastic state-space Z (subspaces $\mathcal{E}^E, \mathcal{E}^{In}, \mathbb{R}^d$) for the inelastic behaviour of a multi-component material.

macroscopic material body are obtained from the "micro-domain" and "meso-domain" operators in terms of their corresponding distributions as shown below. In most cases of practical applications, it is sufficient to establish the expected or mean value of these operators.

In general, one can formally express a material functional for the elastic response of a polycrystalline solid, by:

$$M = M\{E, \mathbf{a}, \rho, \phi, \underline{Q}, \theta, t \dots\} \tag{2.99}$$

in which E is the modulus of elasticity of the solid, $\mathbf{a}$ the lattice parameter, ρ the dislocation density, ϕ the bond potential between microelements, $\underline{O}$ an orientation matrix, θ the temperature and t the time, etc. Evidently, it is extremely difficult to assess all of these parameters, some of which may not be required in the description of a particular phenomenon.

In order to discuss the type of operators involved in the subsequent analysis beginning with the smallest scale, i.e. that of a microelement or that of a microdomain Γ, one can define the "local" material operator as follows:

Def. 1:

$$\left. \begin{aligned} {}^{\alpha}m &: {}^{\Gamma}\Sigma_d \to {}^{\Gamma}\mathcal{U}_d; \;\; {}^{\Gamma}\Sigma_d \oplus {}^{\partial\Gamma}\Sigma_d = {}^{m}\Sigma_d \subset \mathbf{X} \\ &\text{for all } \alpha = 1 \dots N, \text{ or } \Gamma = 1 \dots N; \end{aligned} \right\} \quad (2.100)$$

where the superscripts $\Gamma, \partial\Gamma$ refer to the microdomain of an element or the somewhat larger "microdomain Γ" and their boundaries, respectively. The subscript d indicates the "dense sets" of the stress-space $\Sigma \subset \mathbf{X}$ and deformation space $\mathcal{U} \subset \mathbf{X}$.

On the assumption that in most cases these micro-operators will be linear and bounded one can also use an inverse operator, i.e.:

Def. 2:

$$({}^{\alpha}m)^{-1}: {}^{\Gamma}\mathcal{U}_d \to {}^{\Gamma}\Sigma_d; \; {}^{\Gamma}\mathcal{U}_d \oplus {}^{\partial\Gamma}\mathcal{U}_d = {}^{m}\mathcal{U}_d \subset \mathbf{X} \,. \qquad (2.101)$$

Hence the constitutive map in this case is expressed by:

$$ {}^{\alpha}\mathbf{u} = {}^{\alpha}m\,{}^{\alpha}\sigma; \;\; {}^{\alpha}\sigma \in {}^{\Gamma}\Sigma_d, \; {}^{\alpha}\mathbf{u} \in {}^{\Gamma}\mathcal{U}_d \,, \qquad (2.102)$$

in which the "microstress ${}^{\alpha}\sigma$" is considered to be a "generalized stress" acting on an element of the structure (see also [23]) and ${}^{\alpha}\mathbf{u}$ is the associated local deformation.

Considering the next scale, i.e. that of a "meso-domain" which is formed by an ensemble of microelements or by an ensemble of micro-domains Γ, the resulting operator can be obtained from the mean value of the micro-operators as follows:

Def. 3:

$$\left. \begin{aligned} {}^{M}m(\alpha) &= E\{{}^{\alpha}m\} = \sum_{\alpha=1}^{N} {}^{\alpha}m\Delta\mathcal{P}\{{}^{\alpha}m\} \\ &\text{and similarly} \\ {}^{M}m(\Gamma) &= E\{m(\Gamma)\} = \sum_{\Gamma=1}^{N} m(\Gamma)\Delta\mathcal{P}\{m(\Gamma)\} \end{aligned} \right\} \quad (2.103)$$

Again the "meso-operator" can be considered for the interior of the meso-domain $\mathcal{D}_M$ and on its boundary $\partial \mathcal{D}_M$ so that:

$$\left. \begin{aligned} &^M m: {}^M \Sigma \to {}^M \mathcal{U}; \quad {}^{\mathcal{D}_M} \Sigma \oplus {}^{\partial \mathcal{D}_M} \Sigma = {}^M \Sigma \subset X \\ &\text{and} \\ &(^M m)^{-1}: {}^M \mathcal{U} \to {}^M \Sigma; \quad {}^{\mathcal{D}_M} \mathcal{U} \oplus {}^{\partial \mathcal{D}_M} \mathcal{U} = {}^M \mathcal{U} \subset X. \end{aligned} \right\} \quad (2.104)$$

It is apparent that the constitutive map for the macroscopic material domain is thus given by:

Def. 4:

$$\mathcal{M}: {}^M \Sigma \to {}^M \mathcal{U}; \quad {}^M \Sigma \text{ and } {}^M \mathcal{U} \subset X , \qquad (2.105)$$

in which $^M \Sigma$, $^M \mathcal{U}$ are recognized as the union of non-intersecting disjoint sets of the state-space or probabilistic function space, i.e.:

$$^M \Sigma = \cup_{p=1}^P {}^M \Sigma_p \text{ and } {}^M \mathcal{U} = \cup_{p=1}^P {}^M \mathcal{U}_p. \qquad (2.106)$$

The macroscopic material functional or operator "$\mathcal{M}$" is an element of the space of linear continuous operators, so that:

$$\mathcal{M} \in \mathcal{L}({}^M \Sigma, {}^M \mathcal{U}) \qquad (2.107)$$

(see also J.J. Moreau [135, 136]). The above chosen subspaces of the state-space $Z \equiv X$ pertaining to a structured solid are assumed in general to be Banach spaces. Thus, if X and hence Z is a Banach space, a proper norm can be introduced for its topology. Considering, for instance the micro-deformations of a structured solid and the subspace $\mathcal{U}$ as the space of all $\mathcal{P}^u$-regular measurable and bounded functions discussed previously, which are discrete, the expected value of $^\alpha \mathbf{u}$ will be given by:

$$< {}^\alpha \mathbf{u} >= \sum_{\alpha=1}^N {}^\alpha \mathbf{u} \mathcal{P}\{E({}^\alpha \mathbf{u})\} = \sum {}^\alpha \mathbf{u} \Delta \mathcal{P}^u. \qquad (2.108)$$

Thus in an experiment, where $< {}^\alpha \mathbf{u} >$ is equal to zero, although individual micro-deformation $^\alpha \mathbf{u}$ may still exist, $| < {}^\alpha \mathbf{u} > |$ will be a semi-norm. In this case the deformation space $\mathcal{U}$ will be a Fréchet space with this semi-norm. However, if in the same experiment one is rather interested in finding the average value of the magnitude of the micro deformations $^\alpha \mathbf{u}$, a similar definition to relation (2.108) would satisfy the properties of a norm, that is:

$$\| {}^\alpha \mathbf{u} \| = < |{}^\alpha \mathbf{u}| >= \int_\mathcal{U} |{}^\alpha \mathbf{u}| d\mathcal{P}^u \qquad (2.109)$$

in which $\mathcal{U}$ is now a Banach space. Using the standard deviation of $\mathbf{u}$ one obtains in accordance with (2.108):

$$D^u = \{\sum |^\alpha \mathbf{u}_s - <{}^\alpha \mathbf{u}_s>|^2 \Delta \mathcal{P}^u\}^{1/2} \tag{2.110}$$

where D^u again satisfies the properties of a semi-norm. If $\mathcal{U}$ is a locally convex space with this semi-norm and if it is completely metrizable with respect to $<|^\alpha \mathbf{u}|>$, it is then a Fréchet space. Although the analysis in a Fréchet space is analogous to that in a Banach space, the choice of the proper space will depend on the type of phenomenon and the medium under consideration.

These subspaces as well as explicit forms of the material operator will be considered in the applications of the stochastic mechanics theory in Chapt. 4. Finally, one can regard the deformations occuring in a discrete solid as stochastic processes in which the above operators map at any instant of time, the elements of the stress-space into elements of the deformation-space and thus perform the basic role of "constitutive relations".

Since the macro-operator is also time-dependent, in general, it can be designated more generally by $\mathcal{M}(t)$. From the statement given in (2.107) it connects then the macroscopic stress and deformation fields, induced by external loads. It is to be noted however, that there are mathematical restrictions on this operator such as measurability, convergence, etc. In practical applications it is more convenient to consider the macroscopic maps in terms of the distribution of the meso-operators as will be shown by various examples in Chapt. 4. The operator $\mathcal{M}(t)$ and its distribution with time on the basis of Markov theory will be further discussed in the following chapter dealing with the random evolution of discrete media.

2.4 Markov processes and stochastic differential equations.

In view of the importance of the theory of Markov processes and their application in modelling the behaviour of discrete media, the fundamentals given in section 1.6 of Chapt. 1 will be somewhat extended in this section. More specifically, the concepts of "transition probabilities" and "states" of the medium become important in the construction of probabilistic models. Further, the relation of Markov processes and the Brownian motion as well as the associated stochastic differential equations shall be briefly reviewed.

(i) Markov processes:

As mentioned earlier stochastic processes may be regarded as a collection or a family of random variables on the same probability space $[X, \mathcal{F}, , \mathcal{P}]$ usually denoted by $x_t = \{x(t), t \in T\}$, where the parameter t belongs either to a set of non-negative integers $\mathbb{Z}_+$ or to the index set T of the real line $\mathbb{R}^1$. Since t is usually considered as the time, it is often called an "epoch" (see for instance Gihmann and Skorohod [137], Lipster and Shiryayev [138] and others). Since most of the material models are based on processes with a "finite state-space" or on "finite stochastic processes" with the Markov property, it implies that it is sufficient to know the state of the process at time t (epoch) in order to determine the probability distribution of $x(u), u > t$ for a given interval $(s < t < u) \in \tau \subset T$ and where the information about $x(s), s < t$ is not required. Recalling the definition of "independent and identically distributed (i.i.d.)" random variables (section 1.5 of Chapt. 1), a sequence of such variables $\{x_n, n \in \mathbb{Z}_+\}$ will be identically distributed, if for every $n \in \mathbb{Z}_+$ one has:

$$\left. \begin{array}{l} P\{x_n = i\} = p_i; i = 0, 1, 2 \ldots ; p_i \geq 0, i \in \mathbb{Z}_+ \\ \sum_{i=0}^{\infty} p_i = 1. \end{array} \right\} \quad (2.111)$$

They will be independently distributed, if for every $n \geq 2$ one has:

$$\left. \begin{array}{l} P\{x_1 = i_1 \ldots x_n = i_n\} = \prod_{k=1}^{n} P\{x_k = i_k\}, \\ \qquad\qquad\qquad \text{for all } i_1 \ldots i_n \in \mathbb{Z}_+ \end{array} \right\} \quad (2.112)$$

The above characteristics show that the random variables can be regarded as a sequence of discrete variables taking on values in a finite subset of $\mathbb{R}^1$. They form a "Markov chain" with state-space Z, if for every $n \geq 1; i_0 .. i_r .. i_{n+1} \in Z$, one has:

$$P\{x_{n+1} = i_{n+1} | x_0 = i_0, \ldots x_n = i_n\} = P\{x_{n+1} = i_{n+1} | x_n = i_n\} \quad (2.113)$$

as given in section 1.5, but where the initial condition is given for $n = 0$ rather than $n = 1$, since in practical applications n being identified with t the process starts at $n_0 = t_0 = 0$. The above representation also permits the grouping of "states". Thus, if a state "i" goes to "j" there exists an $n \in \mathbb{Z}_+$ such that $p_{ij}^{(n)} > 0$ and every state leads to itself as $p_{ii}^{(0)} = 1$. Two states i, j are said to "communicate", if they lead to each other. A "Markov chain" is said to be "irreducible", iff its state-space Z is minimal closed, (a closed set $\overline{Z}$ is minimal closed, if no proper subset of Z is closed). If for a state i, $p_{ii} = 1$, then i is said to be an "absorbing state". Hence for an absorbing state, $p_{ij}^{(n)} = 0$ for all $j \neq i$ and all $n \in \mathbb{Z}_+$. If i is a fixed state and Z_i the set of states which communicate with i, it forms a "class of states". A property of one state in the class, implies that all states in that class have the same property.

A classification of states of a Markov chain can also be given in terms of the probability of its "return to its original state". Thus, if $p_{ij} = \pi_{ij}^{(1)}$, then:

$$\left. \begin{aligned} \pi_{ij}^{(n)} &= P\{x_n = j, x_r \neq j; (0 < r < n)|x_0 = i\} \\ &= P\{x_{m+n} = j, x_{m+r} \neq j; (0 < r < n)|x_m = i\}; n \geq 2. \end{aligned} \right\} \quad (2.114)$$

It is to be noted that, if $\pi_{ii}^{(n)}$ is the probability of the first return of the Markov chain to the state "i" at an epoch "n" and since these events are mutually exclusive, one has:

$$\sum_{n=1}^{\infty} \pi_{ii}^{(n)} = \pi_{ii} \, ,$$

or the probability of at least one return of the chain to the state "i". One says that the state "i" is "persistent", if $\pi_{ii} = 1$ and "transient", if $\pi_{ii} < 1$. It has been mentioned earlier that a stochastic process $(x_t, 0 < t < \infty)$ is a process with "stationary and independent increments", if for all $n \geq 1, 0 = t_0 < t_1 < \ldots < t_n$, the increments $x(t_1) - x(0)$, $x(t_2) - x(t_1), \ldots x(t_n) - x(t_{n-1})$ and their joint distribution is the same as for the increments or:

$$x(t_1 + h) - x(h), x(t_2 + h) - x(t_1 + h) \ldots x(t_n + h) - x(t_{n-1} + h)$$

for all $h > 0$ and further, if they are independently distributed for all real $x_1 \ldots x_n$ so that:

$$\left. \begin{aligned} &P\{x(t_1) - x(0) \leq x_1 \ldots, x(t_n) - x(t_{n-1}) \leq x_n\} \\ &= \prod_{i=1}^{n} \{x(t_i) - x(t_{i-1}) \leq x_i\}. \end{aligned} \right\} \quad (2.115)$$

It is apparent that by choosing the state-space for the process $\{N(t), 0 \leq t < \infty\}$ or the set of non-negative integers, where $N(t)$ counts the number of occurrences of the events in $(0, t)$, that $N(s) \leq N(t)$ whenever $s \leq t$. Hence it can be called a "counting process' that will be considered later in the text. Its relation to a special Markov or the Poisson process can be readily shown by letting $\{N(t), 0 \leq t < \infty\}$ be a process with stationary and independent increments such that:

$$\left. \begin{aligned} &\text{(i) } P\{N(0) = 0\} = 1 \\ &\text{(ii) } P\{N(t) = k\} = \exp(-\lambda t)(\lambda t)^k / k! \end{aligned} \right\} \quad (2.116)$$

indicating that indeed it is a Poisson process with the rate λ. A generalization of equation (2.113) can easily be carried out, which then shows that the Poisson process is also Markovian. Considering now the time-continuous parameter case and the process $x_t = \{x_t, 0 \leq t < \infty\}$ defined on a complete probability space $[X, \mathcal{F}, \mathcal{P}]$, it is said to be a "discrete Markov process", if its state-space is either a finite or countably infinite subset of $\mathbb{R}^n$.

In the subsequent modelling of various discrete media mainly discrete Markov processes will be of interest. The latter are also referred to as "finite Markov processes" [83], [145].

An important aspect of Markov processes is the transition probability function of the process briefly indicated in chapt. 1. Thus one can give the following definition of a discrete Markov process. It has "stationary transition probabilities" or alternatively it is a time homogeneous process, if its transition probability function for all states $i, j \in Z, s > 0$ is given by:

$$P\{x(t) = j|x(0) = i\} = P\{x_{s+t} = j|x_s = i\} \tag{2.117}$$

(compare with eqns. 1.179, 1.181, 1.187) of section 1.5).

Analogously by the usual designation of Markov chains with the initial probability:

$$P_{ij}(0) = \delta_{ij} = \left\{\begin{array}{ll} 1, & \text{if } i = j \\ 0 & \text{otherwise} \end{array}\right. \left.\begin{array}{l} \\ \\ \end{array}\right\} \tag{2.118}$$

one can then define a Poisson process in terms of the transition probability function by:

$$P_{ij}(t) = \left\{\begin{array}{ll} \exp(-\lambda t)(\lambda t)^{j-i}/(j-i)!, & \text{if } j \geq i, i = 0, 1, 2 \ldots \\ 0 & \text{otherwise.} \end{array}\right\} \tag{2.119}$$

There are several ways of defining a Markov process, which are equivalent and are based on the usual definition of the Markov property of the stochastic process $\{x_t\}$. However, due to the subsequent utilization of a wider class of Markov processes known as "jump Markov processes" the "strong Markov property" should be briefly considered. It becomes necessary for this purpose to discuss in general the notions of the "history, measurability, stopping times" etc. of Markov processes.

As stated in section (1.5), if $[X, \mathcal{F}]$ is a measurable space a "history $\mathcal{F}_t, t \geq 0$" on that space is the family of sub-σ-fields of $\mathcal{F}$ such that for all $0 \leq s \leq t$: $\mathcal{F} = \{\mathcal{F}_t^x, t \geq 0\}$ and $\mathcal{F}_s \subset \mathcal{F}_t$. Thus a history is an increasing function of sub-σ-fields of $\mathcal{F}$ indexed by the non-negative real numbers. It is also called a "filtration" of the space. One defines usually by $\mathcal{F}_\infty = \cup_{t \geq 0} \mathcal{F}_t$. Hence the family of sub-σ-fields $\{\mathcal{F}_t\}$ can be regarded as describing the history of some phenomenon in which $\mathcal{F}_t$ is the σ-field of the events occuring before the time t. It is to be noted that:

$$\mathcal{F}_{t_+} = \cap_{s > t} \mathcal{F}_s, \ t > 0 \text{ and } \mathcal{F}_{t_-} = \cup_{s < t} \mathcal{F}_s. \tag{2.120}$$

The filtration $\{\mathcal{F}_t\}$ is said to be "right continuous", if $\mathcal{F}_t = \mathcal{F}_{t_+}$ ($\mathcal{F}_{t_+}$ is always right-continuous). If the time index set is either continuous in time (i.e. $T = [0, \infty)$) or discrete: $n = \{0, 1, 2 \ldots\}$, then for $\tau = n$, one has:

$$\mathcal{F}_{n_+} = \mathcal{F}_{n+1} \text{ and } \mathcal{F}_{n_-} = \mathcal{F}_{n-1} \, , \qquad (2.121)$$

and the filtration is constant. If $[X, \mathcal{F}]$ is a measurable space with the filtration $\{\mathcal{F}_t\}, t \in T$, a random variable $\tau \colon X \to \tau$ is called a "stopping time", if for every $t \in T \colon \{\tau \leq t\} \in \mathcal{F}_t$. Another significant notion in the present context is that of an "adapted process".

Thus, if $\{\mathcal{F}_t\}, t \in T$ is a filtration of the measurable space $[X, \mathcal{F}]$ and $\{x_t\}$ a process defined in $[X, \mathcal{F}]$ with values in $[Z, \mathcal{F}^z]$, then $\{x_t\}$ is said to be "adapted to $\{\mathcal{F}_t\}$", if x_t is $\mathcal{F}_t$-measurable for each $t \in T$. Thus, supposing $\{x_t\}, t \in T$ is a stochastic process defined in $[X, \mathcal{F}, \mathcal{P}]$, then $\{x_t\}$ is adapted to the filtration $\{\mathcal{F}_t\}$, where

$$\mathcal{F}_t = \sigma\{x_s \colon s \leq t\} \, ,$$

or the σ-field on X generated by all random variables $x_s, s \leq t$.

In general this filtration may not be complete. It can be completed however by adding all $\mathcal{P}$-null sets of $\mathcal{F}$, where $\mathcal{F}^{\mathcal{P}}$ designates the completion of $\mathcal{F}$ and $\{\mathcal{F}_t^{\mathcal{P}}\}$ is a filtration on $[X, \mathcal{F}, \mathcal{P}]$. The Borel fields of the events E (defined by 2.87) can be recognized by letting $X \equiv Z$ to be a topological space. The latter in most applications in stochastic mechanics is either $Z \equiv \mathbb{R}^1$ or $Z \equiv \mathbb{R}^n$. By specifying the σ-field generated by the events E as the Borel field $\mathcal{B}(Z)$ in Z, then $\mathcal{B}(Z) = \mathcal{B}(\mathbb{R}^1)$ and similarly for $Z = \mathbb{R}^n, \mathcal{B}(Z) = \mathcal{B}(\mathbb{R}^n)$. If $Z = \mathbb{R}_+$, then $\mathcal{B}(Z) = \mathcal{B}(\mathbb{R}_+)$ and $\overline{\mathcal{B}}$ is also $\mathcal{B}(\overline{\mathbb{R}})$, where $\overline{\mathbb{R}} = [-\infty, +\infty]$ is the extended real line. The topologies chosen in this manner are then the Euclidean topologies $\mathbb{R}^n$ and $\mathbb{R}_+$ and the extended Euclidean topology of $\overline{\mathbb{R}}$. For any subset A of the events in the topological space Z one can define:

$$\mathcal{B}(A) = \{C | C = A \cap B, B \in \mathcal{B}(Z)\}.$$

It is evident that the open sets in $\mathbb{R}$ may be defined in terms of "open intervals", i.e. open, semi-open or closed. A more comprehensive discussion on filtration, measurability as well as on "martingales" can be found among others in Delacherie [139], Brémaud [47], Elliott [48].

In view of the above statements the "strong Markov property" of a stochastic process can be defined as follows. Let $\{x(t), t \in \mathbb{R}_+\}$ be a process with the Markov property with respect to the right-continuous increasing family $\{\mathcal{F}_t, t \in \mathbb{R}_+\}$ of sub-σ-fields. The process will have the strong Markov property with respect to $\{\mathcal{F}_t, t \in \mathbb{R}_+\}$, if for any Borel set $E \in \mathcal{F}_t$:

$$P\{x_{(\tau+u)} \in E | \mathcal{F}_t\} = P\{x_{(\tau+u)} \in E | x_\tau\} \, , \qquad (2.122)$$

for all positive random variables τ, which are "stopping times" with respect to $\{\mathcal{F}_t, t \in \mathbb{R}_+\}$. A finite Markov process has this strong property with respect to the sub-σ-fields (see also Chung [140]). Generally, the process $x_t \equiv \{x_t\}$, when it is Z-valued, defined on the probability space $[X, \mathcal{F}, \mathcal{P}]$ and "adapted" to some history $\mathcal{F}_t$ can be regarded as a $(\mathcal{P}, \mathcal{F}_t)$-Markov process, if for all $t \geq 0$:

$$\sigma(x_s, s \geq t) \text{ and } \mathcal{F}_t \text{ are } \mathcal{P}\text{-independent given } x_t. \tag{2.123}$$

Conventionally, since $P \equiv \mathcal{P}$ the Markov process x_t is an $\mathcal{F}_t^x$-Markov process as in the case considered earlier, one can then express the expected or mean value of the process by:

$$E\{f(x_t)|\mathcal{F}_s\} = E\{f(x_t)| \sigma(x_s)\}, \text{ for all } 0 \leq s \leq t \tag{2.124}$$

and all measurable functions or mappings $f \colon Z \to \mathbb{R}$.

Using the definition of the transition functions already given in relation (1.124) one has:

if $[Z, \mathcal{F}^z]$ is a measurable space, then for each $0 \leq s \leq t$ the transition probability function is given by:

$$P(z, E), z \in Z \subset X, E \in \mathcal{F}^z \text{ as a function from } Z \times \mathcal{F}^z \to \mathbb{R}_+ \text{ such that :}$$

(i) $E \to P(z, E)$ is a probability on $[Z, \mathcal{F}^z]$ for all $z \in Z$

(ii) $x \to P_{s,t}(z, E)$ is in $\mathcal{B}_+/\mathcal{F}^z$ for all $E \in \mathcal{F}^z$

(iii) for all $0 \leq t$; $t \leq u \leq s$, all $z \in Z$ and all events $E \in \mathcal{F}^z$ $\left.\right\}$ (2.125)

$$P_{s,t}(z, E) = \int P_{s,u}(z, dy) P_{u,t}(y, E) ,$$

where relation (iii) is the Chapman-Kolmogorov equation. The function $P_{s,t}(x, E)$ in accordance with section (1.6) is called the Markov transition function on $[Z, \mathcal{F}^z]$. As stated earlier, if $P_{s,t}(x, E) = P_{t-s}(x, E)$, in the homogeneous case, hence one can also define a Markov process as follows:

if x_t is a Z-valued $(P, \mathcal{F}_t)$-Markov process and $P_{t,s}(x, E)$ the Markov transition function on that space and if

$$E\{f(x_t)|\mathcal{F}_s\} = \int P_{s,t}(x_s, dy) f(y) , \tag{2.126}$$

for all $0 \leq s \leq t$ and all bounded measurable $f \colon Z \to \mathbb{R}$. One says that the $(P_t, \mathcal{F}_t)$-Markov process x_t "admits the Markov transition function $P_{s,t}(x, E)$". The latter can also be expressed by:

$$P_{s,t}(x_t, E) = P\{x_t \in E|\mathcal{F}_s\} \tag{2.127}$$

showing that $P_{s,t}(.)$ is the probability that starting from a point x at time s the process x_t ends at time t in the event set E, conditionally to $\mathcal{F}_s$ and $x_s = x$.

Hence for x_t to be a "state-valued" stochastic process defined in the probabilistic function space $[X, \mathcal{F}, \mathcal{P}]$ and an adapted history $\mathcal{F}_t; (\mathcal{F}_\infty)$ the following statements are equivalent:

(i) x_t is a $(P, \mathcal{F}_t)$-Markov process

(ii) for each $t \geq 0$ and each bounded $\sigma(x_s, s \geq t)$-measurable random variable y:

$$E\{y|\mathcal{F}_t\} = E\{y|\sigma(x_t)\} \,, \tag{2.128}$$

(iii) for all $0 \leq s \leq t$ and all bounded measurable mappings $f: Z \to \mathbb{R}$

$$E\{f(x_t)|\mathcal{F}_s\} = E\{f(x_t)|\sigma(x_t)\}. \tag{2.129}$$

For a homogeneous Markov process x_t admitting a homogeneous transition function $P_{t-s}(x, E)$, one can define for each $t \geq 0$ the "Markov operator P_t" mapping onto itself the set of bounded measurable functions $f: Z \to \mathbb{R}$ by:

$$P_t f(x) = \int_Z P_t(x, dy) f(y). \tag{2.130}$$

It is evident from the Chapman-Kolmogorov eqn (2.125 (iii)) that in terms of the transition operator P, one has:

$$P_t P_s = P_{t+s} \,, \tag{2.131}$$

in which the family $(P_t, t \geq 0)$ is known as the transition semi-group associated with the stochastic process $x_t \equiv \{x_t\}$.

In general given a probability space $[X, \mathcal{F}, \mathcal{P}]$ as a measure space any linear operator $P: L^1 \to L^1$ which satisfies:

$$\left. \begin{array}{l} \text{(i) } Pf \geq 0, \text{ for } f \geq 0, f \in L^1 \,, \\ \text{(ii) } \| Pf \| = \| f \|, \text{ for } f \geq 0, f \in L^1 \,, \end{array} \right\} \tag{2.132}$$

is referred to as a Markov operator and where the elements of L^1, i.e. f, Pf are functions that can only differ on a set of measure zero. Thus for such functions the properties $f \geq 0$ and $Pf \geq 0$ hold almost everywhere. Markov operators have certain properties. For example, if $f, g \in L^1$, then:

$$Pf(x) \geq Pg(x), \text{ whenever } f(x) \geq g(x). \tag{2.133}$$

An operator with the above characteristic is said to be "monotone" (see also J. v. Neumann [141] and [142]). Thus one can give a definition of a monotone operator A on a real Hilbert space for instance such that $A: H \to 2^H$ is monotone, if

(i) $< y_1 - y_2, x_1 - x_2 > \geq 0$, whenever $x_1, x_2 \in H$ and $y_1 \in Ax_1, y_2 \in Ax_2$,

(ii) the operator is "strictly monotone", if $<y_1 - y_2, x_1 - x_2> > 0$ whenever $x_1, x_2 \in H, x_1 \neq x_2, y_1 \in Ax_1, y_2 \in Ax_2$.

The notion of a "monotone mapping" is in fact a generalization of a non-decreasing function $f : \mathbb{R}^1 \to \mathbb{R}^1$, since $f(x_1) \geq f(x_2)$ for all $x_1 \geq x_2$, iff $[f(x_1) - f(x_2)](x_1 - x_2) \geq 0$ for all $x_1, x_2 \in \mathbb{R}^1$. There exist certain inequalities, that are satisfied by the Markov operator, the most significant one being the contractive property. Thus the inequality:

$$\| Pf \| \leq \| f \| ,$$

or the property given in (ii) of (2.132) is satisfied by the Markov operator P and is called a "contraction". Another characteristic of interest is that for some functions $f \in L^1, Pf = f$. Here f is called a "fixed point" of the operator P. In the stochastic mechanics of discrete media the Markov operator is often defined by a "stochastic kernel". Considering the probability space $[X, \mathcal{F}, \mathcal{P}]$ and a kernel $G: X \times X \to \mathbb{R}$ to be a measurable function such that:

(i) $G(x, y) \geq 0$

(ii) $\int_X G(x, y)dx = 1; (dx = P(dx))$ $\left.\right\}$ (2.134)

is called a "stochastic kernel". By using the integral operator P, then:

$$Pf(x) = \int_X G(x, y)f(y)dy; \text{ for } f \in L^1. \tag{2.135}$$

Evidently, since it is linear and non-negative this operator is also a Markov operator. It is of interest to note, that if X is a finite space and $\mathcal{P}$ a counting measure characterizing a Markov chain, the operator $\mathcal{P}$ is a stochastic matrix. Two Markov operators P_1, P_2 with their corresponding stochastic kernels G_1, G_2, P_1, P_2 is also a Markov operator and an integral operator, where:

$$G(x, y) = \int_X G_1(x, z)G_2(z, y)dz . \tag{2.136}$$

Designating a "composite kernel" by $G = G_1 \circ G_2$, it has the properties:

(i) $G_1 \circ (G_2 \circ G_3) = (G_1 \circ G_2) \circ G_2$ (associative law)

(ii) Any kernel formed by the composition of stochastic $\left.\right\}$ (2.137)

kernels is also stochastic

A more detailed treatment of stochastic kernels and Green functions are given among others by Lasota and Mackey [143], Adomian [144], Doob [50]. Since one of the main aims of the stochastic mechanics is the formulation of the evolution of the microstructural characteristics during the deformation and or flow of discrete media, the concept of "dynamical semi-groups" of the transition operators is rather important. Thus, identifying the relevant vector-valued field variables by the state vector $\mathbf{z} \in Z \subset X$, it is convenient to use either Z or other subspaces of X to represent the motion of the discrete medium. For this purpose, it is assumed that the σ-algebra $\mathcal{F}^z$ can be identified in Z and that an appropriate measure on Z, i.e. $\mathcal{P}^z$ can be established.

Hence the entire probability space $[Z, \mathcal{F}^z, \mathcal{P}^z]$ for any particular time t will be characterized by a set of function spaces or a "product space". In particular, if $\mathbb{R}_+ = [0, \infty)$ where for each $t_r \in \mathbb{R}_+, r = 1 \ldots N$, there corresponds a triple $[Z, \mathcal{F}^z, \mathcal{P}^z]$, the N-fold product of such spaces forms a product space in which $\mathbf{z}(X, t)$ becomes a measurable function and where $\mathbf{X}$ designates the position vector to the C.M. of an individual microelement of the structure.

For convenience the product space can be designated by $Z_\infty, \mathcal{F}^z_\infty, \mathcal{P}^z_\infty$, in which case $\mathbf{z}(Z, t)$ can be regarded as a time-continuous random function. Of greater interest however is the alternative representation by the "one-parameter" family of transformations L_t such that:

$$L_t : Z \to Z \text{ for all } t \in \mathbb{R}_+ = [0, \infty]; \ \mathbf{z} \in Z . \tag{2.138}$$

Hence a deformation process for a discrete solid for instance can be analytically defined by the "automorphism L_t" for all $t \in \mathbb{R}_+$. However a deformation process will be formed generally by a family of the L_t's and thus by an "endomorphism" of transformations. In an analogous manner one can think of a discrete fluid with the velocity field $\mathbf{v}(\mathbf{r}, t)$, where $\mathbf{r}$ is the position vector to the C.M. of a molecule, to be an element of the velocity space, i.e. $\mathbf{v}(\mathbf{r}, t) \in \mathcal{V} \subset X$ and the flow will then be described by the set $\{L_t\}$, each L_t representing a unique history of the possible motion.

Since the σ-algebra $\mathcal{F}_\infty$ of the product space has a countably finite number of Borel sets: $E_1(\mathbf{z}) \times E_2(\mathbf{z}) \times \ldots E_N(\mathbf{z})$ corresponding to the sequence $t_1, t_2 \ldots t_N \in \mathbb{R}_+$ and where each Borel set is obtained from the other by L_t or its inverse, it follows that:

$$\left. \begin{array}{l} L_{\Delta t_r} E_r(\mathbf{z}) = E_{r+1}(\mathbf{z}); \ r = 1, 2 \ldots N - 1, \\ \Delta t_r = t_{r+1} - t_r. \end{array} \right\} \tag{2.139}$$

It can readily be shown that if $\mathcal{P}^z$ is a regular measure on E_r and L_t is defined so as to satisfy (2.139), E_{r+1} is also $\mathcal{P}^z$-regular measurable. This can be generalized to any Borel sets $E_r(\mathbf{z}); r = 1, 2 \ldots N$ in $\mathcal{F}^z$. Hence, it may be concluded that at any time during a deformation or flow process the random variables $\mathbf{z}$ are $\mathcal{P}^z$-regular measurable. In view of the modelling of the behaviour of discrete media by means of the Markov theory, it is more fundamental to consider "conditional probabilities". Thus to each automorphism $L_{\Delta t_r}$ there corresponds a conditional probability measure: $\mathcal{P}\{E_{r+1}|E_r\}$ such that whenever

$$L_t E_r(\mathbf{z}) = E_{r+1}(\mathbf{z})$$

holds, one has:

$$\mathcal{P}^z\{E_{r+1}\} = \mathcal{P}^z\{E_{r+1}|E_r\}\mathcal{P}^z\{E_r\}, \tag{2.140}$$

which can be generalized to:

$$\mathcal{P}^z\{E_N\} = \mathcal{P}^z\{E_1\} \prod_{r=1}^{N-1} \mathcal{P}^z\{E_{r+1}|E_r\} \tag{2.141}$$

for any sequence $E_1 \supset E_2 \supset \ldots E_{N-1} \supset E_N$ corresponding to the time sequence $t_1 < t_2 < \ldots < t_N$ and a set of conditional probabilities $\mathcal{P}\{E_{r+1}|E_r\}$; $r = 1, 2 \ldots N - 1$.

It is of interest to note that eqns (2.140, 2.141) are very significant in stochastic mechanics, since they allow the determination of probability distributions of certain field variables at any time during the deformation or flow process, if the distribution at any other time or the initial one, is known. Such distributions can be established from appropriate experimental observations of the microstructure of a given medium and the use of "geometrical probabilities". The latter will be discussed in Chapt. 3. To recognize this, one can write expression (2.140) in the following form:

$$\mathcal{P}^z\{E_{r+1}, t_{r+1}\} = \mathcal{P}\{E_{r+1}, t_{r+1}|E_r, t_r\}\mathcal{P}^z\{E_r, t_r\} \tag{2.142}$$

where E_r corresponds to $t_r \in \mathbb{R}_+$ and E_{r+1} to $t_{r+1} > t_r$. Denoting the space of all measures $\mathcal{P}^z\{E_N\}$ by $L(0,1)$ or L^1, it is seen that this conditional probability in $L^1 \to L^1$ is a contraction, i.e. $\| \mathcal{P}(\cdot) \| \leq 1$. Recalling the notation of the transition probability function of section (1.6), the function $P(s, \xi; t, E)$ with probability one will be a transition function for $(s, t) \in T$, if:

$$P(s, \xi; t, E) = \mathcal{P}\{z(t) \in E | z(s) = \xi\} \tag{2.143}$$

with the following properties:

$$\left.\begin{array}{l}
\text{(i) } P(s, \xi; t, E) = \begin{cases} 1 & \text{for } \xi \in E \\ 0 & \text{for } \xi \notin E \end{cases} \\[2ex]
\text{(ii) } P(s, \xi; t, E) \leq 1 \\[1ex]
\text{(iii) for fixed time instants } (s, t) \in T \text{ and } \xi \in Z, \ P(s, \xi; t, E) \\
\qquad \text{is a probability measure on } Z \\[1ex]
\text{(iv) for fixed } (s, t) \text{ and } E \in \mathcal{F}^z, P(s, \xi; t, E) \\
\qquad \text{is a } z\text{-measurable function of } \xi \in Z \\[1ex]
\text{(v) } P(s, \xi; t, E) = \int_Z P(s, \xi, \tau, d\eta) P(\tau, \eta; t, E), s \leq \tau \leq t
\end{array}\right\} \tag{2.144}$$

in which (v) is again the Chapman-Kolmogorov relation. The transition function always exists, if the space Z is contained in or identified with X and is separable, and the probability distributions given by:

$$\mathcal{P}_t\{E\} = \mathcal{P}\{z(t) \in E\}, E \in \mathcal{F}^z \tag{2.145}$$

are perfect measures. The topology of Z induced by the bounded events is how-ever a "weak topology" (see also Guz [146], Silverstein [147]). If the Markov process is homogeneous, i.e. representing for instance a "steady-state" defor-mation of a structured solid or the uniform flow of a simple fluid, the transition function reduces to:

$$P(t, \xi, E) \equiv P(0, \xi; t, E) \tag{2.146}$$

and hence the property in (v) becomes:

$$P(t + s, \xi, E) = \int_Z P(t, \xi, d\eta) P(s, \eta, E). \tag{2.147}$$

For the discrete state-space $Z \subset X$ one obtains accordingly the matrix equation:

$$\underline{P}(t + s) = \underline{P}(t)\underline{P}(s) \tag{2.148}$$

showing that in the temporally homogeneous process the Markov operator ex-hibits the semi-group property. By parametrizing P with time t, one has:

$$P_t[f(\xi)] = \int_{Z \subset X} f(\eta) P(t, \xi, d\eta); \ f \in \subset B(Z) , \tag{2.149}$$

where $B(Z)$ denotes in general the Banach space of all bounded continuous functions $f(\xi)$ on Z with the norm:

$$\| f \| = \sup_{\xi \in Z} |f(\xi)|.$$

Thus it follows from (ii) and (v) of (2.144) that $\{P_t, t \geq 0\}$ is a contraction semi-group of operators on Z, or:

$$\left.\begin{array}{l} \text{(i) } P_{t+s}[f(\xi)] = P_t P_s[f(\xi)] \\ \text{(ii) } P_0 = I \text{ (Identity operator)} \\ \text{(iii) } \| P_t \| \leq 1. \end{array}\right\} \tag{2.150}$$

The above semi-group is often referred to as a "dynamical semi-group" (see also Kossakowski [126], Mackey [127], Guz [146]).

It can be seen from the above properties of the P_t-operator that the con-dition of the transition function $P(t, \xi, E)$ to be stochastically continuous is equivalent to the condition that the semi-group of operators P_t be continuous, i.e.:

$$s\text{-}\lim_{t \to t_0} P_t[f(\xi)] = P_{t_0}[f(\xi)]; \ f(\xi) \in B(Z). \tag{2.151}$$

In the context of Markov theory, one can use the "weak-limit" that corresponds to the weak topology of $B(Z)$, since in this case the two limits are equivalent (see also Butzer and Berens [39]). Another significant concept concerning semi-groups in a Banach space is the "infinitesimal generator A" for the operator $P_t : Z \to Z$, which is given by:

$$Az = s\text{-}\lim_{t \downarrow t_0} t^{-1}(P_t - I)z. \tag{2.152}$$

Of further interest here is a theorem by Yosida [24] in connection with the contracting C_0-class of semi-groups that can be stated as follows:

"if Z is a Banach space and $\{P_t\}$ a one-parameter contracting C_0-class of semi-groups of linear operators in Z, the infinitesimal generator A of this semi-group has the following properties:

(i) A is a linear operator in Z with the domain:

$$\mathcal{D}(A) = \{z \in Z \colon s\text{-}\lim t^{-1}(P_t - I)z \text{ exists in} Z$$

(ii) if $z \in \mathcal{D}(A), t \mapsto P_t z$ is a strongly differentiable function of the variable, $t \geq 0$ such that:

$$\frac{d}{dt}(P_t z) = AP_t z = P_t Az; \ t \geq 0 \tag{2.153}$$

with the solution: $P_t = e^{At}$".

Considering now a continuous Markov process $\{x_t, t \geq 0\}$, which is characterized by the transition function in a given subspace of X, i.e.:

$$P(s, \xi; t, \eta) = \mathcal{P}\{x(t) < \eta | x(s) = \xi\}; \ t > s \tag{2.154}$$

then $P(.)$ is a continuous function of t for a fixed s and ξ. It is also a conditional distribution function in η satisfying the following condition:

$$\lim_{\eta \to -\infty} P(s, \xi; t, \eta) = 0 \text{ and } \lim_{\eta \to \infty} P(s, \xi; t, \eta) = 1. \tag{2.155}$$

Assuming that the corresponding probability density is given by:

$$p(s, \xi; t, \eta) = \frac{\partial P(s, \xi; t, \eta)}{\partial \eta}, \tag{2.156}$$

one can derive the so-called "backward" Kolmogorov equation on the assumption that the transition function in (2.154) will satisfy for a small time interval "Δt" the following conditions:

(i) $\displaystyle \lim_{\Delta t \to 0} \frac{1}{\Delta t} \int\limits_{|\eta - \xi| > \epsilon} P(s, \xi; s + \Delta t, d\eta) = 0$

(ii) $\displaystyle \lim_{\Delta t \to 0} \frac{1}{\Delta t} \int\limits_{|\eta - \xi| \le \epsilon} (\eta - \xi) P(s, \xi; s + \Delta t, d\eta) = a(s, \xi)$ $\qquad (2.157)$

(iii) $\displaystyle \lim_{\Delta t \to 0} \frac{1}{\Delta t} \int\limits_{|\eta - \xi| \le \epsilon} (\eta - \xi)^2 P(s, \xi; s + \Delta t, d\eta) = d^2(s, \xi)$

where $\epsilon > 0$ is a positive number and $a(s, \xi)$ the "drift coefficient" characterizing on the average the evolution of $x(s)$ in the small time interval from $s \to s + \Delta t$ under the condition that $x(s) = \xi$.

The quantity $d(s, \xi)$ determines the mean square deviation of $x(s)$ from its expected value. It is known as the "diffusion coefficient". Assuming that the above limits are satisfied uniformly with respect to s and by introducing a "test function ϕ" that is bounded and continuous, one has:

$$\phi(s, \xi) = \int \phi(\eta) P(s, \xi; t, \eta); \ t > s \qquad (2.158)$$

and further on the assumption that the derivatives of ϕ : $\frac{\partial}{\partial \xi}\phi(s, \xi)$ and $\frac{\partial^2}{\partial \xi^2}\phi(s, \xi)$ are bounded and continuous and ϕ has also a derivative with respect to s, the test function will satisfy the following differential equation:

$$\frac{\partial}{\partial s}\phi(s, \xi) = -a(s, \xi)\frac{\partial}{\partial \xi}\phi(s, \xi) - \frac{1}{2}d^2(s, \xi)\frac{\partial^2}{\partial \xi^2}\phi(s, \xi) \ , \qquad (2.159)$$

with the limiting condition: $\lim_{s \to t} \phi(s, \xi) = \phi(\xi)$.

Using the transition probability density function (2.156) which is assumed to be continuous with respect to s together with its first and second derivative with respect to ξ, it will represent a fundamental solution of the following differential equation:

$$\frac{\partial}{\partial s}p(s, \xi; t, \eta) = -a(s, \xi)\frac{\partial}{\partial \xi}p(s, \xi; t, \eta) - \frac{1}{2}d^2(s, \xi)p(s, \xi; t, \eta) \ , \qquad (2.160)$$

referred to as the "backward Kolmogorov" equation. To obtain the "forward equation", which is the adjoint of the above relation, on the assumption that the transition probability density of (2.156) exists and has derivatives continuous with respect to s and η, and by introducing a non-negative continuous function (see also [137], Bharucha-Reid [44] and Sharpe [148]) satisfying:

$$\begin{aligned} &\phi(\eta) = 0 \text{ for } \eta < \eta_1 \text{ and } \eta > \eta_2; \eta_1 < \eta_2 \\ &\phi(\eta_1) = \phi(\eta_2) = \phi'(\eta_1) = \phi'(\eta_2) = \phi''(\eta_1) = \phi''(\eta_2) = 0 \end{aligned} \qquad (2.161)$$

as well as a second bounded function $\phi(\xi)$ gives upon expanding it about η the density function $p(s, \xi; t, \eta)$ representing a fundamental solution of the following differential equation:

$$\left.\begin{array}{l} \dfrac{\partial}{\partial s}p(s, \xi; t, \eta) = -\dfrac{\partial}{\partial \eta}[a(t, \eta)p(s, \xi; t, \eta)] \\[2ex] \qquad\qquad +\dfrac{1}{2}\dfrac{\partial^2}{\partial \eta^2}[d^2(t, \eta)p(s, \xi)p(s, \xi; t, \eta)] \end{array}\right\} \quad (2.162)$$

which is known as the "forward Kolmogorov" or the "Fokker-Planck" equation. Both the forward and backward relations will be further discussed in applications of the stochastic mechanics theory in Chapt. 4. The above scalar representation can be readily extended to the vector-valued Markov process, which consists of several independent random functions $a_i(t)$, i.e. $\mathbf{a}_N(t) = \{a_1(t), \ldots a_N(t)\}$. The probability density then satisfies a multi-dimensional Fokker-Planck equation of the form:

$$\frac{\partial p}{\partial t} = -\sum_{i=1}^{N} \frac{\partial}{\partial a_i}[A_i(\mathbf{a}_N)p] + \sum_{i,j=1}^{N} \frac{\partial^2}{\partial a_i \partial a_j}[B_{ij}(\mathbf{a}_N)p]. \qquad (2.163)$$

The existence and uniqueness of the solution of the Kolmogorov equations have been studied extensively. Thus theorems concerning these solutions in terms of the semi-group theory were given, amongst others, by Feller [64], Hille [149] and Yosida [24]. It has been stated earlier (section 1.6) that for a homogenous Markov chain, the transition probability densities always satisfy the inequality:

$$\sum_{i,j} q_{ij} \leq q_i \; ; \; (i = 1, 2 \ldots) \qquad (2.164)$$

and under these conditions the transition probabilities $P_{ij}(t)$ also satisfy the backward and forward Kolmogorov relations. Recalling the earlier given limiting conditions (1.185) for the transition probabilities of a time-continuous Markov chain and if this chain has the state space $Z = \mathbb{Z}_+$ then for each $t \geq 0$, the transition probabilities $P_{ij}(t), i, j \in \mathbb{Z}_+$ are characterized by a matrix or:

$$P_{ij}(t) = P\{x_{t+s} = j | x_s = i\}$$

with the continuity condition:

$$\lim_{t \downarrow 0} P_{ij}(t) = \left\{\begin{array}{ll} 0 & \text{for } i \neq j \\ 1 & \text{for } i = j \, . \end{array}\right\} \qquad (2.165)$$

The following limits exists (Prohorov and Rozanov [42]):

$$\lim_{t \downarrow 0} \frac{(1 - P_{ii}(t))}{t} = q_i \leq \infty; \; \lim_{t \downarrow 0} \frac{P_{ij}(t)}{t} = q_{ij} < \infty. \qquad (2.166)$$

If Z is finite, the second of the conditions in (2.166) as well as:

$$q_i = \sum_{i,j} q_{ij} ,$$

are always satisfied. For a homogeneous Markov process $\{x_t, t \geq 0\}$ with a countable state-space and a transition semi-group satisfying (2.131) the transition probabilities $P_{ij}(t)$ will satisfy the forward Kolmogorov relation in the form of:

$$P_{ik}^*(t) = -P_{ik}(t)q_k + \sum_{j \neq k} P_{ij}(t)q_{jk}; \ i, k \geq 0 \tag{2.167}$$

and correspondingly the backward relation:

$$P_{ik}^*(t) = -q_i P_{ik}(t) + \sum_{j \neq i} q_{ij} P_{jk}(t); \ i, k \geq 0. \tag{2.168}$$

An application of the above relations will be given in Chapt. 4 (see also Gihman and Skorohod [137], Lipster and Shirjayev [138]). It is of interest to note, that by summing over i, one obtains the distribution of x_t at time t $(P_k(t) = P[x_t = k])$, where:

$$P_k^*(t) = -P_k(t)q_k + \sum_{j \neq k} P_j(t)q_{jk}; \ k \geq 0 \tag{2.169}$$

and by denoting the initial distribution by $P_0(k) \equiv \pi_k$, then:

$$\pi_k q_k = \sum_{j \neq k} \pi_j q_{jk}, \ k \geq 0$$

which holds for the representation of the "steady-state" behaviour of a discrete medium with $P_k(t) = \pi_k$ for all $t \geq 0$. Thus one may regard π_k as an "equilibrium distribution" of the process x_t (see also [23]). Although this distribution may not be unique, it is however in many cases experimentally accessible. In this context the notion of "reversibility" of a Markov chain is of interest. Thus a stochastic process $\{x_t, t \in \mathbb{R}\}$ is reversible, if for all τ and all $t_1, t_2 \ldots t_n \in \mathbb{R}$, the distributions of $(x_{t_1} \ldots x_{t_n})$ and $(x_{\tau - t_1}, \ldots x_{\tau - t_n})$ are the same. Hence if x_t is a "stable process" as in the case of the "steady-state deformation" or flow of a discrete medium, which can be represented by a conservative Markov chain, an instantaneous equilibrium is expressed by:

$$P\{x_t = i\} = \pi(i) ,$$

so that a necessary and sufficient condition of reversibility can be stated analytically as:

$$\pi(i)q_{ij} = \pi(j)q_{ij} ; \ i, j \in \mathbb{R}_+. \tag{2.170}$$

The above form is due to Kolmogorov [150] and is known from statistical mechanics as a "detailed balance" relation. It describes the idealized behaviour of real discrete media, but will not always hold, since their behaviour cannot in general be represented by a conservative Markov chain.

The Brownian motion or "Wiener process" has been briefly mentioned in section (1.6). It is of great importance in certain applications of the stochastic mechanics theory and, hence, some additional remarks may be indicated. One can define a continuous parameter stochastic process $\{\beta_t, t \geq 0\}$ as a Brownian motion or a "Wiener process $\{w(t), t \geq 0\}$, (1-dimensional) on the probability space $[X, \mathcal{F}, \mathcal{P}]$, if:

$$\left.\begin{array}{l} \text{(i) } \beta_{t=0} = 0, \\ \text{(ii) } \{\beta_t, t \geq 0\} \text{ has stationary increments,} \\ \text{(iii) } \{\beta_t\} \text{ for every } t \geq 0 \text{ is normally distributed,} \\ \text{(iv) the increments } (\beta_t - \beta_s) \text{ have a Gaussian distribution} \\ \qquad \text{with: } E\{(\beta_t - \beta_s)\} = 0 \text{ and a variance: } D\{(\beta_t - \beta_s)\} \\ \qquad\qquad\qquad\qquad\qquad\qquad\qquad\qquad = \sigma^2 |t - s|. \end{array}\right\} \quad (2.171)$$

It follows from these definitions that a "standard" Brownian motion process (1-dimensional) has the following properties:

$$\left.\begin{array}{l} \text{(i) the expected value: } E\{\beta_t\} = 0 , \\ \text{(ii) the covariance function: } cov(\beta_s \beta_t) = E\{\beta_s \beta_t\} = min(s,t) \end{array}\right\} \quad (2.172)$$

giving a probability distribution (see also section (1.6)) of the form:

$$P(\beta_t \leq x) = \frac{1}{\sqrt{2\pi t}} \int\limits_{-\infty}^{x} e^{-y^2/2t} dy. \qquad (2.173)$$

By considering the sub-σ-field of β_t on $[X, \mathcal{F}, \mathcal{P}]$ or:

$$\mathcal{F}_t^\beta = \sigma(\beta_s, s \leq t),$$

the Brownian motion is seen to be a martingale relative to $\mathcal{F}_t^\beta, 0 \leq t \leq T$, where:

$$\left.\begin{array}{l} E\{\beta_t | \mathcal{F}_s^\beta\} = \beta_s; \ t \geq s \text{ and } E\{(\beta_t - \beta_s)^2 | \mathcal{F}_s^\beta\} = t - s , \\ \qquad\qquad\qquad\qquad\qquad\qquad\qquad\qquad t \geq s . \end{array}\right\} \quad (2.174)$$

Since a process with independent increments is Markovian, one can by introducing conditional probabilities for any measurable function $f(x) < \infty$ (see also Doob [50]) write that:

$$E\{f(\beta_{t+s}) | \mathcal{F}_t^\beta\} = E\{f(\beta_{t+s}) | \beta_s\}, a.s.; \ s \geq 0 \qquad (2.175)$$

and in particular for any Borel set $E \in \mathcal{F}$ on $\mathbb{R}^1$:

$$P\{\beta_t \in E | \mathcal{F}_s^\beta\} = P\{\beta_t \in E | \beta_s\}, a.s.; \quad s \geq 0. \tag{2.176}$$

The Brownian motion process also shows a "strong" Markov property. Thus by extension of (2.175) the Brownian motion process $\{\beta_t\} \equiv \beta_t, 0 \leq t \leq T$, for any Markov time τ relative to $(\mathcal{F}_t^\beta, 0 \leq t \leq T)$ with the probability $P(\tau \leq T) = 1$, can be expressed by:

$$E\{f(\beta_{s+\tau}) | \mathcal{F}_\tau^\beta\} = E\{f(\beta_{s+\tau}) | \beta_\tau\} \tag{2.177}$$

(see also Lipster and Shirjayev [138]). Furthermore, if the initial process $\{\beta_t\} \equiv \beta_t$ is defined for all $t \geq 0$ and all Markov times $\tau \in T$ with respect to $(\mathcal{F}_t^\beta), t \geq 0$ and $P(\tau < \infty) = 1$, the process: $\beta_t^* = \beta_{t+\tau} - \beta_\tau$ is also a Brownian motion process and independent of the events of the σ-algebra $\mathcal{F}_\tau^\beta$.

There exist other distributions of interest that are related to the Brownian motion process $\{\beta_t\}, t \geq 0$. The probability density function for instance given by equation (2.156) can be written in an abbreviated form as:

$$p(s, \xi; t, \eta) = \frac{\partial P_{s,\xi}(t, \eta)}{\partial \eta} \, ,$$

where $P_{s,\xi}(t, \eta) = P\{\beta_t \leq \eta | \beta_s = \xi\}$ or the conditional probability distribution. By using the form given for the standard Brownian motion with $\sigma^2 = 1$, this density becomes:

$$p(s, \xi; t, \eta) = \frac{1}{\sqrt{2\pi(t - s)}} e^{-(\eta - \xi)^2 / 2(t - s)} \, ,$$

which can be readily verified. It will satisfy the backward and forward Kolmogorov relations so that:

$$\frac{\partial p(s, \xi; t, \eta)}{\partial \xi} = -\frac{1}{2} \frac{\partial^2 p(s, \xi; t, \eta)}{\partial \xi^2}; \quad s > t \tag{2.178}$$

$$\frac{\partial p(s, \xi; t, \eta)}{\partial t} = \frac{1}{2} \frac{\partial^2 p(s, \xi; t, \eta)}{\partial \xi^2}; \quad s > t. \tag{2.179}$$

Another class of Markov processes are processes in the "wide sense", which are of considerable interest in the random evolution of discrete media. They will be considered in the following chapter.

(ii) Stochastic Differential equation:

The description of the deformation and flow of discrete media in stochastic mechanics occurs via dynamical models that are based on equations of motion of the following types:

(i) an equation of motion representing the development of a dynamic variable x_i or a set $\{x_i\}$, which are the components of a defined "state-vector" $\mathbf{z} \in Z \subset \mathrm{X}$.

(ii) or an equation describing the evolution of the probability distribution $P\{x_i\}$ of the variable x_i or of the set $\{x_i\}$.

In classical mechanics these groups are well-known and correspond to the Newtonian or the Hamiltonian description (i) and the second group (ii) to the description in terms of Liouville's equation. In the present analysis the first group is of considerable interest and is associated with the Langevin equation, while the second is associated with the Fokker-Planck equation, discussed above. Thus it becomes necessary to remark briefly on "stochastic differential equations" in this section. For a more comprehensive study of the extensive field of stochastic differential equations and the methods of solution reference is made to the various texts given in the reference list. With reference to the Brownian motion, Einstein [151] concluded that this motion is:

(i) too complex and hence must be described in a statistical manner.

(ii) requires the use of a small time interval $\Delta \tau$, which is large compared with the time of collision with other particles, but small in relation to the "macroscopic" observation times. Thus the motion is considered in terms of two consecutive time intervals that are mutually independent.

Einstein further showed that the probability $P(\mathbf{r}, t)$ of finding a particle at a position $\mathbf{r} \in C$ (configuration subspace of X) at time t satisfies the diffusion equation, i.e.:

$$\frac{\partial P(\mathbf{r}, t)}{\partial t} = D\nabla^2 P(\mathbf{r}, t) , \qquad (2.180)$$

where D is the diffusion coefficient, which is related to the "friction coefficient ζ" of the fluid particle by:

$$D = \kappa^T / \zeta ,$$

κ being Boltzmann's constant, T the temperature and ζ is defined as the proportionality constant between the external force $\mathbf{f}$ and the terminal velocity $\mathbf{v} \in \mathcal{V} \subset \mathrm{X}$ of a fluid particle or:

$$\mathbf{f} = \zeta \cdot \mathbf{v} .$$

The vectors $\mathbf{f}, \mathbf{v}$ are taken at the same instant of time. Conventionally the force acting on a Brownian particle in the fluid is assumed to be expressible as the "sum of two terms", i.e. a "systematic frictional force", which is linearly related to the velocity $\mathbf{v}$ and a second term related to a "random force or disturbance" so that:

$$\mathbf{f} = -m \int_0^t \beta(t-s)\mathbf{v}(s)ds , \qquad (2.181)$$

where $\mathbf{v}$ is the velocity at time s, m the mass of the particle and $\beta(t-s)$ a friction function. Obviously for the steady state motion in a viscous fluid the linear relation holds. In the non-steady case however, the friction force depends also on the "history of the motion" due to the disturbance in the fluid by the moving particle itself. In the hydrodynamic approximation one can express this friction function by:

$$\beta(t) = 6\pi\eta a \Delta t - \frac{3}{2}(\pi a^4 \rho\eta)^{\frac{1}{2}}t^{-\frac{3}{2}} ,$$

where a is the radius of the "spherical" particle, ρ the density of the fluid and η its viscosity. An extensive study of the Brownian motion can be found among others in Nelson [61], C. Itzykson and J.M. Drouffe [119].

The significance of the "state-space" representation in modelling of discrete media has already been stressed in section (2.3). From a thermodynamics point of view however, the description of transitions from one state of the medium to another adjacent one is usually considered in terms of a family of potentials $\phi(x_i, \vartheta)$ which depends on "i-state variables" and a set of "control parameters" $\{\vartheta\} = \Theta \subset X$. The latter are very important for internal structural changes, i.e. the evolution with time of the microstructure of a specific medium. In the continuum mechanics sense these potentials may be regarded to describe the "local equilibrium" and the stability at various points in the medium such that:

$$\left.\begin{array}{ll} \text{(i)} & \dfrac{\partial\phi(x_i, \vartheta)}{\partial x_i} = 0 \ ; \ x_i \in \mathbb{R}^n \\[4mm] \text{(ii)} & \dfrac{\partial^2\phi(\cdot)}{\partial x_i \partial x_j} > 0 \ , \ \text{for all } \{\vartheta\} \in \mathbb{R}^k; \{\vartheta\} \equiv \Theta \subset X. \end{array}\right\} (2.182)$$

These criteria can also be used in the stochastic approach to the critical behaviour of discrete media.

It is to be noted that the state-vector $\mathbf{z}_t \in Z$ in the state-space representation is a random variable, which contains intensive and extensive variables. Hence a family of such vectors represents then a stochastic process $\{x_t\}$ or $\{\mathbf{z}_t, t \geq 0\}$ in the state-space $Z \subset X$. In order to include the possible interactions between the elements of the microstructure, it becomes necessary to regard the extensive variables as the "controlling" ones so that the state-vector $\mathbf{z}_t$ or $\mathbf{x}_t \in X$ will be an element of $\mathbb{R}^n \otimes \mathbb{R}^m$.

In the general formulation of an equation of motion, one may consider first a "deterministic" state model describing the evolution of the state variables $\{\mathbf{x}_i(\mathbf{r}, t)\}$ belonging to $\mathbf{z}_t$ so that:

$$\frac{\partial \mathbf{z}_t}{\partial t} = \mathbf{f}[\mathbf{z}_t, t]; \ t \geq t_0, \ \mathbf{z}_t \in Z, \tag{2.183}$$

and in the stochastic approach which is influenced by a control parameter ϑ or a set $\{\vartheta\}$ of such parameters one has:

$$\frac{\partial \mathbf{z}_t}{\partial t} = \mathbf{f}_\vartheta[\mathbf{z}_t, t, \vartheta]; \ t \geq t_0, \ \mathbf{z}_t \in Z, \tag{2.184}$$

in which the state-variables are the components of the state vector $\mathbf{z}$ in its evolution with time, and $\mathbf{f}, \mathbf{f}_\vartheta$ can be regarded as "system functionals", respectively (see also Axelrad [156]). These relations characterize the "local change" of the variables $\{\mathbf{x}_i\}$ with time.

The state variable $\mathbf{x}_i$ may denote a deformation, strain, velocity, chemical composition etc. In general the system functionals may have also derivatives with respect to these quantities in the domain of α or Γ, respectively. They may also be non-linear due to the effect of interactions between elements. In order that these relations constitute a well-posed problem, it is further necessary to establish the values of the involved parameters and the boundary conditions on the respective surface of α or Γ. Usually the boundary conditions are of the "Dirichlet" type resulting in fixed values of n-state-variables $\{x_i\}$ on α, Γ, or correspondingly on a particular meso-domain M of the macroscopic material body. More often they are of the "Neumann" type or for fixed values of the fluxes on the respective surfaces. Hence the control parameter and the boundary conditions of a subsystem or the system as a whole impose constraints on these domains from the outside.

In a continuum model of a discrete medium in an unconstrained motion, the future motion is uniquely determined by the actual value of the state-vector so that a change of the latter can be expressed by a difference equation or a set of such equations as follows:

$$\frac{d}{dt}\{\mathbf{x}_i\}_{t+1} = \frac{d\mathbf{z}_{t+1}}{dt} \text{ or } \frac{d\mathbf{x}_{t+1}}{dt} = \mathbf{f}[\mathbf{x}_t, t], \mathbf{z}_t \in Z \text{ or } x_t \in X \tag{2.185}$$

in which $d\mathbf{x}_{t+1}$ is uniquely determined from $d\mathbf{x}_t \in X$ at time t independently of how $\mathbf{z}_t$ or $\mathbf{x}_t$ was reached.

In a stochastic model however the cause-effect relation of continuum mechanics is extended and thus requires the use of the probability distribution of the variables $\{\mathbf{x}_i\}$ for future times, if the value at the current time is known. Hence (2.185) becomes:

$$dx_{t+1} = \mathbf{g}[\mathbf{x}_t, t]dt + d\mathbf{w}[\mathbf{x}_t, t], \ t \in T \tag{2.186}$$

where $\mathbf{g}$ is the conditional mean value m_x of $\mathbf{x}_{t+1}$ given $\mathbf{x}_t$, and $\mathbf{w}$ a "disturbance" vector random variable with zero mean value. This relation also implies that the conditional probability distribution of $\mathbf{w}_t$ given $\mathbf{z}_t$ does not depend on $\mathbf{z}(s)$ or $\mathbf{x}(s)$, $s < t$. In analogy to eqn. (2.185) one can express (2.186) in an abbreviated form as follows:

$$\frac{d\mathbf{x}_{i+1}}{dt} = \mathbf{f}[\mathbf{x}_i, \mathbf{w}_{i+1}, t_i] \ ; \ i = 0, 1 \ldots \tag{2.187}$$

in which it is assumed that the non-linear n-vector functional is continuously differentiable with respect to its arguments. This relation represents a non-linear stochastic vector difference equation. In the terminology of system theory [157–160] the variable $\mathbf{w}_i$ is called a "random noise" input to the system. The sequence $\{\mathbf{w}_i, i = 1, 2 \ldots\}$ is a random vector sequence and so is $\{\mathbf{x}_i, i = 0, 1 \ldots\}$ for which the initial condition is $\mathbf{x}_0$. The latter can either be a given constant vector or a random vector with a specified initial distribution. In absence of such a distribution one obtains for relation (2.187) an ordinary difference equation with $\mathbf{x}_i$ as its solution. With the inclusion of the random distribution it is however necessary to consider the probability density $p\{\mathbf{x}_i\}$. The latter permits then the computation of the expected value $E\{\mathbf{x}_i\}$, the variation $\mathrm{Var}\{\mathbf{x}_i\}$ and so on, thus giving the statistics of the state $\mathbf{x}_i$ or $\mathbf{z}_i \in Z$. For an arbitrary distribution of $\{\mathbf{w}_i\}$ there is no unique solution of eqn. (2.187). It is usually assumed therefore that the disturbances are equivalent to a "white Gaussian noise" for which a solution of (2.187) can be obtained. If eqn. (2.187) represents a linear stochastic dynamical system one obtains the following form:

$$\frac{d\mathbf{x}_{i+1}}{dt} = \underline{f}_{i+1,i} d\mathbf{x}_i + \underline{g}_{i+1} d\mathbf{w}_{i+1} \ ; \ i = 0, 1 \ldots \tag{2.188}$$

in which $\mathbf{x}_i$ designates the n-dimensional state-vector, $\underline{f}_{i+1,i}$ an $n \times m$ non-singular transition matrix, $\underline{g}$ the $n \times m$ disturbance (noise) matrix and $\{\mathbf{w}_i; i = 1, 2 \ldots\}$ the white noise m-dimensional random vector sequence. If in particular $\{\mathbf{w}_i; i = 1, 2 \ldots\}$ is a Markov sequence, then:

$$p\{\mathbf{w}_i | \mathbf{w}_j\} = p\{\mathbf{w}_i\} \ ; \ i > j \tag{2.189}$$

where all $\mathbf{w}_i$'s are mutually independent. As mentioned above this sequence is completely random and hence unpredictable. One can approximate it by using the additivity property of white noise and the observation that a large number of small, independent "random effects" are often Gaussian. Hence, using the standard white Gaussian noise for the representation of the disturbances, then the sequence $\{\mathbf{w}_i; i = 1, 2 \ldots\}$ will be characterized by the mean value $m_\mathbf{w}$ and the covariance matrix $\underline{C}_\mathbf{w}$, i.e.:

$$\left.\begin{array}{l} m_{\mathbf{w}} = E\{\mathbf{w}_i\} \text{ for all } i \geq 1 \\ \underline{C}_{\mathbf{w}} = E\{(\mathbf{w}_i - E\{\mathbf{w}_i\})(\mathbf{w}_j - E\{\mathbf{w}_j\})^T\} \text{ for all } i, j \geq 1. \end{array}\right\} \quad (2.190)$$

The above considerations can be extended for the white noise in a time-continuous parameter system, where the white noise process is then defined by $\{\mathbf{w}_t, t \in T\}$ as a Markov process so that:

$$p\{\mathbf{w}_t | \mathbf{w}_s\} = p\{\mathbf{w}_s\} \; ; \; t > s \in T \tag{2.191}$$

in which the $\mathbf{w}_t$'s are mutually independent for all $t \in T$. If the $\mathbf{w}_t$'s are normally distributed for each $t \in T$ one has again a white noise Gaussian process for which the covariance function is defined by:

$$\underline{C}_{\mathbf{w}_t} = E\{(\mathbf{w}_t - E\{\mathbf{w}_t\})(\mathbf{w}_s - E\{\mathbf{w}_s\})^T\} = \underline{Q}(t)\delta(t - s) \tag{2.192}$$

in which $\underline{Q}(t)$ is a positive semi-definite covariance matrix and $\delta(t-s)$ the Dirac delta function. Hence this process is often referred to as a "delta-correlated" process (see also Aström [152], Horsthemke and Lefever [153] and others).

Returning to the equation of motion for a Brownian particle that is subjected to a random force or disturbance and recognizing the latter as a white noise force, one obtains for the velocity of the particle with $\mathbf{v} \in \mathcal{V}$ (velocity subspace of X):

$$\dot{\mathbf{v}}(t) = -\beta \mathbf{v}(t) + \mathbf{f}(t) \tag{2.193}$$

which has the solution:

$$\mathbf{v}(t) = \mathbf{v}_0 e^{-\beta} + \int_0^t e^{-\beta(t-s)} \mathbf{f}(s) ds \tag{2.194}$$

in which $\mathbf{f}(s)$ corresponds to the white noise term and where the Brownian motion can be identified with the Wiener process (section 1.6), i.e.:

$$\mathbf{w}(t) = \frac{1}{q} \int_0^t \mathbf{f}(s) ds \; . \tag{2.195}$$

However since $\mathbf{w}(t)$ is nowhere integrable (see Doob [50], Brémaud [47] and others), the solution in (2.194) is not well defined. It becomes necessary to consider the "stochastic integral $\mathcal{J}$ (Itô)" and the concept of a "non-anticipating" stochastic process. Considerations to a stochastic state-valued process $\mathbf{x}_t$ which is defined on the probabilistic function space $[X, \mathcal{F}, \mathcal{P}]$ with the adapted history $\{\mathcal{F}_t\} \subset \mathcal{F}$ have been given in section (2.4(i)). For the motion of a "non-anticipating" stochastic process (1-dimensional), one can also consider a family of σ-algebras $\{\mathcal{F}_t\}$, $(a \leq t \leq b) \in T$ contained in the σ-algebra $\mathcal{F}$, which will be non-anticipating, if the following conditions are satisfied:

(i) $\mathcal{F}_s \subset \mathcal{F}_t$ for $s \leq t$, where $\mathcal{F}_t$ increases as t
increases.

(ii) $\mathcal{F}_t \supset \mathcal{F}(w(s): a \leq s \leq t)$, hence $w(s), a \leq s \leq t$
is measurable with respect to $\mathcal{F}_t$.

(iii) $w(t + \epsilon) - w(t)$ is independent of $\mathcal{F}_t$ for $\epsilon \geq 0$ so that all
pairs of sets E_1, E_2 with $E_1 \in \mathcal{F}_t, E_2 \in \mathcal{F}\{w(t + \epsilon) - w(t)\}$
are independent.

$$\left. \right\} \quad (2.196)$$

Assuming that the process $w(t)$ and a family of non-anticipating σ-algebras $\{\mathcal{F}_t\}$ are given, then the stochastic process $\{x_t\}$; $(a \leq t \leq b) \in T$ is "non-anticipating" with respect to $\{\mathcal{F}_t\}$, if:

$$\mathcal{F}_t \supset \mathcal{F}_s\{x(s): (a \leq s \leq t)\} \text{ so that } x(s) \text{ is measurable} \\ \text{with respect to } \mathcal{F}_t. \qquad \left. \right\} \quad (2.197)$$

The above definitions permit to define an "Itô sum s" for every random process $\{x(t)\}$; $(a, b) \in T$ in the form of:

$$s = \sum_{i=1}^{k} x_{t_{i-1}}[w(t_i) - w(t_{i-1})]. \qquad (2.198)$$

This definition has been introduced by Itô [154, 155] in terms of characteristic functions of the process $\mathbf{x}(t)$, with $[a, b] \in T$ or $(0 \leq a \leq b \leq t) \in T$, where the stochastic integral is defined by:

$$J = \int_a^b \varphi(t) dw(t) \equiv w(b) - w(a). \qquad (2.199)$$

Now considering any function $f(t)$ as a "step function" in the interval $[a, b] \in T$, one has therefore:

$$f(t) = \sum_{i=0}^{i-1} f(t_i) \varphi(t)_{[t_i, t_{i+1}]}. \qquad (2.200)$$

If $f(t)$ is a random function of $w(t)$, it will be independent of the increments $w(t_{i+1}) - w(t_i) = \delta_i w$ and hence from earlier statements, it will be a non-anticipating function. Thus a sequence of Itô sums $\{s_k\}$ for the process $x(t)$ on $[a, b] \in T$ leads to the Itô integral:

$$J = \int_a^b x(t) dw(t), \qquad (2.201)$$

in which $\mathcal{J}$, if considered as a random variable is the strong-limit of the sequence $\{s_k\}$ that always exists for every continuous non-anticipating process. Using the above definitions of an Itô integral, one can now rewrite the stochastic differential equation (2.186, 2.188) for the 1-dimensional case as follows:

$$dx_t = f(x_t,t)dt + g(x_t,t)d\beta_t \; ; \; t \in [0,T] \tag{2.202}$$

where $\{\beta_t, t \geq 0\}$ is a scalar Brownian motion process with the variance equal to unity. The corresponding integral equation is therefore:

$$x_t - x_0 = \int_0^t f(x_s,s)ds + \int_0^t g(x_s,s)d\beta_s \; ; \; t \in [0,T] \tag{2.203}$$

in which the first integral is a Riemann integral in the mean square sense, and the second the above defined Itô integral. It has been shown by Jazwinski [161] that (2.203) has a solution on $[0,T]$ in the mean square sense, if x_o or x_{t_0} is a random variable with $\{|x_0|^2\} < \infty$ and independent of $\{d\beta_t, t \in [0,T]\}$. The solution of the process $\{x_t\}$ has the following characteristics:

(i) x_t is mean square continuous in $[0,T]$

(ii) $E\{|x_t|^2\} < \infty$ for all $t \in [0,T]$

(iii) $\displaystyle\int_0^T E\{|x_0|^2\}dt < \infty$

(iv) $x_t - x_0$ is independent of $\{w\}$ or $\{d\beta_s, s \geq t\}$ for every $t \in [0,T]$.

$$\left.\begin{array}{c}\\\\\\\\\\\\\end{array}\right\} \text{(2.204)}$$

The $\{x_t\}$-process is also a Markov process and is uniquely determined in the mean square sense by the initial condition x_0. The above can be readily extended to the vector valued case, then eqn. (2.203) becomes:

$$d\mathbf{x}_t = \mathbf{f}(\mathbf{x}_t,t)dt + \underline{g}(\mathbf{x}_t,t)d\beta_t \; , \tag{2.205}$$

in which $\mathbf{x}_t$ and $\mathbf{f}$ are n-dimensional vectors, g an $(n \times m)$ matrix, β_t the m-dimensional vector of an independent Brownian motion each with a variance parameter equal to unity and where $\mathbf{x}, \underline{g}$ have the norms:

$$|\mathbf{x}| = \left(\sum_{i=1}^n x_i^2\right)^{\frac{1}{2}} = [\mathbf{x}^T\mathbf{x}]^{\frac{1}{2}}$$

$$|\underline{g}| = \left(\sum_{i=1}^n \sum_{j=1}^m g_{ij}^2\right)^{\frac{1}{2}} = [tr(gg^T)]^{\frac{1}{2}}.$$

Since the process $\{x_t, t \in [0,T]\}$ generated by the Itô stochastic differential equation (2.205) is Markovian, it can also be characterized by the probability density function $p\{\mathbf{x}_t\} = p(\mathbf{x}, t)$ for all $t \in [0, T]$. Hence the transition probability density function can be written as:

$$p\{\mathbf{x}_t|\mathbf{x}_s\} = P_{x_t|x_s}(\mathbf{x}|\mathbf{y}) \equiv p\{\mathbf{x}, t; \mathbf{y}, s\} \text{ for all } (t > s) \in [0, T].$$

In the 1-dimensional case, the evolution of these densities leads to the earlier given Kolmogorov relations, which now take the form:

$$\left.\begin{aligned}
\frac{\partial p[x, t; y, s]}{\partial t} &= -\frac{\partial [p(x, t; y, s) f(x, t)]}{\partial x} \\
&\quad + \frac{1}{2} \frac{\partial^2 [p(x, t; y, s) g^2(x, t)]}{\partial x^2}
\end{aligned}\right\} \quad (2.206)$$

where this partial differential equation conforms with that of section (2.1) (eqn. 2.162) and represents the Kolmogorov "forward" equation. Thus the process is a diffusion process. Similarly to eqn. 2.160 of section (2.1) one can derive the "backward" relation, viz.:

$$-\frac{\partial p[x, t; y, s]}{\partial s} = f(y, s) \frac{\partial p[x, t; y, s]}{\partial y} + \frac{1}{2} g^2(y, s) \frac{\partial^2 [p(x, t; y, s)]}{\partial y^2} \quad (2.207)$$

which is adjoint to eqn. (2.206). Again in the vector case of the Itô stochastic differential equation, one obtains analogously the evolution for the probability density function of the process with $\mathbf{x}_t, \mathbf{f}$ being the n-dimensional vectors, g the $n \times m$ matrix from before and $\{\beta_t\}$ the m-dimensional Brownian motion process $(E\{d\beta_t d\beta_t^T\} = \underline{Q}(t))$ the forward Kolmogorov equation, or:

$$\frac{\partial p}{\partial t} = -\sum_{i=1}^{n} \frac{\partial (p f_i)}{\partial x_i} + \frac{1}{2} \sum_{i,j=1}^{n} \frac{\partial^2 [p(g\underline{Q}g^T)_{ij}]}{\partial x_i \partial x_j} . \quad (2.208)$$

This relation can also be written in another form by using the diffusion operator: $\mathcal{L} = dp/dt$, as follows:

$$\mathcal{L}(\cdot) = -\sum_{i=1}^{n} \frac{\partial (\cdot f_i)}{\partial x_i} + \frac{1}{2} \sum_{i,j=1}^{n} \frac{\partial^2 [\cdot(g\underline{Q}g^T)_{ij}]}{\partial x_i \partial x_j} . \quad (2.209)$$

This operator will be further considered in Chapt. 4. The above brief discussion on stochastic differential equations clearly shows that these equations have no proper differentials and correspondingly are not integrable in the usual sense. They therefore require certain rules of the Itô calculus for their solutions. In this context, one has to be careful about the meaning of the solutions. Thus with regard to eqn. (2.202) and its solution being the process x_t, it is a solution only if $\mathbf{x}_t$ is Markovian and iff the disturbances are equivalent to "white noise". To illustrate this consider for simplicity a physical process η_t in form of a scalar differential equation or:

$$\dot{\eta}_t = f(\eta_t, t) + g(\eta_t, t)m_t , \qquad (2.210)$$

in which m_t characterizes a random Gaussian disturbance with $\{m_t\}$ as a zero-mean exponentially correlated stationary process. The correlation function is then:

$$C_\rho(t + s, t) = E\{m_{t+s} m_t\} = \sigma^2(\rho/2)e^{-\rho|s|} , \qquad (2.211)$$

where ρ is the correlation coefficient. If ρ is large, $\{m_t\}$ is called a "wide-band" process that approximates white noise. If $\{m_t\}$ is correlated then the process $\{\eta_t\}$ is not Markovian. However the above correlation function is integrable and $\{\eta_t\}$ differentiable so that equation (2.210) is a well-defined differential equation for the sample functions of the process. The process $\{m_t\}$ is often referred to as "coloured noise", that can be obtained from a linear stochastic differential equation, whose forcing function is white noise, i.e.:

$$dm_t = -\rho m_t dt + \rho b d\beta_t , \qquad (2.212)$$

where $\{\beta_t\}$ is a Brownian motion process with unit variance and ρ, b fixed constants. Taking both equations (2.210 and 2.212) into consideration, it is readily seen that the composite system can be represented by a vector differential equation or:

$$\mathbf{x}^T = [\eta_t, m_t] \qquad (2.213)$$

in which $\mathbf{x} \in Z$ (state-space) and where the process $\{x_t\}$ is a Markov process. However the state-space in this case has been enlarged from the 1-dimensional to a 2-dimensional one. In general, the coloured noise problem shows that only the "pair-process" $[x_t, \vartheta_t]$ consisting of the state-variables of the system and the noise process, is Markovian. With reference to the stochastic differential equation (2.202), it is to be noted that there are strong and weak solutions of it. Thus, if $[X, \mathcal{F}, \mathcal{P}]$ is a probability space and by letting for simplicity $T = 1$, the sub-σ-algebras $\{\mathcal{F}_t\}$, $t \leq 1$ form a non-decreasing family of such algebras and $\mathrm{w} = \{\mathrm{w}_t, \mathcal{F}_t\}$, $t \leq 1$ is a Wiener process. Denoting by $[\mathcal{C}_1, \mathcal{F}_1]$ the measurable space of the continuous functions $x = \{x_t, 0 \leq t \leq 1\}$ on $[0, 1]$ and the σ-algebras: $\mathcal{F}_1 = \sigma\{x : x_s, s \leq 1\}$, $\mathcal{F}_t = \sigma\{x : x_s, s \leq t\}$, then by assuming that the functions $f(x, t)$, and $g(x, t)$ are measurable non-anticipating functions, i.e. $\mathcal{F}_t$-measurable for each $t, (0 \leq t \leq 1)$, the random process x_t is a "strong solution" of the stochastic differential eqn. (2.202). It has the $\mathcal{F}_t$-measurable initial condition x_0, if for each t the variable x_t are $\mathcal{F}_t$-measurable, where:

$$\left. \begin{array}{c} P\{\int_0^1 |f(x,t)|dt < \infty\} = 1 \\ \text{and} \\ P\{\int_0^1 g^2(x,t)dt < \infty\} = 1 \end{array} \right\} \qquad (2.214)$$

so that with probability one for each t, $(0 \leq t \leq 1)$:

$$x_t = x_0 + \int\limits_0^t f(x,s)ds + \int\limits_0^t g(x,s)dw_s. \tag{2.215}$$

It is often necessary to consider the "weak solution" of eqn. (2.202) with the initial condition x_0 having a prescribed distribution function $P\{x\}$. Thus a solution in the weak sense exists, if there are: a probability space $[X, \mathcal{F}, \mathcal{P}]$, a non-decreasing family of sub-σ-algebras $\{\mathcal{F}_t\}$, $t \leq 1$, a continuous random process $\{x_t, \mathcal{F}_t\}$ and a Wiener process $\{w_t, \mathcal{F}_t\}$ such that relation (2.214, 2.215) are satisfied and the probability $P\{w: x_0 \leq x\} = P\{x\}$ is the prescribed distribution function of x.

The solution of the stochastic differential equation concerning the vector valued process $\mathbf{x}_t$ (eqn. 2.205) is more difficult to obtain. A rigorous discussion on the solution is given among others by Lipster and Shiryayev [138]. Stochastic differential and integral equations can also be analyzed by the Green's function method as dealt with by Adomian [144] and in terms of a random operator formalism has been discussed by Bharucha-Reid [129] and references therein.

3 RANDOM EVOLUTION AND GEOMETRIC PROBABILITIES

3.1 Wide-sense Markov processes

This chapter is mainly concerned with enlarging the scope of the stochastic analysis to the description of the "random evolution" of discrete media and the application of geometric probabilities. In the preceding sections (1.5, 2.3) most of the models used in the phenomenology of discrete media were based on homogeneous Markov processes in which the transition probabilities were time independent and displayed the semi-group properties. In many applications however, the "distribution of states" is also required in the representation of the deformational or flow behaviour of the media, in particular, if interaction effects between structural elements are to be included. Evidently, this necessitates the use of a wider class of Markov processes usually involving "jump processes", partially observed jump processes and also "semi-Markov processes".

(i) Wide-sense Markov processes:

In order to discuss these random processes, it may be indicated first to remark on the homogeneous Markov processes with a countable set of states and on the differentiability of the transition probabilities as well as some limiting conditions. In the simple case when the state-space Z or X contains a countable number of points (states) one can also identify X with the set of all integers $\mathcal{J}$ or a subset of this set. Hence using the index set $\mathcal{J}$ instead of X, the elements of $\mathcal{J}$ correspond then to the states $i, j, \ldots$ Since the transition probabilities for processes with a countable number of states can be defined by a set of functions: $p_{ij}(t) = P\{t, i, \{j\}\}$, where $\{j\}$ denotes the singleton containing j and $(i, j) \in \mathcal{J}$; these functions will satisfy the following conditions:

$$\left.\begin{array}{l}
\text{(i) } 0 \le p_{ij}(t) \le 1 \\[6pt]
\text{(ii) } \sum_{i \in \mathcal{J}} p_{ij}(t) \le 1 \\[10pt]
\text{(iii) for all } i,j \text{ in } \mathcal{J} \text{ and } 0 < t, 0 < s, \\[4pt]
\qquad p_{ij}(t+s) = \sum_{k \in \mathcal{J}} p_{ik}(t) p_{kj}(s) \\[10pt]
\text{(iv) } \lim_{t \downarrow 0} p_{ij}(t) = p_{ij}(0) = \delta_{ij} \; ; \; \delta_{ij} = \begin{cases} 1, & \text{if } i = j \\ 0, & \text{if } i \ne j. \end{cases}
\end{array}\right\} \tag{3.1}$$

For "non-cut off" processes [137], it is apparent that the inequality in (3.1(ii)) becomes an equality or a weaker condition. Condition (3.1(iii)) merely restates the Chapman-Kolmogorov relation of section (1.6). If the condition (3.1(iv)) is satisfied, the transition probability and the process are said to be "stochastically continuous". In the earlier given classification of states (section 1.6) the quantities "q_{ij}" were introduced. It was also mentioned that they have a finite limit. Thus for a subset $\mathcal{J}_1 \subset \mathcal{J}$ not containing the state "i" the following limit exists:

$$\lim_{t \to 0^+} \frac{1 - p_{ii}(t)}{t} \ge \sum_{j \in \mathcal{J}_1} \frac{p_{ij}(t)}{t} \tag{3.2}$$

and consequently $-q_{ii} \ge \sum_{j \in \mathcal{J}_1}$ and $\sum_{j \ne i} q_{ij} \le -q_{ii}$.

A state "i" of the process is called "instantaneous", if $q_{ii} = -\infty$, otherwise it is said to be non-instantaneous or "delaying". In the latter case the state "i" is called "regular", if:

$$\sum_{j \ne i} q_{ij} = -q_{ii}$$

is satisfied, otherwise it is called a "non-regular" state. A process in which all states are regular is also said to be "locally regular". If the state "i" is non-instantaneous and the process is separable, then for a countable set "S" (everywhere dense in $[0, \infty)$) there exists a "random interval $(0, \Delta)$" such that:

$$P_i\{x(t) = i, t \in (0, \Delta)\} = 1. \tag{3.3}$$

It can be shown that for a sequence of such sets "S_n" which increases monotonically and where $\cup S_n = [0, t] \cap S$, one obtains:

$$P_i\{x(t) = i, t \in S \cap [0, t]\} = e^{q_{ii} \cdot t} \tag{3.4}$$

(see also Gihman and Skorohod [137]). Thus for a variable time τ_i which denotes the time of the first exit out of the state "i", one has the exponential distribution above with the parameter q_{ii}. If the state is an instantaneous one $P\{\tau_i > 0\} = 0$. In this case the inequality:

$$p_{ii}(t) \geq e^{q_{ii} \cdot t}$$

is always valid. Using the notation introduced in section (1.6), ie. where $q_{ii} = -\lambda_i$ and $q_{ik} = \lambda_i \pi_{ik}$ (for $q_{ii} \neq 0$); $\sum_{k \neq i} \pi_{ik} = 1$, one can readily obtain the following relation:

$$p_{ij}(t) = \int_0^t \lambda_i e^{-\lambda_i \cdot s} \sum_{k \neq i} \pi_{ik} p_{kj}(t-s) ds \qquad (3.5)$$

in which $\lambda_i e^{-\lambda_i s}$ is the probability density function of the random variable τ_i. This indicates that in order that a system or subsystem moves from a state $i \to j$ during the time t, it must first exit from "i" into the state "k" at some time s and then move into the state "j". Analogously one obtains also the relation:

$$p_{ii}(t) = e^{-\lambda_i \cdot t} + \int_0^t \lambda_i e^{-\lambda_i \cdot s} \sum_{k \neq i} \pi_{ik} p_{ki}(t-s) ds. \qquad (3.6)$$

Since the transition probabilities are differentiable with respect to t, by employing the above relations together with the relation between q_{ij}, π_{ij} and λ_i, one obtains:

$$\frac{d}{dt} p_{ij}(t) = \sum_k q_{ik} p_{kj}(t); \quad (j \in \mathcal{J}) \qquad (3.7)$$

which are known as the "first system" of Kolmogorov's equation, if it is also satisfied for all $i \in \mathcal{J}$. This will be the case when $\mathcal{J}$ is finite and all states are regular. For the previously mentioned "locally regular" processes, one obtains similarly the "second system" of Kolmogorov's equation, ie:

$$\frac{d}{dt} p_{ij}(t) = \sum_k p_{ik}(t) q_{kj}. \qquad (3.8)$$

The proof of these statements are given by Gihman and Skorohod, who also showed that, if the numbers q_{ij} satisfy for $(i, j) \in \mathcal{J}$ the conditions:

(i) $q_{ij} \geq 0$ for $i \neq j$;

(ii) $q_{ii} \leq 0$;

(iii) $\sum_j q_{ij} = 0$;

(iv) $\sup_i |q_{ii}| < \infty$

$$\left. \right\} \qquad (3.9)$$

then the system (3.7) has a unique bounded solution and the system (3.8) a unique solution satisfying $\sum_k |p_{ik}(t)| < \infty$ with the initial condition: $p_{ik}(0) = \delta_{ik}$.

These solutions are the transition probabilities with the properties in (3.1). Hence the class of "wide-sense Markov processes" can be regarded as a generalization of the previously treated Markov processes. It is to be noted, however, that it is not necessary for the process to be a time homogeneous one, provided the transition probabilities satisfy certain additional conditions. Thus one condition which is related to (3.1(iv)) is that at the time $s = t$, the transition probability should be continuous or:

$$\lim_{s \downarrow t} P\{x, t; s, E\} = \varphi_E(x) \; ; \; s > t \tag{3.10}$$

where $P\{\cdot\}$ is the transition probability of the Markov process, $(s, t) \in T$, the events $E \in \mathcal{F}$ or the fixed σ-algebra on $[X, \mathcal{F}, \mathcal{P}]$ and φ_E the characteristic function of the process. Further, other additional conditions are as follows:

$$\left.
\begin{aligned}
&\text{(i)} \; \frac{P\{x, t; s, E\} - \varphi_E(x)}{s - t} \to q(x, t, E) \text{ for } s \to t, \\
&\text{(ii) for fixed } (x, E) \text{ the function } q(x, t, E) \\
&\qquad \text{is continuous with respect to } t \in T \text{ and} \\
&\qquad \text{uniformly continuous with respect to } (x, E), \\
&\text{(iii) as a consequence of (i), one also has:} \\
&\qquad |q(x, t, E)| \leq k \text{ for all } t \in T, x \in X, E \in \mathcal{F} \\
&\qquad \text{where } k \text{ is a positive constant.}
\end{aligned}
\right\} \tag{3.11}$$

A Markov process in the wide sense satisfying conditions (i), (ii) above is called a "jump process". It is regular, if the convergence in (3.11(i)) is uniform in (s, x, E) and the function $q(s, x, E)$ is continuous for fixed (x, E) in $s \in [0, t]$, when t is an arbitrary point of the set $\mathcal{J}$ or T.

(ii) Partially observed Markov processes:

The wide sense jump Markov processes can be conveniently employed in the description of the random evolution of discrete media as will be demonstrated in the next paragraph. Since a significant part of the material response frequently encountered in polycrystalline solids for example, is the "transient behaviour", it is often convenient to use the class of "partially observed jump processes" in the description. Such incomplete observed jump Markov processes also form a link between a possible "state estimation" and the statistics of discrete state measurements obtained from appropriate experimental observations. During such transients, changes of states of the microstructure are induced by certain internal mechanisms that are due to some "unobservable" random variables. Consequently the "evolution of states" can be regarded as a partially observed jump process. The latter is well-known in system control theory. In general the evolution of a random phenomenon such as structural changes in a discrete solid that occur immediately upon application of an external force (at $t = 0$) corresponds to a transiency until a "steady-state"' deformation is attained at a certain time t. It can be characterized by a family of random variables $\{x_t\}, t \in [0, T]$ taking on values in the space R^n. This family can be recognized as a stochastic process $\mathbf{x}_t \in X$. It is important to note that a partially observed process is in fact one, that consists of a pair of component random processes in which the "observable component ξ_t" is a Markov process with a finite countable number of states and the "unobservable process η_t" which takes discrete values in a corresponding subspace of X. Hence a partially observed jump process $\mathbf{x}_t$ or $\{x(t)\}; t \in [0, T]$ is defined with values in the product space $\mathbb{R}^{n_1+n_2}$, where:

$$\mathbf{x}_t = (\xi_t, \eta_t); \ \mathbf{x}_t \in X, t \in [0, T] \tag{3.12}$$

in which ξ_t is the vector of the first n_1 components of the "state-vector" $\mathbf{z}_t$ or $\mathbf{x}_t \in X$ and the "unobservable vector η_t" of the last n_2 components of $\mathbf{x}_t$. Hence the σ-field generated by the part of the process $\mathbf{x}_t$ is given by:

$$\mathcal{F}_t = \sigma\{\mathbf{x}(s): 0 \leq s \leq t\} \tag{3.13}$$

and that generated by the past "observations" by:

$$\mathcal{G}_t = \sigma\{\boldsymbol{\xi}(s): 0 \leq s \leq t\}. \tag{3.14}$$

Using the filtering analysis of section (1.6), one can define a functional ϕ on the past of $\mathbf{x}_t$ as a measurable real-valued stochastic process which for each t is $\mathcal{F}_t$-measurable and where the expected value, ie. $E\{|(\phi_t)|\} < \infty$, can be established. It is well-known from system theory that "optimal filtering" will occur, when the conditional expectation of this functional: $E\{\phi(t)|\mathcal{G}_t\}$ can be determined. The optimality conditions for partially observed jump process have been considered amongst others in refs. [162], [163].

Since partially observed jump processes have piece-wise continuous paths with probability one, there will be a sequence of times at which the process $\mathbf{x}_t$ jumps from one state to another and hence a "sequence of states". The latter can be associated with the change in a given spatial arrangement of the microstructure, whereby "phase-transitions" are excluded. To delineate such "structural changes" from others, where phase transitions are admitted, one may refer to these changes as subcritical or "metabatic" changes. By denoting the number of jumps the process $\mathbf{x}_t$ undergoes within the time interval $(0, t) \in T$ by $N(t)$ and by τ_n the time of the nth jump, then $\mathbf{x}_t$ in the 1-dim. case can be characterized in the following manner:

For the initial conditions at $t_0 = t = 0$, then $\xi_0 = \xi(0), \eta_0 = \eta(0)$ and hence:

$$x_{t_0} = (\xi_0, \eta_0), \text{ if } 0 \leq t < \tau_1 \qquad \text{(a)}$$

and for any time t between τ_n, τ_{n+1} one has: $\qquad\qquad\qquad$ (3.15)

$$x_{t_n} = (\xi_n, \eta_n), \text{ if } \tau_n \leq t < \tau_{n+1} \qquad \text{(b)}$$

The above evidently forms a sequence of random elements, which for $n \geq 1$ can be written as:

$$(\xi_0, \eta_0, \tau_1, \xi_1, \eta_1 \ldots \tau_n, \xi_n, \eta_n \ldots) \tag{3.16}$$

including the "jump times" and the "successive states". The probability distribution of this sequence can be determined from the conditional distribution of the "next jump time", given the history up to the "time of the nth jump" and the conditional distribution of the "location of the jump" (experimental observations) given the history up to the nth jump time and the fact that the $(n + 1)$ jump has just occured, viz:

$$\left. \begin{array}{ll} P\{\tau_{n+1} \leq t | \xi_0, \eta_0, \tau_1, \xi_1, \eta_1 \ldots \tau_n, \xi_n, \eta_n\} & \text{(a)} \\ P\{\xi_{n+1}, \eta_{n+1} \in E | \xi_0, \eta_0, \tau_1, \xi_1, \eta_1 \ldots \tau_n, \xi_n, \eta_n, \tau_{n_1}\} & \text{(b)} \end{array} \right\} \tag{3.17}$$

in which the Borel sets $E \in \mathcal{F}$. Abbreviating the sequence of the random elements in (3.16) to X_n, the distribution function can also be expressed by using the conditional density as follows:

$$P\{\tau_{n+1} \leq t | X_n\} = \int\limits_{\tau_n}^{t} p(s | X_n) ds, \qquad \text{on } [\tau_n, t] \in T . \tag{3.18}$$

In a partially observed jump process representing the transient behaviour of a discrete medium in which the internal mechanisms acting on the microstructure are regarded to control the evolution process, it becomes important to define a "controlled jump rate" of the process (see also [164], [165]). Since the event set $E \in \mathcal{F}$, $\mathcal{F}$ being the σ-algebra on $X = \mathbb{R}^{n_1 + n_2}$, one can write for (3.18) an abbreviated form, ie. $\pi\{E | X_n, \tau_{n+1}\}$ so that the jump rate of x_t on $\tau_n \leq t < T$ (T being here the final time of the transiency) is defined by:

$$q(t | X_n) = \frac{p(t | X_n)}{\int_{\tau_n}^{t} p(s | X_n) ds + P\{\tau_{n+1} = T | x\}} \tag{3.19}$$

which upon computation (see also Richel [164]) leads to a conditional probability density of the form:

$$p\{t | X_n\} = q(t | X_n) \exp[-\int\limits_{\tau_n}^{t} q(s, X_n) ds]. \tag{3.20}$$

Since the "current state" of x_t or the sequence X_n at (ξ_n, η_n) corresponds in the earlier notation to the pair of states (i, j), one obtains the conditional jump rate of a "partially observed controlled jump process" by the use of the previously introduced control parameter $\vartheta \in \Theta$ (subspace of X) in the form of:

$$\pi\{(k, \ell) | (i, j), t, \vartheta\}. \tag{3.21}$$

Recalling from (3.14) expressing the past observations $\mathcal{G}_t$, the control of the process can be seen as a family $\{\vartheta(t, \mathcal{G}_t)\}$ of observations and time t, where the control parameters are assumed to take on values in a closed set of the control or subspace $\Theta \subset X$. In this sense one can define the controlled jump process by a family of functions $\{\vartheta(t), \mathcal{G}_t\}$ in which the conditional distribution of the location of the next jump is expressed by:

$$P\{\tau_{n+1} > t | X_n\} = \exp[-\int_{\tau_n}^{t} q(s, i, j, \vartheta(s), \mathcal{G}_s) ds]$$

or

$$P\{(\eta_{n+1}; \xi_{n+1}) = (\ell, m | X_n, \tau_{n+1}\} = \pi\{(\ell, m) | (i, j), \tau_{n+1}, \vartheta(\tau_{n+1}, \mathcal{G}_m)\} \quad (3.22)$$

In the above relation $\mathcal{G}_n$ reflects the measurement or observation history corresponding to the sequence X_n with $(\xi_n, \eta_n) = (i, j)$. Thus for an initial fixed distribution P_{ij}, ie:

$$P_{ij} = P\{\xi_0 = i, \eta_0 = j\},$$

one can construct the finite dimensional distribution of the variables forming X_n, and extend this to a probability measure $\mathcal{P}$ on X defining the controlled process x_t such that:

$$x_t = \begin{cases} (\xi_0, \eta_0), & \text{if } 0 \le t < \tau_1 \\ (\xi_n, \eta_n), & \text{if } \tau_n \le t < \tau_{n+1}. \end{cases} \left. \right\} \quad (3.23)$$

On the assumption that the jump rate q is bounded the above process will have a finite number of jumps in each finite time interval. It should be noted however, that this process depends on the current values of time, state and control, ie. $q = q(t, (i, j), \vartheta)$ and is usually called a "partially observed controlled jump Markov process", although in general it may not be Markovian.

A related class of random processes often employed in the modelling of discrete media are the "semi-Markov" processes. They are of the form: $x_t = [x(t), \xi(t)]$, where $x(t)$ designates the "state" of the process at time t and $\xi(t)$ denotes here the time for which the process remains in this state. Evidently the duration or the time period for $\xi(t)$ up to the time instant t is significant. It is possible to define a semi-Markov process by considering the "constructive definition" of a Markov process as given by Gihman and Skorohod [137]. In this sense, if $[X, \mathcal{F}]$ is a measurable space with $\mathcal{F}$ the σ-algebra of its subsets and assuming that $x(t)$ is a sample function continuous from the right with respect to the discrete topology of X, the semi-Markov process spends a positive amount of time in each state. Then, if τ is the instant of time of the first exit from the initial state and is exponentially distributed with a parameter $\lambda(x)$, the one-step transition probability is given by:

$$P\{x(\tau) \in E\} = \pi(x, E) \; ; \; E \in \mathcal{F} \; . \tag{3.24}$$

Thus, up to the moment of the first accumulation of an infinite number of jumps the process can be characterized as follows:

$$x_0 = x(0), x_1 = x(\tau) \ldots x_n = \gamma_\tau x_{n-1} \tag{3.25}$$

so that the sequence $\{x_n\}$ forms a "Markov chain" with the one-step transition probability: $\pi(x, E)$. Introducing the variables

$$\tau_1 = \tau, \ldots \tau_n = \gamma_\tau \tau_{n-1} \tag{3.26}$$

and letting: $\eta = \sum_{k=1}^{\infty} \tau_k$, one has for $t < \eta$:

$$x(t) = x_n, \text{ if } \sum_{k=1}^{n} \tau_k \leq t < \sum_{k=1}^{n+1} \tau_k. \tag{3.27}$$

Analogously, as shown in section (1.3), one can establish the joint distribution function of the variables. By using the Markov chain $\{x_n\}$ in $[X, \mathcal{F}]$ with the above transition probability and letting the times $\tau_1, \tau_2 \ldots$ be a sequence of random variables such that the "joint distribution" given the values $x_0, x_1 \ldots x_n$ coincides with the distribution of independent random variables that are exponentially distributed, ie. $\lambda(x_0), \lambda(x_1) \ldots$ then the sequence ξ_k of i.i.d. random variables can be established such that:

$$\tau_k = \frac{1}{\lambda(x_k - 1)} \xi_k \; ; \; k = 0, 1, 2 \ldots \tag{3.28}$$

Hence the random process x_t is defined in terms of (3.27) and the variables x_k, τ_k as a Markov process for $t \in [0, \sum_1^{\infty} \tau_k]$.

A more general definition of a semi-Markov process than above can also be given (see Gihman and Skorohod [137], Rosenblatt [83] and others). It is of interest to note that for Markov and semi-Markov processes successive states form a chain. However in a Markov process the time period spent in a given state depends only on this state and is exponentially distributed, whereas in a semi-Markov process this period also depends on the state into which the process moves. This time period is arbitrary. For the evaluation of such processes the computation of marginal distributions is essential, but is rather complex. Reference is made to the discussion on this subject in [162].

(iii) Random evolution of discrete media:

To illustrate the jump Markov processes discussed in the foregoing section, the evolution of a structured solid during the transient response and that of a simple fluid will be considered next.

(a) The transient behaviour of a structured solid:

It is apparent, that immediately upon the application of an external load a specimen of a polycrystalline solid will experience a spatial rearrangement of its microstructure with the constraints imposed by the occuring interaction effects between the elements of the structure. A more detailed analysis of the kinematics and a corresponding general deformation theory will be given in Chapt. 4. In this section the description of the "transient response" of the structure is merely a phenomenological one, ie. interaction effects are excluded. The transiency that takes place after the load application and lasts until a "steady-deformation" of the solids sets in, can be regarded as a random phenomenon that is characterized by a set of random variables $\{x_t\}, t \geq 0$ taking on values in the Euclidean space $\mathbb{R}^n$ or equivalently by a stochastic process $z_t \in Z \subset X$, where the subspace of X is a restricted "state-space Z" as shown earlier in Fig. (11, 12). If the "interaction effects" are considered to control the transient response, then evidently in the "state-space" representation the state-space Z must be augmented by a subspace or the "control space Θ" of X formed by the set of control parameters $\{\vartheta\}$. Hence the stochastic state space and the corresponding subspaces will be as shown in the diagram of Fig. 13 (see also [49, 156]).

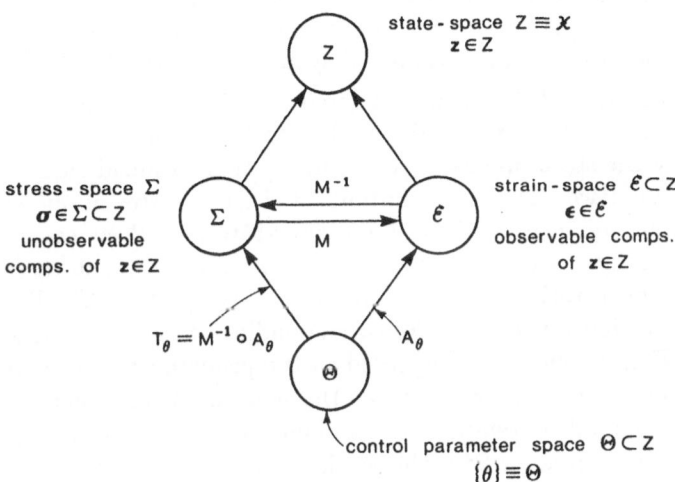

Fig. 13. Stochastic state-space with control-parameter set $\{\vartheta\} \subset Z$.

From a stochastic mechanics point of view the state vector $^\alpha \mathbf{z}$ of an element of the microstructure is an outcome or elementary event E in Z as a result of the statistical experiment α. Hence excluding interactions the subsets $E_n \subset Z$ will represent the states of the solid within a certain range $\Delta \mathbf{z}^n$ of experimental observations, ie. $E_n = \{\mathbf{z}^n < {}^\alpha \mathbf{z} < \mathbf{z}^n + \Delta \mathbf{z}^n\}$; $\cup_N E_n = Z$; $E_n \cap E_k = \emptyset, n \neq k$. Assuming that Z or if extended to X is locally compact, the subsets E_n are also compact and bounded under closure. The elements of E_n of the σ-algebra $\mathcal{F}^z$ are Borel sets with the properties given in (2.88) of Chapt. 2, and $[Z, \mathcal{F}^z]$ is a measurable space.

The measure on these subsets given earlier is then $0 \leq \mathcal{P}^z\{E_n\} \leq 1$; $\mathcal{P}^z\{E_n\} = 0$, if $E_n = \emptyset$ and $\mathcal{P}^z\{Z\} = 1$. This measure in accordance with probability theory represents the distribution of the relevant field quantities. The latter are considered here to be the "observable strains $\epsilon \in \mathcal{E}$" and the "unobservable internal stresses $\sigma \in \Sigma$". It may also be seen from Fig. 13 that the most significant operator is the mapping "M" (or its inverse) linking the subspaces $\mathcal{E}$ and Σ of Z. This mapping depends largely on the strict or strong monotonicity of the operator M on some dense sets of $\mathcal{E}_d, \Sigma_d$, respectively, belonging to Z. It is usually assumed that the weaker condition, ie. the strong monotomicity of M holds. It may also be noted that the control parameter subspace $\Theta \subset Z$ is connected to the strain space $\mathcal{E}$ and stress space Σ by the operators A_θ and $T_\theta = M_0^{-1} \circ A_\theta$, respectively, where the latter is a composition operator, an explicit form of which can be established from the knowledge of the constitutive operator M or M^{-1}. Hence for the derivation of evolution relations of the structure during the transient response one can employ these operators and the abstract dynamical system $[Z, \mathcal{F}^z, \mathcal{P}^z]$.

The description of the behaviour can occur either by using the set of internal variables $\{X_i\}$ and correspondingly the stochastic process $\mathbf{x}_t \in X, t \geq 0$ or equivalently the state process $\mathbf{z}_t, t \geq 0$ in Z. As mentioned earlier in the state-space analysis, two notions are important, ie. the states of the structure and the observable quantities. Thus a specific state $\mathbf{z} \in Z = \{z_0, z_1 \dots\}$ can be characterized by a single observation, ie. the strain ϵ, but more often by a set of such observations or $\mathcal{E}_i = \{\epsilon_0, \epsilon_1 \dots\}$; $\mathcal{E}_i \in \mathcal{E} \subset Z$. Thus given the single observation $\epsilon \in \mathcal{E}_i$ and the corresponding state $z \in Z$, the distribution $P(\epsilon, z, E), E \in \mathcal{F}^z$ can be interpreted as the probability of the "observable ϵ" having values in an event set E, when the structure is in a specific state z. The fundamental condition relating the random process $\mathbf{z}_t$ in Z or $\{z(t_i) = x_i; t_i = 0, 1, 2 \dots\}$ to the measure $\mathcal{P}^z$ is the Markov principle, or $\mathcal{P}^z\{z(t) \in E|z(s) = x\}$. Thus the function $P\{s, x; t, E\}$ will be a transition function with probability one, if:

$$P\{s, x; t, E\} = \mathcal{P}^z\{z(t) \in E|z(s) = x\} \tag{3.29}$$

so that $P\{.,.\} = 1$ for $x = E$ and 0 for $x \neq E$; $P\{.,.\} \leq 1$. For a fixed time interval (s, t) and $x = E, P\{.,.\}$ is the measure $\mathcal{P}^z$ on Z and hence a

Z-measurable function of $x \in Z$ or X, resulting in the Chapman-Kolmogorov relation.

There are two ways in the application of "partially observed jump" processes to the approximate analysis of the structural changes during the transient response of a polycrystalline solid. Thus one can employ the concepts of "control theory" that relate to discontinuous processes and martingale analysis or one can use the "conditional distribution of states" of Markov theory given the "history of the observations". By defining a jump process (one-dimensional) $\{z_t, t \geq 0\}$ in Z, which takes values in X by a countable sequence of random variables, or:

$$\{\tau_0, z_0, \tau_1, z_1 \ldots \tau_n, z_n\}$$

defined on $[Z, \mathcal{F}^z, \mathcal{P}^z]$ with the jump times $\{\tau_i\}$ and the states $\{z_i\}$, the sample path of the random process will be given by:

$$x_t \equiv z_i, \ t \in [\tau_i, \tau_{i+1}]; \ i = 0, 1, 2 \ldots \tag{3.30}$$

Evidently the condition: $0 = \tau_0 < \tau_1 < \ldots \tau_n$, a.s. holds. Using the σ-algebra $\mathcal{F}^z$ or the family of σ-algebras $\{\mathcal{F}(x_s, s \leq t)\}$ for the process $\{x_t, t \geq 0\}$, one can introduce a family of "counting processes" such that:

$$P\{t, E\} = \sum_{\substack{s \leq t \\ x_s \neq x_t}} I_{(x_s \in E)} = \sum_{\tau_i \leq t} I_{(z_i \in E)} \tag{3.31}$$

in which I designates the indicator function for the set E. By using the first form in (3.31) and the "local description" of the jump process x_t with the "base measure $\mathcal{P}^x$" one also has to introduce a pair of measures (see also Prohorov and Rozanov [42]) designated here by $[t, m(t, x_t, E)]$, where the corresponding probability $\overline{P}$ becomes:

$$\overline{P}(t, E) = \int_0^t m(s, x_s, E) ds. \tag{3.32}$$

Hence the jump can be expressed by: $q(t, E) = P(t, E) - \overline{P}(t, E)$ so that it will be an $\mathcal{F}_t$-martingale for each event $E \in X$ and the base measure $\mathcal{P}^x$. The jump process x_t in general can be regarded as a regular step Markov process (see also Davis [162], Blumenthal and Getoor [82], Brémaud [47]).

In the second type of approximation to the transient response during the period $[0, T] \in \mathcal{J}$, the process can be described by:

$$z_t = \{z(t); t \in [0, T] \subset \mathcal{J}\} \tag{3.33}$$

in which the process starts at $t = 0$ and ends at $t = T$ (the onset of a steady-state deformation). Hence, it is defined on the closed time interval $[0, T] \in \mathcal{J}$

with values in the n-dimensional Euclidean space or $\mathbb{R}^n = \Sigma^{n_1} \otimes \mathcal{E}^{n_2}$ with probability one. Since $\mathbf{z}$ has values in the product space $\mathbb{R}^n = \mathbb{R}^{n_1} \times \mathbb{R}^{n_2}$ or $\mathbb{R}^{n_1 + n_2}$, it can be regarded to be a two component process:

$$\mathbf{z}_t = (\boldsymbol{\sigma}_t, \boldsymbol{\epsilon}_t) \tag{3.34}$$

where $\boldsymbol{\sigma}_t \in \Sigma$ is the vector of the 1st n_1-components of the state-vector $\mathbf{z}$ and $\boldsymbol{\epsilon}_t \in \mathcal{E}$ the vector of the last n_2-components of $\mathbf{z}$. Hence $\boldsymbol{\epsilon}$ is the "observable" and $\boldsymbol{\sigma}$ the "unobservable" component of $\mathbf{z}$. Since the process $\mathbf{z}_t$ has piece-wise continuous paths with probability one, it is also a sequence of "subcritical or metabatic" states, that can be expressed by:

$$Z_n := [z_0, \tau_0, \tau_1, z_1, \tau_2, \tau_2 \ldots \tau_n, z_n] \tag{3.35}$$

At time τ_i at which the process jumps from the state $z_{i-1} \to z_i$, one has also a sequence of random elements:

$$F_n := [\sigma_o, \epsilon_0, \tau_1, \sigma_1, \epsilon_1 \ldots \tau_n, \sigma_n, \epsilon_n] \tag{3.36}$$

in which the correspondence between (3.35) and (3.36) is considered to be one-to-one. Thus given the sequence F_n, the state process $\mathbf{z}_t$ in the closed time-interval $[0, T] \in \mathcal{J}$ can also be defined by:

$$\left. \begin{array}{l} \mathbf{z}_0(t) = [\sigma(0), \epsilon(0)] \equiv (\sigma_0, \epsilon_0) \quad \text{for} \ \ 0 \leq t \leq \tau_1; \tau_0 = 0 \quad \text{(a)} \\ \text{and for any time } t \text{ by} \\ \mathbf{z}(t) = [\sigma(\tau_n), \epsilon(\tau_n)] \equiv (\sigma_n, \tau_n) \quad \text{for} \ \ \tau_n \leq t < \tau_{n+1}, \\ \text{if } \tau_n = t = T. \qquad\qquad\qquad\qquad\qquad\qquad\qquad\qquad\ \text{(b)} \end{array} \right\} \tag{3.37}$$

The initial conditions σ_0, ϵ_0 in (3.37a) correspond to the beginning of the transiency for which the constitutive operator M^{-1} can be determined from the observable strain ϵ_0 (at $t = 0^+$) and the unobservable σ_0-stress evaluated by means of the statistical method to be discussed in Chapt. 4. For the description of the state jump process $\mathbf{z}_t$ the two conditional distribution functions (3.17a, b) become now:

$$\left. \begin{array}{l} P_\tau \{\tau_{n+1} \leq t | (\sigma_0, \epsilon_0, \tau_1, \sigma_1, \epsilon_1 \ldots \tau_n, \sigma_n, \epsilon_n\} \qquad\qquad \text{(a)} \\ \text{and} \\ P_n \{\sigma_{n+1}, \tau_{n+1} \in E | (\sigma_0, \epsilon_0, \tau_1, \sigma_1, \epsilon_1, \ldots, \tau_n, \sigma_n, \epsilon_n, \tau_{n+1}\} \quad \text{(b)} \end{array} \right\} \tag{3.38}$$

which by using the conditional probability density on $[\tau_n, T]$, (eqn. 3.18) is:

$$P_\tau(\tau_{n+1} \leq t | F_n) = \int_{\tau_n}^t p(s | F_n) ds; \ [\tau_n, t) = T \in \mathcal{J}.$$

In order to distinguish between the history of the process $\mathbf{z}_t = \{z_t\}$ as a whole, and that of the observations only, one can define the latter by:

$$\mathcal{G}_\epsilon := [\epsilon_0', \tau_1', \epsilon_1' \ldots \tau_n', \epsilon_n'] \tag{3.39}$$

where the sequence of the "observable strains $\{\epsilon_0', \ldots \epsilon_n'\}$" corresponds to the sequence $\{\epsilon_0, \ldots \epsilon_n\}$ of the state-process z_t and where these observations occur in a subset $\{\tau_1' \ldots \tau_n'\}$ of the set of jump times of the process, or $\{\tau_1 \ldots \tau_n\}$; $\{\tau_i\} \in \mathcal{J}$.

On the assumption that the number of jumps of states in $[0, T] \in \mathcal{J}$ is $N(t)$ and that of the strains in $\mathcal{E}$ is $K(t)$ so that when $N(t) = n$, $K(t) = k; k \leq n$, then since $\mathcal{G}_\epsilon$ on $[0, T] \Rightarrow \{\epsilon_0' \ldots \epsilon_k'\}$, eqn. (3.38b) can be written as:

$$P_n\{E|F_n, \tau_{n+1}\} \text{ for any Borel set } E \in \mathcal{F}_t \text{ on } \mathbb{R}^n \text{ or } \Sigma^{n_1} \otimes \mathcal{E}^{n_2} \tag{3.40}$$

It has been shown by Richel [164] that there exists a mapping L of the part history of the observed process and the observation history $\mathcal{G}_\epsilon$ such that:

$$L(F_n) = \mathcal{G}_t$$

and consequently the jump rates of the process can be taken as:

$$q(t, F_n) = q(t, \mathcal{G}_\epsilon, \sigma_n)$$

so that the distribution function is finally expressed by:

$$P_\tau\{E|F_n, \tau_{n+1}\} = P_n\{E|\mathcal{G}_\epsilon, \sigma_n, \tau_{n+1}\} \tag{3.41}$$

showing that in the application of a conditional Markov jump process to the transient behaviour of a structured solid, the jump rate and state jump distribution depend on the observation history $\mathcal{G}_\epsilon$ and the current value of the "unobservable stress σ_n". This component of the state-vector z can be assessed in an analogous manner to that mentioned for the initial condition. It requires however apart from the quantitative measurements of "micro deformations" or strains, an appropriately determined material operator M or M^{-1}, which can only be obtained by the use of stereological methods.

Another approach to the transient response of structured solid, in particular with reference to binary structures will be considered in Chapt. 4.

(b) Evolution relations for simple fluids:

There is a great variety of "complex or structured fluids" which have distinctive properties caused by their large polyatomic structures. They also show a wide variation in their mechanical responses examples of which will be considered in Chapt. 4. In contrast "simple fluids" consist of rather compact molecules that interact strongly with their nearest neighbours. The interaction energy per molecule for these fluids is of the order of the thermal energy (kT) or approximately 10^{-1} eV at ambient temperature. The time scale of their molecular motion is comparable with the time interval of successive molecular collisions.

The kinematics and flow dynamics of simple fluids from the stochastic mechanics point of view will be considered in Chapt. 4.

This section is mainly concerned with the description of the evolution of a simple fluid during a short transient period preceding a "uniform flow" by the application of the theory of Markov processes. However some introductory remarks may be indicated. Recalling from Chapt. 2 the axiomatic definitions of discrete media and particularly those pertaining to molecular fluids, the r-dimensional state vector $^{\alpha}\mathbf{z}$: $^{\alpha}z_i, (\alpha = 1 \ldots N; i = 1 \ldots r)$ represents also the states of an element (molecule) of the structure. The components of $^{\alpha}\mathbf{z}$ are real-valued functions and for a meso-domain of the fluid, which conceptually correspons to the "control volumne" of hydrodynamics one has a set of these vectors or $\{^{\alpha}z_i\}$. Since the stochastic mechanics of fluids concerns the random behaviour of either a single element (molecule) or their "collective" mode of motion, a measure of the events in X is required that must satisfy the conditions:

$$0 \leq \mathcal{P}\{E\} \leq 1; \mathcal{P}\{E\} = 0, \text{ if } E = \emptyset \text{ and } \mathcal{P}\{X\} = 1.$$

There are however other measures necessary for the description of the "distribution of velocities" and that of "configurations" of molecules and their ensembles which will be considered in Chapt. 4. It is usually assumed in dealing with simple fluids that they can be modelled by "hard spheres" and a Lennard-Jones interaction potential. An "idealized" simple fluid covering a meso-domain of the former is schematically shown in Fig. (14a) and its rearrangement of the molecular structure due to an applied shear in Fig. (14b).

As mentioned earlier, it is convenient for the analysis to choose suitable subspaces of X. In this case they are the configuration space $\mathcal{C}$ and the velocity space $\mathcal{V}$. Hence

$$^{\alpha}\mathbf{r}(t) \in \mathcal{C} \subset X; {}^{\alpha}\mathbf{v}(t) \in \mathcal{V} \subset X \tag{3.42}$$

in which $^{\alpha}\mathbf{r}$ is the position vector of an element α (molecule) with respect to a fixed frame. The velocity space $\mathcal{V}$ together with the σ-algebra $\mathcal{F}^{\mathrm{v}}$ of the events and the measure $\mathcal{P}^{\mathrm{v}}$ satisfying the condition of regularity (Doob [50]) form then an abstract dynamical system $[\mathcal{V}, \mathcal{F}^{\mathrm{v}}, \mathcal{P}^{\mathrm{v}}]$ for the analysis of the flow of the molecular fluid. By considering this triple as a function space at any particular time instant $t_r \in \mathbb{R}^1 = [0, \infty]; r = 1 \ldots n$ during the entire flow of the fluid, the n-fold product of these spaces yields a product space in which the velocity vector $\mathbf{v}(^{\alpha}\mathbf{r}, t)$ is a time continuous random function. Alternatively, one can consider $[\mathcal{V}, \mathcal{F}^{\mathrm{v}}, \mathcal{P}^{\mathrm{v}}]$ as a one-parameter family of transformations T_t so that:

$$T_t: \mathcal{V} \to \mathcal{V} \text{ for all } t_r \in \mathbb{R}^1; \mathbf{v}_t(\mathbf{r}) \in \mathcal{V} \subset X . \tag{3.43}$$

Hence the velocity field is described in terms of a random function generated by the automorphism T_t for all $t_r \in \mathbb{R}^1$. As shown earlier by considering the

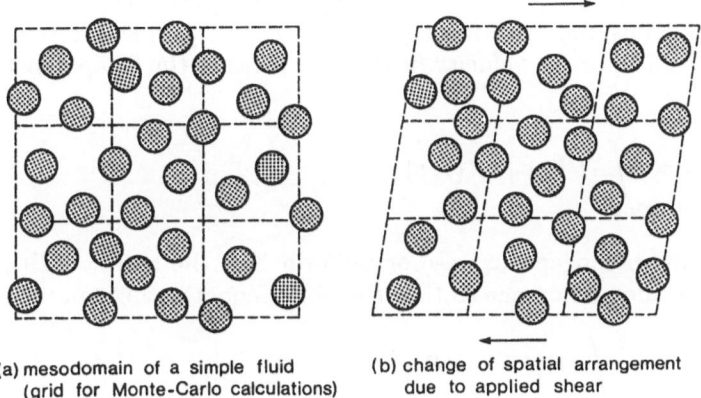

(a) mesodomain of a simple fluid   (b) change of spatial arrangement
    (grid for Monte-Carlo calculations)   due to applied shear

Fig. 14. Microstructure of a simple fluid.

velocity field $\mathbf{v}(^\alpha\mathbf{r}, t)$ and conditional probabilities, then one can define for each automorphism T_t a corresponding probability measure $\mathcal{P}^v\{E_{r+1}|E_r\}$ such that whenever:

$$T_t E_r = E_{r+1}$$

one has

$$\mathcal{P}^v\{E_{r+1}\} = \mathcal{P}^v\{E_{r+1}|E_r\}\mathcal{P}^v\{E_r\} \ . \tag{3.44}$$

This is readily generalized to the set of conditional probability measures, ie:

$$\mathcal{P}^v\{E_n\} = \mathcal{P}^v\{E_1\} \sum_{r=1}^{n} \mathcal{P}^v\{E_{r+1}|E_r\} \tag{3.45}$$

which is valid for any sequence $E_1 \supset E_2 \ldots E_r \supset E_n$ corresponding to the time sequence $t_1 < t_2 < \ldots t_r < t_n$ and a set of conditional probabilities $\mathcal{P}^v\{E_{r+1}|E_r\}, r = 1, 2 \ldots (n-1)$. It can be written more explicitly as:

$$\mathcal{P}^v\{E_{r+1}, t_{r+1}\} = \mathcal{P}^v\{E_{r+1}, t_{r+1}|E_r, t_r\} \tag{3.46}$$

where E_r corresponds to $t_r \in \mathbb{R}^1$ and E_{r+1} to $t_{r+1} > t_r$.

The above relations indicate that the velocity distribution at any time during the uniform flow of the fluid can be established, if the former at any other time or the initial one is known. Denoting the space of all measures $\mathcal{P}^v\{E_r\}$ by $L(0,1)$, it is readily seen that $P(.,.)$ on $L(0,1) \rightarrow L(0,1)$ is a contraction operator: $\| P\{E\} \| = 1$, which is known as the transition probability of the Markov process. Hence by choosing a closed time interval $[t,s] \in \mathbb{R}^1$ subdividing it into smaller intervals and selecting a point $\tau > t$, $\tau \in [t,s]$ shows that the transition probability for the velocity field $\mathbf{v}(\mathbf{r},t)$ satisfies the Chapman-Kolmogorov relation, ie.:

$$P\{t,s\} = \int_{v \subset x} P\{t,\tau\}dP\{\tau,s\} \qquad (3.47)$$

For the time homogeneous case or uniform flow the function $P(.,.)$ depends only on the time difference so that the above equation becomes:

$$P\{t-s\} = \int_{v \subset x} P\{t-\tau\}dP\{\tau-s\} \qquad (3.48)$$

This relation is most significant in the molecular dynamics of discrete fluids since it connects the stochastic theory with the functional analytic representation. Eqn. (3.47) can be written in matrix form:

$$\underline{P}(t+s) = \underline{P}(t)\underline{P}(s) \qquad (3.49)$$

showing the semi-group property of the distributions. Hence the time evolution for the uniform flow can be represented by an one-parameter semi-group of linear operators $T_t(t \geq 0)$, $t \in \mathbb{R}^1$ given in (3.43). These operators satisfy the following conditions:

$$T_t T_s = T_{t+s} \text{ for all } t,s \geq 0 \text{ in } X. \qquad (3.50)$$

If every $T_t(t \geq 0)$ is a bounded operator in X, it has the properties:

(i) $T_0 = I$ (identity operator)

(ii) and by the definition of a strong limit (eqn. (1.31), Chapt. 1)
$$s\text{-}\lim_{t \to t_0} T_t s = T_{t_0} s \text{ for all } t_0 \geq 0 \text{ and all } s \in X. \qquad \left.\begin{array}{c}\\\\\\\end{array}\right\} \quad (3.51)$$

This class of operators is called a C_0-class of semi-groups (Yosida [24]). The linear semi-group of T_t is said to be contracting, if for all $t \geq 0 \| T_t s \| \leq \| s \|$, $s \in X$. For the modelling of the flow by Markov processes the "weak limit" is more significant. It can be defined by:

$$w\text{-}\lim_{t \downarrow t_0} T_t s = s \text{ for every } s \in X \qquad (3.52)$$

(see also eqn. (1.33)). This indicates on the basis of a theorem by Guz [146] and the analysis of Kossakowski [126] that bounded observables or events induce a "weak" topology on X, that is more closely related to Markov processes. Generally, if the state-space $Z \equiv X$ and $P\{t, s, X\}$ a Markov process on it, the family of linear operators T_t in X representing the process, can be written as:

$$[T_t f](s) = \int_{Z \equiv X} P\{t, s, dr\} f dr; f \in X \tag{3.53}$$

in which T_t is a contracting one-parameter semi-group such that $T_t f \geq 0$, whenever the function $f \geq 0$ and $T_t 1 = 1$.

More recently, it has been shown by Pinsky [167] by means of a "random velocity model" based on the characteristics of a "two-state Markov chain", that one can construct the evolution process of the fluid by employing a finite number of contraction semi-groups. Thus from the properties considered earlier, the contraction semi-group of linear bounded operators will have the following characteristics:

(i) $|T_t f| \leq f$; $t > 0, f \in X$ (Banach space)

(ii) $T_{t+s} \cdot f = T_t T_s f$; $(t, s) > 0, \ f \in L$

(iii) $\lim_{t \downarrow 0} T_t f = f$.

$$\left. \right\} \tag{3.54}$$

Assuming that X is the space of the real-valued continuous functions f with the $\lim_{|x| \to \infty} f(x) = 0$, one can also define the semi-group by:

$$T_t \cdot f(x) = f(x + \mathrm{v}t)$$

where the velocity v is a fixed real number and the properties in (3.54) hold. Assuming that X is separable, one can form a direct sum where $X = X_1 + X_2 + \ldots X_n$ consisting of n-vector functions $\mathbf{f} = \mathbf{f_1} + \mathbf{f_2} + \ldots \mathbf{f_n}; \mathbf{f_i} \in X_i$ for $(1 \leq i \leq n)$ with the norm $|\mathbf{f}| = |\mathbf{f_1}| + \ldots |\mathbf{f_n}|$. Hence the "sum semi-group" on X will be given by:

$$\mathbf{T_t} \cdot \mathbf{f} = \begin{bmatrix} T_{1_t} & \mathbf{f_1} \\ T_{2_t} & \mathbf{f_2} \\ \vdots & \vdots \\ T_{n_t} & \mathbf{f_n} \end{bmatrix} \tag{3.55}$$

which is a contraction semi-group on X. Similarly as discussed earlier, the infinitesimal generator A of the semi-group T_t is now defined by:

$$A\mathbf{f} = \lim_{t \downarrow 0} \frac{T_t \mathbf{f} - \mathbf{f}}{t} \tag{3.56}$$

and if it exists the domain of this generator is defined by the function $\mathbf{f} \in X$. For the random velocity model the domain designated by $\mathcal{D}_A$ is thus specified for a set of functions: **f** which for each component f_i, $(i = 1 \dots n)$, the limit exists:

$$\lim_{t \downarrow 0} \frac{f_i(x + \mathbf{v}t) - f_i(x)}{t} = \mathbf{v} f_i'(x) \tag{3.57}$$

in which $\mathcal{D}_A$ is the set of functions $\mathbf{f} \in X$ with a continuous derivative $\mathbf{f}'$ since the $\lim \mathbf{f}(x) = 0 = \lim \mathbf{f}'(x)$ when $x \to \infty$. Hence the generator A of the evolution process becomes:

$$\mathbf{f} \to A\mathbf{f} = \begin{bmatrix} A_1 & \mathbf{f}_1 \\ A_2 & \mathbf{f}_2 \\ \vdots & \vdots \\ A_n & \mathbf{f}_n \end{bmatrix} \tag{3.58}$$

with $\mathcal{D}_A = [\mathbf{f} \in X \colon f_i \in \mathcal{D}_{A_i}; \ (1 \le i \le n)$. This generator A and the contraction group of (3.55) results in an "evolution operator $L(t)$" satisfying the following differential equation:

$$\frac{d}{dt}(L_L \cdot \mathbf{f}) = L_t A \cdot \mathbf{f} \ ; \ \mathbf{f} \in X_1 \cap X_2 \dots \cap X_n \tag{3.59}$$

where the mapping is continuous and differentiable at times $t \ne \tau_1, \tau_2 \dots \tau_n$, the latter being the jump times of the jump Markov process.

3.2 Interaction effects in discrete media:

The interest and research on interfacial phenomena of discrete media has grown dramatically in the past decade. At a boundary or interface separating the elements of a given microstructure various thermodynamic, structural and dynamic phenomena may occur. They have been earlier referred to as "interaction effects". Due to the great variety of heterogeneous media, a classification of such effects appears to be impossible. However at a molecular level at which the stochastic analysis also applies, one may distinguish two types of significant interactions, namely "reactive ones" that lead under certain conditions so the formation or breaking of covalent bonds and "non-reactive" or physical interactions that may result in the formation of new molecular complexes. The latter type are known as "weak" or van der Waals interactions and are of particular interest here. They will be further considered in the applications of the stochastic analysis in Chapt. 4.

It is well-known that externally applied disturbances to a fluid system are dampened due to internal relaxation processes within the molecular structure. In a macroscopic description the occuring phenomena are for instance diffusion, viscous flow and thermal conduction processes, which determine the "transport characteristics" of the fluid. Even without external influences there are at a given temperature "microscopic fluctuations", which are dissipated similarly to the external perturbations. Thus in a fluid thermal fluctuations occur in a natural manner with a distribution of certain wave lengths and frequencies. For long wave length and low frequency fluctuations, the fluid behaves like a continuum and the response can be formulated in terms of the hydrodynamic relations. At wave lengths comparable to the "molecular distances" however, the molecular structure of the fluid becomes rather significant and the description of the flow must be carried out in terms of "ensembles of interacting molecules". This approach is known as "molecular dynamics". Its application on the basis of probabilistic arguments and Markov theory will be further dealt with in Chapt. 4.

The analysis in this section is essentially concerned with "interaction potentials" and "intermolecular forces" required in the stochastic mechanics of discrete media.

(i) Interaction-potentials:

The main information on "interaction potentials" is obtained from experimental observations by the use of "scattering" (molecular beam) techniques, spectroscopic and nuclear magnetic resonance methods, etc. on solids and fluids. The measurements result in empirical forms for the potentials, which do not describe these potentials accurately over a wide range of molecular separation. Hence for their analytical utilization the important parameters are usually adjusted to match the theoretical values leading to "semi-empirical" forms. Nevertheless these forms can be regarded as a good approximation to the interaction potentials of various discrete media. Intermolecular forces and potentials are also considered amongst others by Margenau, Kestner [168], Yvon [117], Buckingham [169], Croxton [98], Hobza and Zaradnik [170], [171].

The most significant interaction effects in the present study arise from the forces and couples that act on "internal surfaces". These forces are in general derivable from the corresponding potentials assumed to act at discrete points (matching points of adjacent element surfaces) or at "clusters" of such points. In a polycrystalline solid for example "internal surfaces" are formed by the two surfaces belonging to two contiguous crystals or grains α, β respectively. Here, one has to deal with an "interaction zone", which although it may be small is still of finite dimensions. Such a zone may therefore serve in the formulation of the "bonding or de-bonding (decohesion)" behaviour between the elements of the microstructure. With regard to the interaction potentials a molecule may be seen as a group of atoms (or a single atom), whose binding energy is large enough to permit an interaction with its surrounding without losing its structural identity.

For instance a "hydrogen molecule" can be classified as a molecule, whilst an "argon molecule" is better considered as a "bound pair of argon atoms". If a molecule is tightly bound, its thermally populated "vibrational and rotational states" have similar structure and properties. Equilibrium and non-equilibrium properties of matter are characterized by the presence of inter-atomic and intermolecular forces. From elementary considerations of liquids for example, it is clear that molecules attract one another when they are sufficiently apart. Since solids and liquids have finite densities of a magnitude, observed under normal conditions, it shows that molecules repel at "short distances" and that there is a balance of forces corresponding to a minimum energy. Hence one obtains an interaction potential between two monatomic "spherical" molecules as indicated in Fig. 15 below, which are held together by central forces.

This figure clearly indicates that the "repulsive force" dominates at small separations of the atoms or molecules. At larger separations an "attractive force" becomes predominant. The balance of these forces occurs at an equilibrium separation "r_e" that would conform to the actual distance between the atoms of a diatomic molecule (Fig. 16).

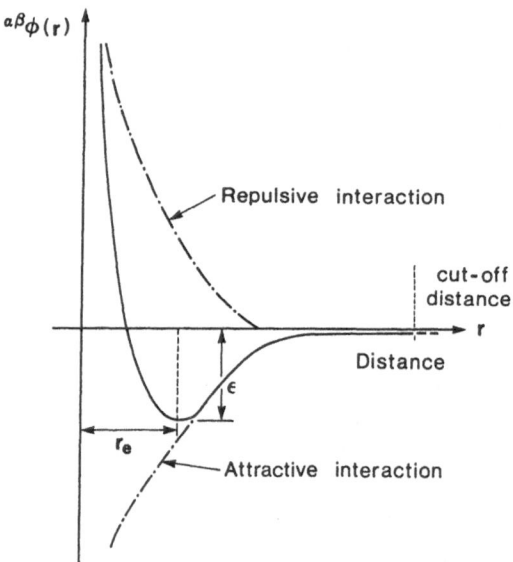

Fig. 15. Schematic form of the "pair-potential" $^{\alpha\beta}\phi(\mathbf{r})$ (central force model)

In a solid "r_e" will not necessarily be equal to the "nearest neighbour separation" denoted by $^{\alpha\beta}r_e$ due to effects caused by the more distant neighbours, but it will be near to it, if the forces are assumed to be "central". If additional non-central forces are present in the system, $^{\alpha\beta}r_e$ may be appreciably different from r_e or the latter may not exist at all. The existence of a potential energy function is based from statistical mechanics on the well-known Born-Oppenheimer approximation [170], whereby the nuclei in a particular configuration are fixed and the potential $\phi(r)$ is considered as the "difference in energy" of the system in that configuration from its value, when the separation "r" $\rightarrow \infty$. The number of variables on which the interaction energy depends increases sharply with an increase in molecular size. Thus, for instance, for an atom or molecule β and a diatomic molecule α as shown in Fig. 16, there are three variables, ie. the separation "r" of β from the C.M. of α, the angle θ between the molecular axis $A_1 - A_2$, the intermolecular separation r and the distance "d" between the nuclei A_1, A_2 in the molecule α. For two diatomic molecules there are six variables, ie. the intermolecular separation r, the angles θ_1, θ_2 between the molecular axis and the centre line, and angles ϑ_1, ϑ_2 between the plane containing r and the molecules, and two "intramolecular distances d_1, d_2". Denoting by n_α, n_β the number of atoms in the molecules α, β respectively, then in general for two interacting non-linear polyatomic molecules, one has an interaction energy that depends on $3(n_\alpha + n_\beta) - 6$ independent variables. Six of these variables are required for the description of the relative position and orientation of the molecules α, β and the remaining $(3n_\alpha - 6) + (3n_\beta - 6)$ are the internal vibrational coordinates of the two molecules. In certain cases these coordinates are of minor or no interest so that a considerable reduction of the variables can be obtained.

By considering a fluid or gas on the basis of a "hard-sphere model", the force in the collision of two monatomic molecules is given from physics to be equal to "$-\partial\phi(r)/\partial r$" (see for instance Yvon [117]). For the interaction occuring in a diatomic molecule and monatomic one (Fig. 16), there is also a torque present, which is equal to "$-\partial\phi(r)/\partial\theta$". It is of interest to note that the above notion of interaction forces does not apply for the determination of forces between ions in an aqueous solution. In such cases, it may be more convenient to use a potential of an "average force", which is equivalent to the Helmholtz free energy representing the "mean interaction energy" of two ions at a fixed separation r averaged over all configurations of all other molecules and ions that are present in the solution.

Intermolecular potentials can be distinguished by two classes of interactions, ie. the short-range and the long-range ones. The former decrease exponentially with increasing intermolecular distances "r" and are essentially due to an overlap of the electronic wave functions describing the isolated molecule. Long-range interactions vary in the form of "r^{-m}", m being a positive integer at large distances r. At long range the electrons become rather indistinguish-

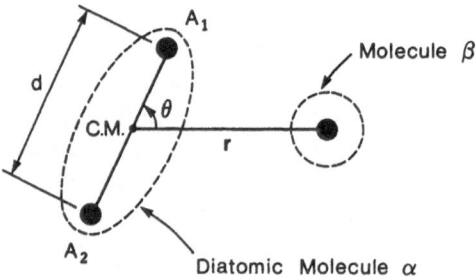

Fig. 16. Interaction between a diatomic molecule α and molecule β. (variables r, θ, d).

able in their exchange between molecules and hence in the determination of interactions one may consider the electrons as if they were associated with one or the other of the molecules. Short-range forces may be attractive or repulsive, but at small separations, they are always repulsive. Long-range forces can also be attractive and repulsive. For spherical atoms the long-range forces are attractive. In this classification of short and long-range interactions, the short-range group consists of "overlap (Coulomb and exchange) forces" which are non-additive. In the long-range interaction group, there are five distinct types of interaction energies and forces, e.g. electrostatic, induction, dispersion, resonance and magnetic. the electrostatic interactions are additive, the dispersion nearly additive and the magnetic only weakly so, the induction and resonance effects however are non-additive (see for instance Hirschfelder [103], Margenau and Kestner [168], Buckingham [169], Schuster[172] and others.

The intermolecular potential of simple fluids may be assumed in a first approximation to be the sum of effective pair-potentials. This potential is then characterized for a real simple fluid by a short-range strong repulsion and a long-range attraction. More recently the theory of pair-potentials has become important as well as the conditions on these potentials for a solid-liquid transition. As classified above, the simple fluids are characterized by spherically symmetric non-saturating interactions. It will be assumed in the subsequent modelling of simple fluids (see also Chapt. 4) that the forces are central, ie. acting through the centre of gravity of the molecule or particle and that they are pair-wise decomposable. In this case the total N-body configurational energy can be represented by the sum of pair-interactions, where the factor $\frac{1}{2}$

prevents counting the interactions $\alpha\beta$ and $\beta\alpha$ as distinct. Hence, one has:

$$\phi(1\ldots N) = \frac{1}{2}\sum_{\alpha}^{N}\sum_{\beta}^{N}\phi(\alpha,\beta) \tag{3.60}$$

which holds strictly for two-body effects only and thus excludes three-body effects. For the latter description a correction term in the form of a triplet potential $\phi(\alpha\beta\gamma)$ would have to be included in the summation: $\phi(\alpha\beta)+\phi(\beta\gamma)+\phi(\gamma\alpha)$. This, however, is usually not done due to the large increase in the calculation of the total interaction potential. It is rather assumed in dealing with simple fluids that the well-known "Lennard-Jones potential" and a number of idealized interactions (hard-spheres, square-well, etc.) can be used in an approximate theory.

So far as crystalline solids are concerned the pair-potential of atoms α, β at separation r, designated in the following by "$^{\alpha\beta}\phi(r)$" can also be taken as $\frac{1}{2}\phi(\alpha,\beta)$ such that the potential energy per atom can be expressed by the lattice sum $\sum\frac{1}{2}\phi(\alpha,\beta)$, which is then taken over all pairs that an atom can form with the remaining ones in the lattice. For short-range interactions only the "nearest neighbour of α" need be included in this sum. It is evident that if $\phi(r)$ is the only potential present in the microstructure of the solid, its form must be such that the lattice is "mechanically stable" and an energy increase can only be ascribed to an arbitrary small perturbation of the atoms from their equilibrium positions. Thus the conditions which $\phi(r)$ has to satisfy for a "mechanically stable" structure to exist will be as follows:

(i) $\phi(r)$ must have a minimum at a finite r,

(ii) for large separations, its magnitude must tend to zero more rapidly than r^{-3}.

These conditions and the finiteness of the lattice energy ensure the stability of the structure with respect to an "infinitesimal" homogeneous tension or compression of the lattice. However, shear deformations are of equal importance for the lattice stability, but will not be considered here. Reference is made to the work of Born [173], Mott [174] among others.

A frequently used form of the potential $^{\alpha\beta}\phi(r)$ in crystalline solids is of the form:

$$^{\alpha\beta}\phi(r) = A(\frac{r_e}{r})^n - B(\frac{r_e}{r})^m, \; n > m , \tag{3.61}$$

in which the first term represents the repulsive part and the second the attractive part of the potential, r_e is the equilibrium separation and A, B are material characteristics. So far as central force models are concerned, it is known that for crystals of the solidified rare gases, the attractive part of the potential originates from van der Waals forces. It can be shown that m in (3.61) is equal to six.

The most appropriate repulsive potential corresponds to values of $n = 10 \div 12$. If the value 12 is chosen the interactions are sometimes called Lennard-Jones forces and the associated interaction potential the L-J.-potential. Although no specific form of the pair-potential can be given for the great variety of polycrystalline solids and particularly for metals, it is frequently assumed that the L-J. potential is a good approximation. An alternative way to the assumption of a force law consists of evaluating the parameters in (3.61) by comparison with experimentally obtained values for a given solid (see Macpherson [106]). It is sometimes preferable to use instead of (3.61) a potential in which the repulsive and attractive parts are expressed in form of exponentials. In this case the potential takes the form:

$$^{\alpha\beta}\phi(r) = D\{\exp[-2\nu(r - r_e)] - 2\exp[-\nu(r - r_e)]\} \tag{3.62}$$

in which D and ν are material constants determined by appropriate experiments. This form is known as the "Morse potential" and has been found useful in the description of interactions in both b.c.c. and f.c.c. metals for instance. A modified form has been employed previously in the stochastic analysis of other heterogeneous media (see [22, 23] and Chapt. 4).

(ii) Stochastic models of the interfacial behaviour in solids:

For a discussion on the "interfacial behaviour" of discrete solids, two important classes of materials, ie. polycrystalline solids and fibrous networks will be considered in this section. In contrast to classical continuum theory which in general does not include interaction effects, the interfacial potential in (3.62) will be employed in the formulation of the interactions occuring at the "internal surfaces". However the use of such a "surface potential" requires for its implementation in the analysis, the concept of a "unit cell" on which it acts and where the latter is repeated over the dimensions of the interface (length × width). It has been shown in refs. [23, 175] that this potential can also be determined by computer simulations and has been evaluated for the microstructures of copper and aluminium in [176]. The interaction effects in polycrystalline solids occur essentially on the grain boundaries and in fibrous structures (cellulose, polymers) they further depend on the "bonding" of individual fibres in the network. There are of course "internal effects" due to defects in the crystals themselves, but attention will be given here mainly to the grain boundary effects. To analyse the latter one can use a Morse-type potential of the form (see also [22, 23]):

$$\phi\{|^{\alpha\beta}\mathbf{d}|\} = \phi_0\{1 - \exp[-b|^{\alpha\beta}\mathbf{d}|]\}^2 \tag{3.63}$$

in which the "distance vector" $^{\alpha\beta}\mathbf{d} = |^{\alpha\beta}\mathbf{d}| \cdot \mathbf{e}$; (e unit vector of the reference frame). To clarify the role of the vector $^{\alpha\beta}\mathbf{d}$ for the occuring interactions at the grain boundaries, it is necessary to consider the basic deformation kinematics

([22], [23]). A schematic drawing of an element α of the polycrystalline solid in the "undeformed configuration" is given in Fig. 17(a) below.

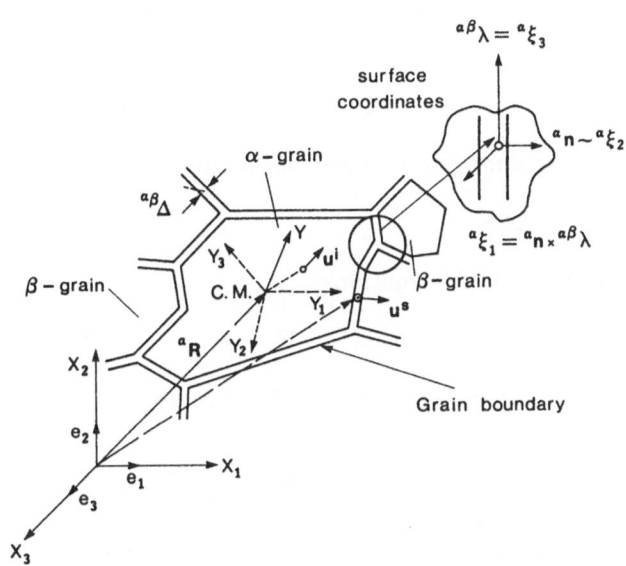

Fig. 17. Schematics of an element α; $(\alpha = 1 \ldots N)$ of a polycrystalline solid (undeformed configuration).

It is well-known form continuum mechanics that the deformation of a continuum is completely described by a set of vectors $\mathbf{u}(\mathbf{X}, t) \in \mathcal{U}$ (deformation-space) as functions of the "undeformed position vector $\mathbf{X}$" of a material point and time t. The formulation in stochastic mechanics is distinctly different since the interactions between material elements are included. Hence with reference to Fig. 17(a,b), the deformation $^{\alpha}\mathbf{u}$ of a microelement or grain α will be described by two random vectors $\mathbf{u}^i(\mathbf{X}, t)$ and $\mathbf{u}^s(\mathbf{X}, t)$ respectively, that are associated with the "internal" and "surface" deformation of this element. Here majuscules designate the "undeformed" and miniscules the "deformed" configurations of the element. One has therefore:

$$\mathbf{u}^i(\mathbf{X}, t) = \mathbf{x}(t) - \mathbf{X} \qquad \text{(a)}$$
$$\mathbf{U}^s(\mathbf{X}, t) = \mathbf{g}(\mathbf{x}, t) - \mathbf{G}(\mathbf{X}) \qquad \text{(b)} \qquad \left.\right\} \quad (3.64)$$

in which $\mathbf{X}, \mathbf{G}$ denote the position vectors of an arbitrary point inside the element and on the surface of the αth-grain, respectively, with respect to the external coordinate frame. The vectors $\mathbf{x}, \mathbf{g}$ are their counter parts in the deformed configuration. Using a "body frame" in accordance with the given Definitions of Chapt. 2 one can also write:

$$\mathbf{X} = \underline{O} \cdot \mathbf{Y} + \mathbf{R} \tag{3.65}$$

where $\mathbf{O}$ is a "random orientation matrix" and $\mathbf{R}$ the random position vector to the C.M. of the microelement. Thus relations (3.64a,b) can be expressed as:

$$\mathbf{u}^i(\mathbf{Y}, t) = \underline{o} \cdot \mathbf{y}(t) - \underline{O} \cdot \mathbf{Y} + \mathbf{r}(t) - \mathbf{R} \tag{3.66}$$

and by introducing a moving surface coordinate frame (see also ref. [22]):

$$\mathbf{H}: \mathbf{G} = \underline{O} \cdot \mathbf{H} + \mathbf{R}$$

one obtains the surface deformations:

$$\mathbf{u}^s(\mathbf{H}, t) = \underline{o} \cdot \mathbf{h}(t) - \underline{O} \cdot \mathbf{H} + \mathbf{r}(t) - \mathbf{R} \tag{3.67}$$

Introducing "generalized deformations $^\alpha \hat{\mathbf{u}}$" of individual microelements, which are functions of $\mathbf{u}^i$ and $\mathbf{u}^s$, one has:

$$^\alpha \hat{\mathbf{u}}(\mathbf{X}_{C.M.}, t) = f[^\alpha \mathbf{u}(\mathbf{X}, t)] = f[\mathbf{u}^i(\mathbf{X}, t) \oplus \mathbf{u}^s(\mathbf{X}, t)] \tag{3.68}$$

or equivalently by using the H-frame (moving coordinate frame):

$$^\alpha \hat{\mathbf{u}}(\mathbf{X}_{C.M.}, t) = f[\mathbf{u}^i(\mathbf{Y}, t) \oplus \mathbf{u}^s(\mathbf{H}, t)]. \tag{3.69}$$

However, the quantities $\mathbf{u}^i, \mathbf{u}^s$ are not continuous functions in the strict sense, but in general may be regarded to be at least piecewise continuous and that derivatives of all orders within some compact support exist. In this sense the "generalized deformation $\hat{\mathbf{u}}(\mathbf{X}, t)$" will be a discrete random function $^\alpha \hat{\mathbf{u}}(\mathbf{X}, t), \alpha \in M$ (meso-domain).

For the solution of problems encountered in practice, the evolution of these deformations is of interest in terms of a general "deformation process". For this purpose one can consider a sequence of deformations $\hat{\mathbf{u}}(\mathbf{X})$ indexed by the time t and designated by $\hat{\mathbf{u}}(\mathbf{X}, t)$ as a "random process". Thus the latter defines a random deformation process in the deformation-space $\mathcal{U} \subset X$ (probability space), where t belongs to the positive half of the real line $\mathbb{R}^+$. The function $\hat{\mathbf{u}}(\mathbf{X}, t)$ is a random function that for a fixed time $t \in \mathbb{R}^+$ is a "random vector $\hat{\mathbf{u}}(\mathbf{X})$" in $[\mathcal{U}, \mathcal{F}^u, \mathcal{P}^u]$. Thus considering the triplet $[\mathcal{U}, \mathcal{F}^u, \mathcal{P}^u]$ for any particular t, the whole "deformation process" is then representable by a set of these function spaces or a product space. In particular, if $\mathbb{R}^+ = [0, \infty)$, where for each time $t_r \in \mathbb{R}^+$, $r = 1 \dots N$, there corresponds the triplet $[\mathcal{U}, \mathcal{F}^u, \mathcal{P}^u]$ one obtains a product space in which $\hat{\mathbf{u}}(\mathbf{X}, t)$ may be regarded as a time-continuous random function. A general stochastic deformation theory of discrete solids will be presented in Chapt. 4.

The above kinematic quantities together with some other variables such as "surface forces" caused by the surface potentials occuring at the grain boundaries represent the basic kinematic parameters involved in the response behaviour of discrete solids. In order to assess the contribution due to the grain boundary effects the simple model of Fig. 17 is extended to the schematics of the grain boundary kinematics of two adjacent grains α, β in Fig. 18.

It has been shown in previous work [22, 23], that one can employ a probabilistic "surface molecular coincidence lattice" model in the formulation of the grain boundary effect. This model is based on the geometrical theory of "coincidence lattices" due to Bollman [177] and Goux [178]. The schematics of a typical "undeformed coincidence cell" on the grain surface is shown in Fig. 18 (b) below. In Bollman's theory grain boundaries are analytically defined in terms of coincidence lattices obtained from the interpenetration of two neighbouring grains and where the "coincidence lattice points" form equivalent groups to the lattice points of the crystals. On the other hand, Goux considers the grain boundary to have a fairly disordered or amorphous structure that separates the two adjacent grain surfaces (boundary zone as indicated in the figure). However, the latter model admits the existence of a certain "distance" between lattice points of two adjacent grains. Evidently, the model suggested in the present theory may be regarded as a combination of these representations, where the "distance vector between two coincidence points" at the grain surfaces of α, β is considered as the fundamental kinematic parameter.

This kinematic variable is here referred to as the "relative displacement **d**" of two adjacent surfaces under the action of an external load. Thus the distance between the crystal lattices in the grain boundary zone or the "calculation zone" for the inherent grain boundary energy [23, 176], in accordance with the previous notation can be stated for the undeformed state of the solid as follows:

$$^{\alpha\beta}\boldsymbol{\Delta}^u = {}^{\beta}\mathbf{G} - {}^{\alpha}\mathbf{G} = {}^{\beta}\underline{Q}{}^{\beta}\mathbf{H} - {}^{\alpha}\underline{Q}{}^{\alpha}\mathbf{H} + {}^{\beta}\mathbf{R} - {}^{\alpha}\mathbf{R} \qquad (a)$$

and for the deformed state, by:

$$^{\alpha\beta}\boldsymbol{\Delta}^d = {}^{\beta}\mathbf{g} - {}^{\alpha}\mathbf{g} = {}^{\beta}\underline{\varrho}{}^{\beta}\mathbf{h} - {}^{\alpha}\underline{\varrho}{}^{\alpha}\mathbf{h} + {}^{\beta}\mathbf{r} - {}^{\alpha}\mathbf{r} \qquad (b)$$

$$\left.\right\} \quad (3.70)$$

so that the random relative displacement is given by:

$$^{\alpha\beta}\hat{\mathbf{d}} = {}^{\alpha\beta}\hat{\boldsymbol{\Delta}}^d - {}^{\alpha\beta}\boldsymbol{\Delta}^u \qquad\qquad (3.71)$$

where the " ^ " sign again refers to the "discrete character" of the parameters. The above kinematic parameters as well as some other physical ones can be established from the corresponding distribution functions. Thus for instance the crystallographic orientations $({}^{\alpha}\underline{Q}, {}^{\alpha}\underline{\varrho})$ can be assessed by an X-ray diffraction technique ([106], [179]). The distribution of "grain sizes" can be obtained by micrography (SEM-observations) and a subsequent stereological evaluation. For small grains the grain boundary effects are also small and a simplification of the analysis is then achieved by neglecting the grain boundary effects. In this case,

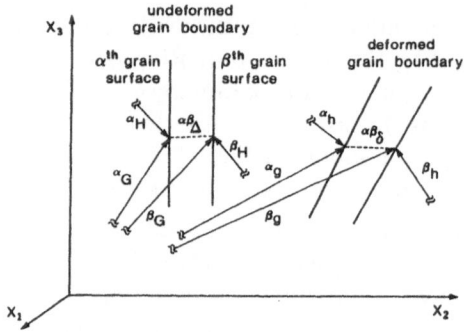

(a) Grain surface displacement

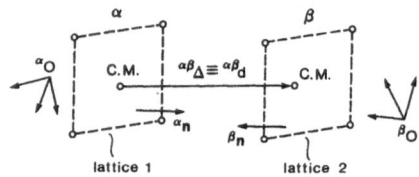

(b) Coincidence or "unit cell" model (area $^{\alpha\beta}$A)

Fig. 18. Grain boundary kinematics of two adjacent crystals α, β.

by assuming that the distributions are Gaussian and statistically homogeneous, non-isotropic and applicable throughout a particular meso-domain, although they may vary from one domain to another, one can employ correlation theory. The distribution functions are then fully determined by their first and second moments. A suitable correlation parameter in this case will be again a distance vector "ν" measured between the C.M. of two adjacent microelements or grains (Fig. 17) so that:

$$\nu = {}^{\beta}\mathbf{R} - {}^{\alpha}\mathbf{R} \qquad (3.72)$$

The expectation and correlation functions for the orientations and grain sizes are then obtained from:

$$\left. \begin{array}{l} E\{{}^{\alpha}\underline{Q}\} = <{}^{\alpha}\underline{Q}>_M ; \ C_0(\lambda) =< {}^{\alpha}\underline{Q}({}^{\alpha}R){}^{\alpha}\underline{Q}({}^{\alpha}R+\lambda) >_M , \\ E\{{}^{\alpha}v\} =< v >_M ; \ C_v(\lambda) =< v({}^{\alpha}\mathbf{R})v({}^{\alpha}R + \lambda) >_M , \\ \qquad\qquad \alpha = 1 \ldots N; M = 1 \ldots p \end{array} \right\} \qquad (3.73)$$

where C with the corresponding suffix is the correlation function of a particular meso-domain M.

With reference to the schematics of the undeformed "coincidence cell" model in Fig. 18(b), the surface between any two crystals in general, will take up a position that the crystals exhibit more or less "optimum matching". Thus, on the assumption that lattice 1 is fixed and lattice 2 is changing, ie. undergoing a translation and rotation, the latter will be translated in such a manner that at least one point coincides with a point on the surface of lattice 1. This point is called a "coincidence site". Due to the inherent periodicity of the two lattices and assuming that they are "ideal", ie. free of any defects, a finite number of such points will exist. These points form, from a purely geometrical point of view, another lattice termed "coincidence lattice".

Hence Bolmann's theory leads to a grain boundary topology in terms of two idealized crystals, that form an interpenetrating mathematical translation lattice, which depends of course on the "misfit angle θ", between the given crystal lattices. As a measure of the "coincidence cell" lattice one may consider the surface of its "unit cells", which in turn also depends on the lattice vector "a" of the specific polycrystalline structure under consideration as well as on the "relative orientation". Since in the present analysis the orientations for two adjacent crystals, ie. $^{\alpha}\underline{Q}, ^{\beta}\underline{Q}$ are taken as random quantities, the "coincidence areas $^{\alpha}\Delta A^q = {}^{\beta}\Delta A^q(q = 1 \ldots p; \ p = $ number of coincidence cells) will be random functions of both $^{\alpha}\underline{Q}$ and $^{\beta}\underline{Q}$ (Fig. 18(b)). Thus, generally the size and shape of the boundary coincidence cells are dependent on the distribution of the misfit angle or $\mathcal{P}(\theta)$ occuring in the grain boundary region. To account for this mismatch angle between individual crystals and the grain boundary effect between them, it is convenient to use a "surface coordinate" system within the "interaction zone" for α and β as indicated in Fig. 17(b). Thus, if $^{\alpha}\mathbf{n}$ denotes the normal to the surface of the αth crystal, then in terms of the surface coordinates one has the following relations:

$$\{^{\alpha}\xi\} = (^{\alpha}\xi_1, {}^{\alpha}\xi_2, {}^{\alpha}\xi_3): {}^{\alpha}\xi_1 = {}^{\alpha}\mathbf{n} \times {}^{\alpha\beta}\lambda, {}^{\alpha}\xi_2 = {}^{\alpha}\mathbf{n}, {}^{\alpha}\xi_3 = {}^{\alpha\beta}\lambda \qquad (3.74)$$

in which the rotation about a common axis is expressed by the eigenvector λ as shown in Fig. 17(b). The results of the computer simulation employed for the determination of grain boundary influences including thermal effects on grain boundaries has been given for instance in [176, 180]. The two scalar quantities ϕ, ϕ_0 in eqn. (3.63) as well as the material parameter "b" can be determined from spectroscopic studies.

The other group of discrete solids, which is of considerable interest in practice, are fibrous structures as shown previously in the micrograph of Fig. 4. Here, the interaction effects between the elements of the structure (polymeric, cellulose networks, etc.) are best illustrated by considering the "bond behaviour" of overlapping fibres (see also [22], [23] and [181–184]). Since in this class of materials the fibres are randomly arranged the most significant parameter that contributes to their strength characteristics will be the "bond dissociation" analogous to a "decohesion" in polycrystalline solids. This bond dissociation with particular reference to "cellulose networks" is predominantly controlled by a partial or complete hydrogen bonding (H-bonding), although other factors such as humidity (water absorption), temperature may also contribute to the overall mechanical response.

For the understanding of the bond dissociation process a few remarks on the molecular properties of hydrogen bonds may be indicated. For a more comprehensive study the reader is referred to a more recent text on the theory of hydrogen bonding [172]. Generally an H-bond occurs, if one hydrogen atom is bonded to more than one other atom say X, Y. If the two bonds of H to X and Y have different strengths, the stronger bond is written X–H (normal X–H bond), whilst the weaker one is designated by $H \ldots Y$ and called H-bond. Evidently in this case the strength of $H \ldots Y$ can be associated with the "dissociation energy" of the molecular complex H–$H \ldots Y$. The H-bond forming the complex can either be symmetrical or unsymmetrical, depending whether the "surface energy" for the proton between atoms X and Y is symmetric or not. The H-bond is "intramolecular" or "intermolecular", if the atoms X and Y belong to the same molecule or not. The term molecule is thus understood in the chemical sense.

From a physical point of view each intermolecular bond forms a larger molecular complex (macromolecule) that exhibits its own "force constant, chemical reactivities", etc. If the complex forming molecules are of the same type as for instance in water, the intermolecular association is referred to as "self-association". If the molecules are of a different type then the association is called a "mixed association". Due to recent developments in the theory of hydrogen bonding, it is possible to define certain contributions to the H-bond energies and to give an estimate to the relative significance of various forces etc. (see [169] [172]). However, apart from these H-bonding characteristics the problem of "intermolecular" interactions must also be considered. In the conventional description of molecular properties on the basis of quantum mechanics, one uses a time-dependent Hamiltonian, which by neglecting any relativistic effects for a system of N-atoms and n-electrons can be expressed by:

$$H = A_N + A_K + A_P \tag{3.75}$$

in which A_N, A_K and A_P are the well-known operators characteristic of the nuclear, electron kinetic and potential energies, respectively. They contain the basic quantities of mass, charge and position of the nuclei.

The quantum theoretical approach is usually restricted to considerations of stationary states and the interaction between molecules can then be represented by a set of equations (stationary Schrödinger equations) that govern the nuclear and electron motion. However an exact solution of these equations is only available for very simple cases and thus approximate methods of solutions are employed [170]. More recently perturbation techniques have come to the forefront. In the latter approach the main interest is on the electron exchange between molecules. Restricting such an exchange to the motion of single electrons, which seems to be an adequate approximation for large molecular distances, one can express the "interaction energy" between molecules essentially by four terms, ie.:

$$\epsilon = \epsilon_C + \epsilon_E + \epsilon_P + \epsilon_D \tag{3.76}$$

in which "ϵ" denotes the energy of complex formation for two molecules α, β, e.g.:

$$\alpha + \beta \rightarrow \alpha\beta \text{ and } \epsilon = {}^{\alpha\beta}\epsilon - ({}^{\alpha}\epsilon^0 + {}^{\beta}\epsilon^0) \tag{3.77}$$

where ${}^{\alpha}\epsilon^0, {}^{\beta}\epsilon^0$ are the total energies of the isolated molecules as obtained from the solution of the Schrödinger equation with the Hamiltonians ${}^{\alpha}H^0$, ${}^{\beta}H^0$, respectively, and

$$ {}^{\alpha}H^0 {}^{\alpha}\psi^0 = {}^{\alpha}\epsilon^0 {}^{\alpha}\psi^0 \; ; \; {}^{\beta}H^0 {}^{\beta}\psi^0 = {}^{\beta}\epsilon^0 {}^{\beta}\psi^0 \tag{3.78}$$

in which ${}^{\alpha}\psi^0, {}^{\beta}\psi^0$ are the corresponding wave functions of the molecules. For a rigorous treatment of the quantum mechanical approach and semi-empirical methods of solution concerning hydrogen bonding see for instance [168] and [170, 185]. It is seen that the interaction energy ϵ is composed to a first approximation of the Coulomb or electrostatic energy of interaction ϵ_C and the exchange energy ϵ_E. The other quantities in (3.76) refer to the polarization and the dispersion energy, respectively. The energy ${}^{\alpha\beta}\epsilon$ is associated with the total energy of the macromolecule $(\alpha\beta)$. The overall Hamiltonian of this molecule can be decomposed into H^0 and the Hamiltonian due to perturbation H', so that:

$$H^0 = {}^{\alpha}H^0 + {}^{\beta}H^0 \; ; \; {}^{\alpha\beta}H = H^0 + H' \tag{3.79}$$

Thus, $H' = {}^{\alpha\beta}\phi(\mathbf{R})$ is the intermolecular potential, which is a function of the distance between molecules and $\mathbf{R}$ the configuration vector, that also contains the nuclear coordinates. Since the main interest here is on the bonding behaviour in hydrogen bonded solids (natural cellulose) as shown in Fig. (9a) and its molecular structure indicated in Fig. 19.

It is to be noted that apart from the "intermolecular bonding" present in this type of bond, an "intramolecular bonding" must also be admitted in order to account for the spatial configuration of the bond. Intramolecular bonding is also important for the consideration of polymer-melts an analysis of which is given in Chapt. 4.

Experimental observations on cellulosic systems (X-ray, etc. [186–190]), revealed that cellulose in natural fibres crystallize either partially or completely within the fibres so that "unit cells" are formed. These cells are repeated in form of a chain. A structural model of such repeating units according to reference [191] is shown in Fig. 20.

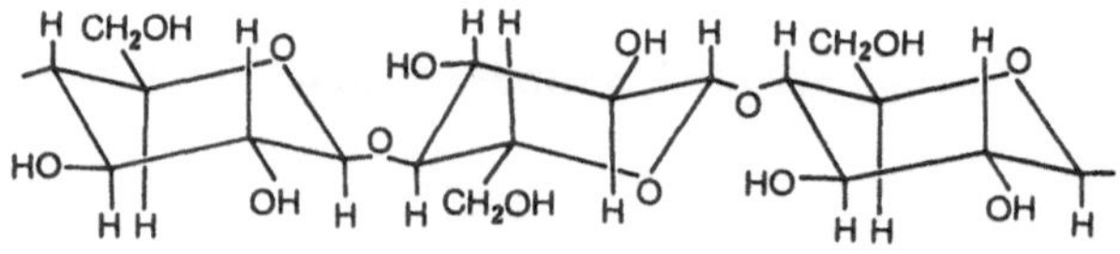

Fig. 19. Molecular structure of "cellulose" (corresponding to Micrograph (Fig. 9a))

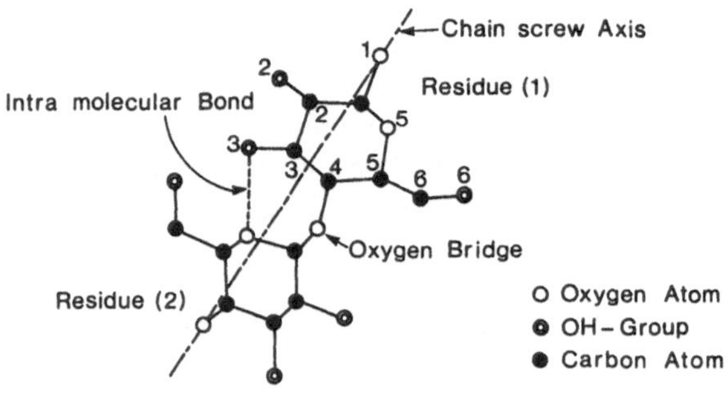

Fig. 20. Repeating unit of natural cellulose (Idealized structure [184]).

It is seen that the unit is composed of two (β–D) glucose residues, which are linked by the oxygen bridge to adjacent residues and are also rotated with respect to one another about a screw-axis so that they form continuous chain segments. There is in addition an "intramolecular bond" between the "Hydroxyl group" of one glucose residue and the oxygen ring of the next residue. In order to form a bond, it is necessary for the neighbouring glucose units to be rotated around the glucose linkage resulting in a "bent" conformation [194]. Structural models of H-bond crystals [191, 192, 193] indicate, that in view of the "monoclinic" character of the unit cell, a set of bonds can be formed in the $(10\bar{0})$ and another in the (101) plane, where the bonds in that plane are "nearer" than those in the $(10\bar{1})$ plane. Since in the present study the mechanical effects on hydrogen bonds are of the main interest, an interfacial bond configuration is visualized as shown in Fig. 21 below. (Fig. 21 a,b).

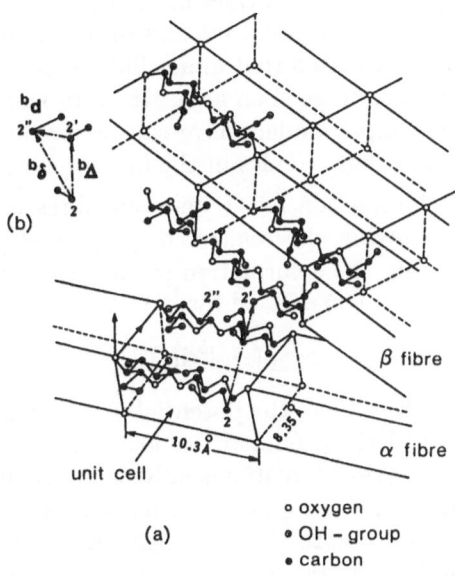

Fig. 21. Fibre-fibre interface model of hydrogen bonds in fibrous structures [22, 184].

This figure represents an "idealized" 3-dimensional configuration of "unit cells" within the interface between two overlapping fibres $(\alpha\beta)$ of the cellulose structure. This model has also been discussed in earlier work [22, 184] with respect to an "effective bonding area" between fibres and for the assessment of a "force-transmission" between the fibres due to bonding. The considerations were based on the previously mentioned geometrical theory of "coincidence lattices" by Bollman. Here, however each lattice is visualized to be formed by the "cellulose unit cells" such that the length of the "repeating units" is in the direction of the fibre-axis. The "coincidence sites" in this case are the OH-group 2 and 2' (Fig. 21(b)) that are initially at a distance $^{b}\Delta$ apart corresponding to the interatomic distance of the bond potential. If an external force is exerted

on the bond in the direction of the fibre axis a "relative displacement $^{\alpha\beta}\mathbf{d}$"
will occur and the OH-group moves to the configuration 2″ Whilst this model
characterizes in the simplest manner the relative motion in the case of "perfect
bonding", the hydrogen bond motion in the actual network is far more complex.
It can be regarded rather as a "dissociation process" occuring in a random
manner, although the significant parameter $^{\alpha\beta}\mathbf{d} \equiv {}^b\mathbf{d}$ of the bond potential is
still retained. As observations by scanning-electron microscopy have revealed
[195], [196], two overlapping fibres will ideally bond over a certain area within
the common surface layers. In these surface layers the bonded areas consist of a
number of individual bonds belonging to it and result in an "effective bonding
area" (see also [22]). It is estimated that in a cellulose structure, there exist six
hydroxyl groups per unit cell in such an area, which provide a perfect bonding
between the fibres. The simple model above can be extended to allow for the
occurrence of bending and shear modes in the bond deformation.

To account for the interaction effect between fibres one can again use a
Morse-type potential involving the relative displacement associated with an
equilibrium position during the deformation so that similar to the form given
in (3.62) this potential can be expressed by:

$$\phi(t) = \phi_0 \{\exp(-2\nu|{}^b\mathbf{d}(t)|) - 2\exp(-\nu|{}^b\mathbf{d}(t)|)\} \tag{3.80}$$

when ϕ_0 is the equilibrium value of the potential, ν a material characteristic of
the fibrous structure and ${}^b\mathbf{d}(t)$ the relative displacement indicated in Fig. 21(b).
It is of interest to note that this potential is only an approximation to the more
exact form, since the terms for the repulsion energy between the hydrogen and
free oxygen atoms and that due to the exchange energy of attraction between
the oxygen atoms have been neglected. However from a stochastic mechanics
point of view, this approximation corresponds to the adopted "microscale" of
postulate 1 (Chapt. 2). In general, the interaction effects will be time-dependent
and the models suggested above as indicated by the form of potential in (3.80)
will have to be extended to include the mechanical relaxation of such materials.
In the case of polycrystalline solids the corrsponding formulation and that
pertaining to fibrous structures will be given in the application of the stochastic
theory in Chapt. 4. The latter uses the concept of a "generalized surface force"
that acts at each coincidence or unit cell between the α and β microelement.

(iii) Markov Models of "bond failure" and fracture:

The interfacial behaviour between grains of a polycrystalline solid and that
of fibres in a fibrous structure has been briefly outlined above. It has been
shown that "perfect bonding" can be expected to occur only at discrete points
(coincidence sites) within a "unit cell" of the interface region. So far as the
"internal bonding" of the given lattice structure is concerned "missed bonds"
or other point defects can be treated by the use of a "random walk" model

(see for instance Montroll and West [197]). In the mechanics of fracture of polycrystalline solids, it is usually assumed that the "fracture zone" consists of two parts, ie. a "cohesive zone" in which the neighbouring crystals still act as completely bonded and a "free zone" in which bonding has ceased to exist. From a phenomenological point of view the free zone is one where the initial crack of an assumed length "a_0" will start a crack propagation towards the cohesive zone. These two zones are characterized in continuum mechanics by smooth non-intersecting curves $z(s)$ of a certain arc length "s" so that both of these zones can be decribed in terms of time-dependent parameters as indicated in Fig. (22) below.

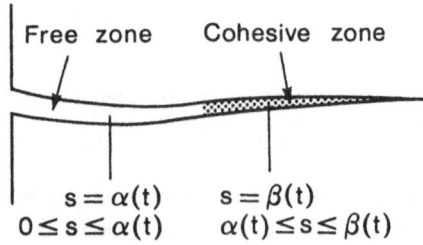

Fig. 22. Continuum Model of crack propagation in a solid.

A continuum mechanics model for the process of crack propagation corresponding to Fig. (22) is due to Gurtin [198]. More recently Maugis [199] has introduced models in the analysis of various fracture phenomena that use interface energies as the significant parameters in the development of fracture. In contrast the stochastic mechanics approach assumes the initial crack formation to be a "random decohesion phenomenon" and the crack propagation to follow rather a path (non-smooth) as shown in Fig. 23.

The crack is assumed in most applications to start at a notch or at some defect on the external surface of the material body and to propagate in a random manner. This phenomenon can therefore be modelled by a random process. Since actual materials have a rather complex microstructure the tendency in modern research is to "simulate" this behaviour by appropriate models. Simulation models in general will be briefly considered later in this chapter.

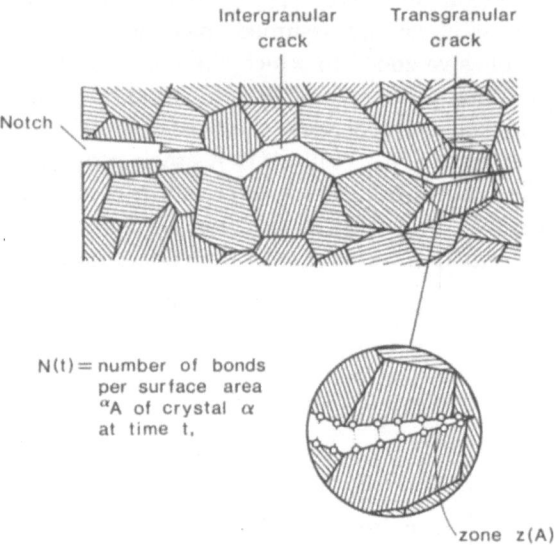

Fig. 23. Discrete Model of crack propagation (non-smooth zone, $z(A)$).

By considering the interactions dealt with in the foregoing section, ie. bonding (cohësion) and de-bonding (decohësion) between two grains α, β of a polycrystalline solid, this process can be characterized by a family of real random variables or the "number of bonds N_t" per "unit cell" of the common interfacial area $^{\alpha\beta}A$ (or A) in form of a stochastic process $\{N_t, t \geq 0\} \in \mathcal{C} \subset X, t \in T$.

The unit cell is based on the "coincidence sites" model mentioned previously, and the grain surface displacement as shown in Fig. 18. If the joint distribution functions of the variables $\{N_{t_1}, N_{t_2} \ldots N_{t_n}\}$ is known for all finite $(n = 1, 2 \ldots)$ and all sets of values $(t_1, t_2 \ldots t_n) \subset T$ and if they are compatible, then there exists a probability triple $[X, \mathcal{F}, \mathcal{P}]$ in which the process $\{N_t, t \geq 0\}$ is defined. The latter may also be defined on any subspace of X, (for instance the configuration space $\mathcal{C} \subset X$, deformation space $\mathcal{U} \subset X$, etc.), for which the measure $\mathcal{P}\{N_{t_1} \leq x_1 \ldots N_{t_n} \leq x_n\}$ is equal to the prescribed distribution for every $(n = 1, 2 \ldots)$; $(t_1 \ldots t_n) \subset T$. In particular the stochastic process $\{N_t\}$ is Markovian, if:

$$\left. \begin{array}{l} \mathcal{P}\{N_t \leq x | N_{t_1} = y_1, N_{t_2} = y_2 \ldots, N_{t_n} = y_n\} \\ \qquad\qquad = \mathcal{P}\{N_t \leq x | N_{t_n} = y_n\} \end{array} \right\} \quad (3.81)$$

for all $t_1 \leq t_2 \leq \ldots t_n; (n = 1, 2 \ldots)$ and all possible values of the random variables. The process is uniquely determined by its initial and conditional distribution functions as follows:

$$\left.\begin{array}{l} \mathcal{P}\{N_0 \leq x\} = P(0,x); \ \ N_0 = N_t \text{ at } t = 0 \\ \mathcal{P}\{N_t \leq x | N_{t_n} = y_n\} = p\{s,y; \ t,x\}; \ \ (s \leq t) \in T \end{array}\right\} \quad (3.82)$$

where $p\{.,.\}$ are the transition probabilities. Since $\mathcal{P}\{N_t \leq x\} = P(t,x) = \int_{-\infty}^{\infty} p\{0,y; \ t,x\}d_y P(0,y)$, it follows from the theorem of total probability that:

$$p\{x,y; \ t,s\} = \int_{-\infty}^{\infty} p\{u,z; \ t,x\}d_z p\{s,y; \ u,z\} \ ; \ \ (s \leq u \leq t) \subset T \quad (3.83)$$

which is the Chapman-Kolmogorov equation. The stochastic process of the formation of bonds and the breaking of bonds can be considered however as a special Markov process, e.g. one with independent increments. In this case the transition probabilities can be regarded to depend on $(x - y)$ apart from t and s. This corresponds to the assumption that only "one event" will occur, namely the formation or the breaking of a bond. Hence for a closed time interval $[0,t] \in T$ the probability of such an event is given by:

$$\mathcal{P}\{N_t \equiv N\} = P_N(t) \text{ for } t \geq 0. \quad (3.84)$$

However the initial condition N_0 is important here. By considering the crack propagation model of Fig. 23, $N_0 = 0$ at $t = 0$, ie. no bonds exist. For further simplification of the analysis by taking the process to be time homogeneous, then the above form becomes:

$$\mathcal{P}\{N_t - N_s \equiv N\} = P_N(t - s) \ ; \ \ t > s \quad (3.85)$$

For this process there exists a positive constant $\lambda > 0$ so that:

$$P_N(t) = \frac{(\lambda t)^N}{N!}e^{-\lambda t} \ ; \ \ (N = 1,2\ldots). \quad (3.86)$$

Hence the distribution function $P_N(t)$ and its evolution with time is determined from the given initial one and the following system of differential equations:

$$\frac{dP_N(t)}{dt} = -\lambda P_N(t) + \lambda P_{N-1}(t) \ ; \ \ (N = 1,2\ldots) \quad (3.87)$$

in which according to (3.86), the transition probability is a step function of x or:

$$p\{s,y;t,x\} = \sum_{N=0}^{(x-y)} \frac{\lambda^N(t-s)^N}{N!}e^{-\lambda(t-s)} \ ; \ t \geq s, \quad (3.88)$$

if y is a non-negative integer, $y \leq x$ and $s \leq t$. Thus this process is a "Poisson process". More specifically by using the number of bonds $^{\alpha\beta}N(t)$ per "unit cell" of the interfacial area $^{\alpha\beta}A$ between the grains α, β, the cohesion and decohesion process can be seen as a stochastic process $\{^{\alpha\beta}N_t, t \geq 0\}$. By identifying $^{\alpha\beta}N_t$ at a given instant of time with a "micro-state ξ" at the interface (α, β), the transition probability in (3.88) can be expressed in terms of the conditional distribution, ie:

$$\mathcal{P}\{\xi_t = n | \xi_s = m\} = p_{m,n}(s, t) ; \tag{3.89}$$

and

$$p\{s, x; t, y\} = \sum_{m \leq x, n \leq y} p_{m,n}(s, t) \tag{3.90}$$

Hence the Chapman-Kolmogorov relation in matrix form becomes:

$$P_{mn}(s, t) = \sum_{\ell} P_{m\ell}(s, u) P_{\ell n}(u, t) ; \; (s \leq u \leq t) \subset T \tag{3.91}$$

It has been shown by Gihman and Skorohod [137], Lipster and Shiryayev [138], that the evolution of the transition probabilities can be obtained from the above relation by means of Kolmogorov's first and second system of differential equations. It is readily seen that in general the cohesion-decohesion process can be characterized by a special Markov or the Birth-Death (BD) Poisson process. Thus for simplicity, by considering the one-dimensional case only and let $^{\alpha\beta}N_t$ to be associated with a micro-state ξ_t at time t, then for one more bond to occur within the time interval $(t, t + \Delta t)$ one has a probability of "formation" given by $\lambda \Delta t + 0(\Delta t), \lambda > 0$. Similarly for decohesion to occur in this time interval the probability will be $\mu \Delta t + 0(\Delta t), \mu > 0$, independently of the other existing bonds. This leads then to the evolution of the probability distribution $P_N(t)$ (with $^{\alpha\beta}N \equiv N$) of the form:

$$\frac{dP_N(t)}{dt} = -N(\lambda + \mu)P_N(t) + (N - 1)\lambda P_{N-1}(t) + (N + 1)\mu P_{N+1}(t) \tag{3.92}$$

and the initial conditions:

$$P_N(0) = \begin{cases} 1 & \text{for } N = 1 \\ 0 & \text{for } N \neq 1. \end{cases}$$

The solution of the above relation can readily be obtained by the method of generating functions. The above relation is based on the assumption that there exists a continous function $c(t)$ holding uniformly in t so that the transition of the interaction, ie. the bonding or debonding between the α, β grains within $(t + \Delta t)$ occurs also with $c(t) + 0(\Delta t)$ (see also [138]). On the assumption that within $(t + \Delta t)$ the function:

$$c(t) = \text{const.} = c_N \equiv N(\lambda + \mu)$$

one can distinguish between bonding to occur, when:

$$C_N p_{N,N+1} = N\lambda \qquad\qquad\qquad\qquad\text{(a)}$$

or de-bonding to occur, when:

$$C_n p_{N,N-1} = N\mu. \qquad\qquad\qquad\qquad\text{(b)}$$

$$\left.\begin{matrix}\end{matrix}\right\} \quad (3.93)$$

However in the propagation problem the stochastic intensities are themselves time dependent, or $\lambda(t), \mu(t)$ and restricted by limiting conditions due to relations (3.89, 3.91). These limiting conditions can be stated as follows:

$$
\begin{aligned}
\text{(i)}\quad &\lim_{\Delta t \to 0} \frac{p_{N,N+1}(t, t+\Delta t)}{\Delta t} = \lambda_N(t) \text{ for } N \geq 0,\\[2mm]
\text{(ii)}\quad &\lim_{\Delta t \to 0} \frac{P_{N,N-1}(t, t+\Delta t)}{\Delta t} = \mu_N(t) \text{ for } N \geq 1,\\[2mm]
\text{(iii)}\quad &\lim_{\Delta t \to 0} \frac{1 - p_{N,N}(t, t+\Delta t)}{\Delta t} = \lambda_N(t) + \mu_N(t) \text{ for } N \geq 0
\end{aligned}
\qquad (3.94)
$$

for all values of $t \geq 0$ and $\mu_0(t) = 0$. Hence it is seen that a "pure birth or B-process" will be characterized by:

$$\mu_N(t) = 0 \text{ for all } t \text{ and } N \qquad\qquad \text{(a)}$$

and a "pure death or D-process" by:

$$\lambda_N(t) = 0 \text{ for all } t \text{ and } N \qquad\qquad \text{(b)}$$

$$\left.\begin{matrix}\end{matrix}\right\} \quad (3.95)$$

If for simplicity of the analysis of crack propagation only, a pure B-process is assumed, one obtains the following probability evolution relation:

$$\frac{dP_{n,m}(t)}{dt} = -\lambda_n P_{n,m}(t) + \lambda_{m+1} P_{n,m+1}(t) \qquad (3.96)$$

where the stochastic intensities λ are considered as one-step transitions from a micro state $n \to m$ eqn. (3.89). This simplified model leads however to considerable deviations of the theoretical values of λ from those obtained from experimental observations. This has been shown to be the case for various polycrystalline solids in the analysis of the "fatigue failure" brought about by the application of cyclic stresses (see Provan [200, 201]. These deviations may be ascribed, to a large extent, to neglecting the decohesion effect in the interfacial regions of the polycrystalline solid, or to the decoupling of the process of decohesion from that of the crack propagation. As is well known from the fundamentals of "fracture mechanics", a crack can only advance, if the available energy is larger than the work needed to break the bonds. It is evident therefore, that in order to obtain a more realistic model, "surface energetics" should be included in the modelling of debonding and crack propagation. It is further apparent that the stochastic intensities λ are not only time-dependent but also functions of the intrinsic "energy release rate" associated with the potential and interfacial energies of the polycrystalline structure. Consideration to surface energetics of binary structures will be given in Chapt. 4. More recently simple stochastic models of material failure have been considered by Meakin et al. [202], and some interfacial phenomena were treated by Maugis [199, 203].

So far as the behaviour of fibrous structures is concerned, the breakdown of bonding leading to a complete failure of the microstructure is briefly discussed below. The main interest here is the effect of bonding, ie. particularly "hydrogen bonding" as indicated earlier by the structure in Figs. (19–21) (see also refs. [23, 182]). As a consequence of the applied external load a "unit cell" (Fig. 21) and correspondingly the coincidence site (hydroxyl groups 2, 2′) will experience a relative displacement which is the argument in the bond potential given in eqn. (3.80). This important parameter in general is time-dependent and characteristic of an "individual bond". As before, it is convenient for the representation of bonding effects to use the number of bonds $^{\alpha\beta}N(t) \equiv N(t)$ per "elemental area $^{\alpha\beta}A$" of the "actual bonding area ^{b}A. The latter is indicated in Fig. 24 showing the kinematics of a "fibre segment". Since stochastic mechanics requires the definition of a "microelement", such an element is defined by the "unsupported fibre segment of α or β", which is anchored at one half of the "actual bonding area ^{b}A" belonging to the crossing of the fibres α, β. Thus this definition ensures that only one bonding area is counted at each crossing of the fibres α, β (Fig. 24).

In order to account for the effect of the hydrogen bonding in the random network and assuming that the number of bonds at time $t = 0$ is N_0 and is experimentally accessible, then by using the general deformation theory given in earlier work [22], [23], the occuring deformations can be represented by a Markov process. The characteristics differential equation for the "bond-deformation process" is the Chapman-Kolmogorov equation or:

$$\frac{dP_{ij}(t)}{dt} = \sum_k q_{ik} P_{kj}(N(t)) \; ; \; P_{ij}(0) = \delta_{ij} \; ; \; q_{ik}(t) \in Q_{ik}(t) \qquad (3.97)$$

where the transition probability $P_{ij}(N(t))$ designates the transition of an "individual bond" from the state i at time t to another state j at time $t + \Delta t$ in one step and where $q_{ik} \in Q_{ik}(t)$ or the stochastic matrix or the relative transition probability.

If the state i is characterized by the number of bonds $N(t)$ forming the event i and the state j is an adjacent event, it may be seen that the above relation is in fact the stochastic analogue to a deterministic "rate equation". Alternatively, relation (3.97) can be expressed by the probability measure $\mathcal{P}_N(t)$ describing the probability distribution of $N(t)$ at time t. It is important to find an explicit expression to depend only on the state of the process and if the bond system is initially in the state i at time t goes in one step to the state j ; $(j = i + 1)$ at $(t + \Delta t)$, the transition probability can be designated by $\lambda_i \Delta t$. The probability of remaining in the state i will then be $(1 - \lambda_i \Delta t)$ and

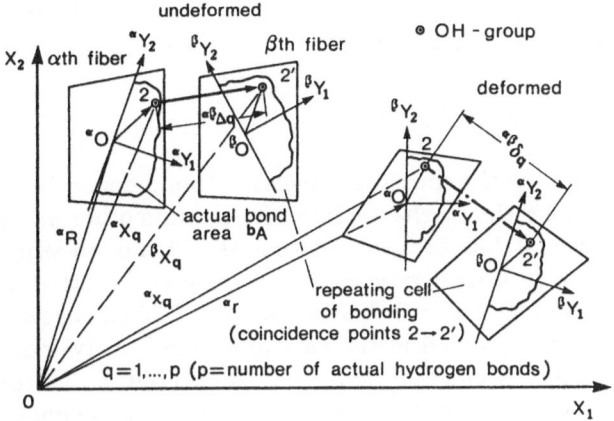

Fig. 24. Kinematics of bond area (coincidence sites, matching points 2–2' of the fibre-fibre interface).

the probability of changing to another state j will be $0(\Delta t)$. Hence one obtains the same form of differential equation as in (3.96), ie.:

$$\frac{dP_{ij}(t)}{dt} = -\lambda_j P_{ij}(t) + \lambda_{j+1} P_{i,j+1}(t)$$

which characterizes a linear B or D-process. It can be recognized that the "dissociation process" of hydrogen bonding is a simple "D-process", which in terms of the above probability measure can be expressed as follows:

$$\frac{dP_N(t)}{dt} = -\lambda_N P_N(t) + \lambda_{N+1} P_{N+1}(t) \ . \tag{3.98}$$

For a linear process one may replace $\lambda_N(t)$ by $\lambda.N(t)$ (see also [137]) so that:

$$\frac{dP_N(t)}{dt} = -\lambda N P_N(t) + \lambda(N+1) P_{N+1}(t) \ ; \ 0 \le N \le N_0. \tag{3.99}$$

The solution of this differential equation is of the binomial form, ie:

$$P(t) = \binom{N_0}{N} e^{-N_0 t} (e^{-\lambda t} - 1)^{N_0 - N} \tag{3.100}$$

from which the mean value of $N(t)$ and the variance are obtained as follows:

$$E\{N(t)\} =< N(t) >= N_0 e^{-\lambda t} \qquad \text{(a)}$$
$$\sigma^2(t) = N_0 e^{\lambda t}[1 - e^{-\lambda t}] \ . \qquad \text{(b)} \qquad \Big\} \ (3.101)$$

Considering the change in the number of bonds in terms of the mean values only, e.g. identifying "$< N(t) >\equiv N(t)$" it is seen that (3.101a) is the solution of a deterministic rate equation, viz.:

$$\frac{dN(t)}{dt} = -\lambda N(t) \ ; \ N(t)|_{t=0} = N_0. \qquad (3.102)$$

Hence one can regard the simplest models for "bond dissociation" to be a linear D-process. However the stochastic intensity permits other interpretations than that given above. It has been found that the "bond breakage" phenomenon frequently occurs in a rather "cooperative manner", ie. that the bonds break in groups. To account for such behaviour Nissan [204] suggested the use of a "cooperative index γ, which in the case of cellulosic networks is based on the hydrogen bond model of Frank and Wen [205]. From a deterministic point of view the bond breakage process may be regarded to depend on the rate of change of the bond ratio $N(t)/N_0$, if N_0 bonds are available for bonding in the initial stage, or before the application of an external influence. Hence

$$\frac{d}{dt}(N(t)/N_0) = -(N(t)/N_0)^\gamma \qquad (3.103)$$

where γ is the overall cooperative index. Values of this index for cellulose structures has been given in ref. [204]. To construct a corresponding stochastics model, one can introduce a non-linear stochastic intensity of the form: $\lambda_N = \lambda N^\gamma$ into equation (3.98), which changes to:

$$\frac{d\mathcal{P}_N(t)}{dt} - \lambda N^\gamma \mathcal{P}_N(t) + \lambda(N+1)^\gamma \mathcal{P}_{N+1}(t) \ , \qquad (3.104)$$

representing a non-linear D-process. However the solution of this relation for $\gamma \geq 2$ is not available. In order to obtain a solution some simplifications are needed. Thus for instance, letting:

$$-\lambda_N = -\lambda^{1+\gamma/2\gamma_N} = -\lambda_1 N \ ; \ \gamma \geq 1 \qquad (3.105)$$

yields a linearized process with the intensity λ_1 so that the distribution is given by:

$$\mathcal{P}_N(t) = \binom{N_0}{N} e^{-\lambda_1 N_0 t}[e^{\lambda_1 t} - 1]^{N_0 - N} \qquad (3.106)$$

and the expected value of $N(t)$ by:

$$N(t) =< N(t) >= N_0 e^{-\lambda_1 + \gamma/2\gamma \cdot t} = N_0 e^{-\lambda_1 t} \ . \qquad (3.107)$$

Alternative forms than above using the index γ also permit a linearization of the non-linear D-process. Hence by using the solution of the differential equation (3.103), given by:

$$N(t) = N_0[1 + \lambda t(\gamma - 1)]^{\frac{1}{1-\gamma}} \qquad (3.108)$$

expanding it in a Taylor series and retaining the first and second terms only results in:

$$\left.\begin{aligned}
N(t) &= N_0[1 - \lambda t + \frac{1}{1!}(\frac{1}{1-\gamma} - 1)\lambda^2 t^2 (\gamma - 1)^2 + \ldots] \\
&= N_0[1 - \lambda t + \gamma \frac{\lambda^2 t^2}{2!} + \ldots]
\end{aligned}\right\} \qquad (3.109)$$

Equivalently, one can use the index of the linearized model of (3.105) with λ_1. An expansion by considering two terms only, yields then:

$$N(t) = N_0[1 - \lambda_1 t + \frac{\lambda_1^2 t^2}{2!} + \ldots]. \qquad (3.110)$$

A comparison of the above non-linear and linear forms (with $\gamma = 3$) indicates that a somewhat larger change in the number of bonds $N(t)$ is the result of the model in terms of a non-linear D-process. In general, the bond breaking may occur in such a manner that energy is released, activating "bond formation" on other sites within an elemental area of the bonding area ${}^bA$. In this case the transition intensity λ_N will depend not only on the number of bonds, but also on the number of "broken bonds" $(N_0 - N)$. Hence the evolution of the distributions of $N(t)$ will then be of the form:

$$\left.\begin{aligned}
\frac{dP_N(t)}{dt} &= -\lambda N_0 P_{N_0}(t) \\
\frac{dP_N(t)}{dt} &= -\lambda N_0 (N_0 - N + 1)P_N(t) + \lambda(N+1)P_{N+1}(t).
\end{aligned}\right\} \qquad (3.111)$$

The second equation in (3.111) may be solved with the initial condition:

$$P_N(0) = \delta_{NN_0}$$

so that the solution (see [44]) is given by:

$$P_N(t) = \lim \frac{1}{2\pi i} \int_{\alpha - i\beta}^{\alpha + i\beta} \overline{P}_N(s)e^{st}ds \qquad (3.112)$$

in which $\overline{P}_N(s)$ is the Laplace transform of $N(t)$, which can be expressed by:

$$\overline{\mathcal{P}}_N(s) = \frac{N_0!(N_0 - N)!}{N!}\lambda^{(N_0 - N)}\prod_{i=1}^{(N_0 - N + 1)}[s + i\lambda(N_0 - i + 1)]^{-1} \qquad (3.113)$$

This form has been investigated by Bailey [206], who has shown that for $N > \frac{N_0}{2}$ and an even N, the form (3.113) can be reduced to:

$$\overline{\mathcal{P}}_N(s) = \sum_{i=1}^{(N_0 - N + 1)}\frac{^\alpha N_i}{[s + i\lambda(N_0 - N + 1)]} \qquad (3.114)$$

in which the coefficients $^\alpha N_i$ take the following form:

$$^\alpha N_i = \frac{(-1)^{i-1}(N_0 - 2i + 1)!N_0!(N_0 - N)!(N - i - 1)!}{N!(i-1)!(N_0 - i)!(N_0 - N - i - 1)!}. \qquad (3.115)$$

The evaluation of these coefficients and the inversion of $\overline{\mathcal{P}}_N(s)$ from (3.14) yields then the distribution $\mathcal{P}_N(t)$ as well as the mean value of $N(t)$ for this process (see also Haskey [207]).

In certain polymer structures an often occuring phenomenon is that of the simultaneous occurrence of bond dissociation and formation, where the former is predominant. Such a process can be formulated by the 'Birth and Death" or BD-process of Markov theory. However, in order to describe this process for the case of hydrogen bonding, it becomes necessary to distinguish between the two stochastic intensities λ and μ (3.93, 3.94). Thus, if the bond system is in a state i at time t, the probability of transition to the state j; $(j = i + 1)$ at $t + \Delta t$ is $\lambda_N\Delta t$ (formation of bonds), but the probability of transition to j; $(j = i - 1)$ at $t + \Delta t$ will be $\mu_N\Delta t$ (dissociation of bonds). Hence the probability of remaining in the state i is given by: $(1 - [\lambda_N + \mu_N]\Delta t)$ and the probability of change to any other state than $j = (i - 1)$ or $(i + 1)$ will be $0(\Delta t)$. The differential equation expressing these transitions becomes therefore:

$$\frac{d\mathcal{P}_N(t)}{dt} = \lambda_{N-1}\mathcal{P}_{N-1}(t) - (\lambda_N + \mu_N)\mathcal{P}_N(t) + \mu_{N+1}\mathcal{P}_{N+1}(t). \qquad (3.116)$$

and for the bond dissociation only:

$$\left.\begin{array}{l} \dfrac{d\mathcal{P}_N(t)}{dt} = \mu_{N+1}\mathcal{P}_{N+1}(t) \ ; \ N = 1, 2\ldots \\[2mm] \text{with the initial condition: } \mathcal{P}_N(0) = \delta_{NN_0} \end{array}\right\} \quad (3.117)$$

Similarly as shown earlier a linearization can be carried out by letting the intensities λ_N, μ_N be equal to "λN" and "μN", respectively, so that (3.116) for the linear case can be written as:

$$\frac{d\mathcal{P}_N(t)}{dt} = \lambda(N - 1)\mathcal{P}_{N-1}(t) - N(\lambda + \mu)\mathcal{P}_N(t) + \mu(N + 1)\mathcal{P}_{N+1}(t) \quad (3.118)$$

or in briefer notation for the dissociation only:

$$\frac{d\mathcal{P}_N(t)}{dt} = \mu_1 \mathcal{P}_1(t).$$

The solution of this differential equation has been discussed by Bharucha-Reid [44] and is of the form:

$$\mathcal{P}_N(t) = [1 - \alpha(t)][1 - \beta(t)][\beta(t)]^{N-1}$$
$$\mathcal{P}_0(t) = \alpha(t) \tag{3.119}$$

in which the functions $\alpha(t), \beta(t)$ are given as follows:

$$\alpha(t) = \frac{\mu[e^{(\lambda-\mu)t} - 1]}{\lambda e^{(\lambda-\mu)t} - \mu} \; ; \; \beta(t) = \frac{\lambda[e^{(\lambda-\mu)t} - 1]}{\lambda e^{(\lambda-\mu)t} - \mu} \; . \tag{3.120}$$

After some calculations the mean value and variance of the process $N(t)$ in terms of the transition intensities λ, μ and the initial number of bonds N_0 can be assessed.

It is of interest to note that the energy release per bond denoted by $\epsilon(r)$ for a "maximum cut-off distance r" in the bond potential (Fig. 15) can also be estimated from experimental observation. This energy release per bond is proportional to $^{\alpha\beta}\phi(r_0) - {}^{\alpha\beta}\phi(r_{max})$ and where $^{\alpha\beta}\phi$ is the interaction potential. From a phenomenological point of view, one can express the rate of the total energy release to occur within a certain bonding area ^bA of the interface by:

$$\frac{dE(t)}{dt} = \epsilon(r)\frac{dN(t)}{dt} = -\epsilon\lambda(t)N(t). \tag{3.121}$$

Using this result of the simple Markov model, it is possible to interpret the stochastic intensity λ of the process. Letting $E(t) = \epsilon N(t)$ and from eqn. (3.101a): $N(t) = N_0 e^{-\lambda t}$, so that: $E(t) = \epsilon N_0 e^{-\lambda t}$ one has:

$$\lambda \equiv t^{-1} \ell n \frac{\epsilon N_0}{E(t)}. \tag{3.122}$$

This intensity is experimentally accessible, if the energy release averaged over a test sample can be established by appropriate experimental measurements.

(iv) Interfaces in fluids:

One of the more complex problems in the molecular dynamics of fluids is undoubtedly the description of "interfaces" that may occur in simple fluids due to external effects or on the boundaries with other phases (solid, gaseous, etc.). Evidently a comprehensive analysis of the interface characteristics is beyond the scope of the present volume and thus only a brief outline of the stochastic modelling is given here. A more detailed study however will be supplied in section 4.3 of the following chapter, which is concerned with the flow dynamics of simple fluids on the basis of the theory of Markov processes. A significant part of the analysis is the relation between the particle (molecule) distribution mentioned earlier in eqn. (2.22) and the occuring interactions between pairs of molecules. A rigorous derivation of such a relation is due to Kirkwood, (see also [208]) in the form of an "integral equation". Many attempts have been made to approximate this relation. For a general reference the reader is referred to Abraham [209], Evans [210], Percus [211], Croxton [98], Rowlinson and Widom [212] among others.

For the analysis of interfaces in general, one may distinguish between "intrinsic local variables" of the fluid and the "interfacial or surface variables" of the fluid structure. So far as the interaction effects are concerned a fundamental relation is the "decomposition" of the pair-interaction potential $^{\alpha\beta}\phi(\mathbf{r})$, (see also Fig. 15) due to van der Waals, where:

$$^{\alpha\beta}\phi(\mathbf{r}) = \phi_1(\mathbf{r}) - \phi_2(\mathbf{r}) \tag{3.123}$$

where $\phi_1(\mathbf{r})$ designates the short-range or "core-part" and $\phi_2(\mathbf{r})$ the long-range or attractive part of $^{\alpha\beta}\phi(\mathbf{r})$. This separation contains however a certain "optimality condition" (see for instance Croxton [98], Temperley and Travena [213]). In the analysis of interfaces and the microdynamics of fluids that are either in or "near" thermodynamic equilibrium, the use of time-correlation functions is often convenient (see also Chapt. 2). In terms of the "local thermodynamics" or rather its idealization, where it is assumed that the fluid density "$\rho(\mathbf{r})$" varies so slowly that it can be taken as "locally uniform", one obtains the "Helmholtz free energy ψ" of a grand ensemble in the form of:

$$\psi = \int [\epsilon_1(\rho(\mathbf{r})) + \{v(\mathbf{r})\}]\rho(\mathbf{r})d^3\mathbf{r} \tag{3.124}$$

in which ϵ_1 is the free energy density per particle of the "core fluid" and $\{v(\mathbf{r})\}$ is the function space corresponding to the velocity field. Since the energy of the latter can be taken as $\int \{v(\mathbf{r})\}\rho(\mathbf{r})d^3\mathbf{r}$, one obtains by subtraction from (3.124) the "bulk" free energy of the fluids as follows:

$$\psi_B = \int \epsilon_1[\rho(\mathbf{r})]\rho(\mathbf{r})d^3\mathbf{r} \tag{3.125}$$

which allows to generate complete direct correlation functions (see Lebowitz and Percus [211]). Since generally the Helmholtz free energy ψ of an inhomogeneous fluid in terms of the local free energy is expressed by:

$$\psi = \int \epsilon d^3\mathbf{r} \tag{3.126}$$

Yang, Fleming and Gibbs [215] derived on the basis of the theory of "density-functionals" an exact form of ψ for a one-component fluid as follows:

$$\psi = \int [\epsilon^h(\rho(\mathbf{r})) - \frac{1}{2}\rho(\mathbf{r}) \int [R(\mathbf{r},\mathbf{r}')\rho(\mathbf{r}')d^3\mathbf{r}']d^3\mathbf{r} \qquad (3.127)$$

where $\rho(\mathbf{r})$ is the local density of the fluid, $\rho(\mathbf{r}) \equiv \rho(^\alpha\mathbf{r}), \rho(\mathbf{r}') \equiv \rho(^\beta\mathbf{r})$, $\epsilon^h(\rho)$ the free energy density of the homogeneous fluid (in the ideal gas state) and the pair-correlation function is given by:

$$R(^\alpha\mathbf{r},{}^\beta\mathbf{r}) \equiv R(\mathbf{r},\mathbf{r}') = 2\int_0^1 t dt \int_0^1 dt'(R(\mathbf{r},\mathbf{r}';\{tt'\rho\})). \qquad (3.128)$$

The pair-correlation function $R(\cdot)$ for the inhomogeneous fluid is however also a functional of the "density-distribution" of the fluid. It is generally assumed that in the "low density" limit, $R(\mathbf{r},\mathbf{r}';\{\rho\})$ is independent of the density and thus becomes equal to its homogeneous value, ie. $R_0(|\mathbf{r}-\mathbf{r}'|)$. In this limit one obtains for the free energy in (3.127) the form:

$$\psi = \int \{\epsilon^h(\rho(\mathbf{r})) + \frac{kT}{4}\int (\rho(\mathbf{r}') - \rho(\mathbf{r}))^2 \times R_0(|\mathbf{r}'-\mathbf{r}|)d^3\mathbf{r}'\}d^3\mathbf{r} \qquad (3.129)$$

and

$$\epsilon^h(\rho(\mathbf{r})) = \epsilon_0^h(\rho(\mathbf{r})) - \frac{kT}{2}[\rho(\mathbf{r})]^2 \times \int R_0(|\mathbf{r}'-\mathbf{r}|)d^3\mathbf{r}' . \qquad (3.130)$$

These relations are strictly valid only at low densities, ie. at densities that permit the evaluation of "ϵ" from terms quadratic in ρ. To make relation (3.129) valid for any density, an approximation can be made by introducing a "mean density" $\bar{\rho} = \frac{1}{2}(\rho(\mathbf{r}) + \rho(\mathbf{r}'))$, assumed to be close to the actual density in the vicinity of the correlated pair $(\mathbf{r},\mathbf{r}')$. Another approximation based on van Kampen's [216] mean field approach is known as the "modified van der Waals" or "MVDW" model [217]. Both of these models are extensions of the theoretical relations and hence will only be valid within certain limits. The MVDW-model can be regarded as accurate in the limit of long-ranged attractive and short-ranged repulsive forces compared to the scale of the density variation of the fluid, while the ADF-model (approximate density functional) yields accurate results for fluids at sufficiently low densities. A minimization of the free energy of the system leads to a "density profile $\rho(\mathbf{r})$" of the fluid. Some remarks on the thermodynamic relations used in other approximations may be indicated here.

Recalling from section 2.2 of Chapt. 2 in the discussion on statistical models, the "configurational integral Z_N" for an N-particle system on the assumption that the latter are indistinguishable has been given as:

$$Z_N(V,T) = \int_V \exp[-\beta U_N(^N\mathbf{r})]d^N\mathbf{r}$$

where the integration extends over each position vector $^\alpha \mathbf{r}; (\alpha = 1 \ldots N)$ of a canonical ensemble contained in the volume V. The corresponding partition function $Q_N(V, T)$ is related to Z_N by:

$$Q_N(V, T) = \frac{Z_N}{N! \lambda^{2N}}$$

with $\lambda = [\frac{2\pi h^2}{\alpha \mu}]^{1/2}$ and $\beta = [k_B T]^{-1}$.

Since Q depends on the various values of the separation $^{\alpha\beta}\mathbf{r}$ for all possible configurations of the particles (molecules), the effect of "interactions" can be assessed in terms of an "equation of state" of the fluid. It is known from equilibrium thermodynamics that:

$$\psi = -k^T \ell n Z; \quad (\frac{\partial \psi}{\partial T})_{V,A} = -S; \quad (\frac{\partial \psi}{\partial V})_{T,A} = -P \tag{3.131}$$

in which T is the temperature, S the entropy, P the pressure and A the total interface area. The "internal energy U" of the fluid is thus defined by:

$$U = \frac{kT^2}{Z}(\frac{\partial Z}{\partial T})_V \tag{a}$$

$$U = \psi + TS = \psi - T(\frac{\partial \psi}{\partial T})_V = -T^2 \frac{\partial}{\partial T}(\frac{\psi}{T})_V \tag{b}$$

$$\left.\right\} \tag{3.132}$$

where equation (3.132b) represents the well-known Gibbs-Helmholtz relation. For a fluid consisting of molecules that interact by "two-body central forces" the internal energy of the MDVW-model for instance can be expressed by:

$$U = \int \{u^h[\rho(\mathbf{r})] + \frac{1}{2} \int \rho(\mathbf{r})\rho(\mathbf{r}') \times R_0(|\mathbf{r}' - \mathbf{r}|; \bar{\rho})u(|\mathbf{r}' - \mathbf{r}|)d^3\mathbf{r}'\}d^3\mathbf{r} \tag{3.133}$$

where $u^h(\rho)$ is the internal energy density (local) of the homogeneous fluid (in the ideal gas state), $u(|\mathbf{r}' - \mathbf{r}|)$, is the internal energy corresponding to the pair potential $^{\alpha\beta}\phi$, $R_0(\cdot)$ the pair-correlation function of the homogeneous fluid at density $\rho(\mathbf{r})$ and $\bar{\rho}(\mathbf{r})$ is the quantity introduced as the mean density (see also McCoy and Davis [217]).

For the consideration of the fluid interfaces some other themodynamic relations may also be of interest. In particular those which relate work and energy at the fluid interface. The latter is usually assumed to be planar for simplification of the analysis. In a more general form the Helmholtz free energy of a classical "multi-component" fluid with a planar interface of area A and a molar composition $n_1 \ldots n_\nu$ constituents can be expressed by:

$$\psi = f\{T, V, A, n_1 \ldots n_\nu\}.$$

However, the free energy represents the total amount of work according to a given "thermodynamic state" of the system and is to be distinguished from the

"Gibbs free energy" that excludes the amounts of work due to a volumetric expansion and a possible area extension.

An infinitismal change of the Helmholtz free energy caused by small changes in temperature, volume, area and composition can be written as a total differential ie.:

$$dψ = [\frac{∂ψ}{∂T}]_{V,A \atop n_1...n_ν} dT + [\frac{∂ψ}{∂V}]_{T,A \atop n_1...n_ν} dV + [\frac{∂ψ}{∂A}]_{T,V \atop n_1...n_ν} dA + \sum_1^ν (\frac{∂ψ}{∂n_ν})_{T,V,A \atop n_1...n_n;n_ν} \quad (3.134)$$

in which the subscripts indicate that the partial differentials are to be taken at constant volume, temperature etc. and the sign ";" means exclusion of $n_ν$. These partial differentials are also related by noting (3.131) to the following thermodynamic quantities:

$$\left. \begin{array}{l} [\frac{∂ψ}{∂T}]_{V,A \atop n_1...n_ν} = -S; [\frac{∂ψ}{∂V}]_{T,A \atop n_1...n_ν} = -P; [\frac{∂ψ}{∂A}]_{T,V \atop n_1...n_ν} = γ; \\ \text{and} \\ [\frac{∂ψ}{∂n_ν}]_{T,V,A} = μ_ν \end{array} \right\} \quad (3.135)$$

where $γ$ designates here the "surface tension" in the interface region (area A) and $μ_ν$ the chemical potential. It is readily seen that the differential of the Helmholtz free energy gives the familiar relation of classical thermodynamics, ie.:

$$dψ = -SdT - PdV + γdA + \sum_ν μ_ν dn_ν \quad (3.136)$$

and in particular for a constant composition, temperature and volume one obtains the well-known formula:

$$\frac{dψ}{dA} = γ \quad (3.137)$$

that relates the free energy to the "surface tension" in the interfacial region. On the continuum assumption, that the "bulk" properties of the fluid system remain the same even so an extension of the interface area A may occur, leads to the fundamental Gibbs relation, ie:

$$dψ_s = -SdT + γdA + \sum_ν μ_ν dn_ν \quad (3.138)$$

where the subscript s refers to the interface region of area A. If it is further assumed that the free energy is a homogeneous function of degree one in the volume, area A and the composition, one obtains from (3.134) the free energy of the surface region as follows:

$$\psi_s = \sum_\nu \mu_{\nu_s} n_{\nu_s} - PV_s + \gamma_s A. \tag{3.139}$$

By neglecting the effect of pressure in this region this relation reduces to:

$$\psi_s = \sum_\nu \mu_{\nu_s} n_{\nu_s} + \gamma A . \tag{3.140}$$

Although the above Gibbs relation (3.140) may be regarded as fundamental for the interfacial region, it is based on the assumption that this region has no volume, ie. is a planar surface so that the "PdV_s" term can be neglected. A more detailed analysis of this matter including liquid surfaces pertaining to binary and multi-phase fluid structures is given by Croxton [98].

It is to be noted that the quantities ψ_s and n_{ν_s} depend also on the location of the interface. The latter may be chosen in accordance with the Gibbs notion of an "equimolar surface" for which the sum $\sum_\nu(\cdot)$ is zero. Hence for a simple one component (pure) liquid relation (3.140) reduces to:

$$\psi_s = A\gamma_s; \quad d\psi = \gamma_s dA + \psi d\gamma_s \rightarrow \frac{d\gamma_s}{dT} = -S_s \tag{3.141}$$

where S_s is the entropy of the interface. The "intrinsic" structure of fluid interfaces has been studied extensively by employing for instance the Ising model (section 2.2 of Chapt. 2). Although these investigations indicated that two free energy models can be used which are formally different, it has been shown by McCoy and Davis [217] that for a "liquid-vapour" interface in a Lennard-Jones fluid (ie. where $^{\alpha\beta}\phi$ is of the form of a 6–12 potential) these models yield approximately the same surface tension. The Ising model approach has also been discussed by Bricmont, Lebovitz and Pfister [218], and has become important in dealing with "melting phenomena". The Ising model with regard to surface tension has been considered by Bricmont, Fontaine and Lebovitz [219].

In this context, of particular interest in the analysis of a structured fluid is the surface energy and the nearest neighbour separation. Evidently in order to create an interface or new fluid surface some of the nearest bonds involving a certain number of molecules will have to be broken. It is usually assumed that one can consider an "interface region" to consist of a sequence of parallel epipedes or cylinders $\Gamma \subseteq \mathbb{Z}^d$. In terms of a "$d$-dimensional" Ising model, one can define a cylinder by: $\Gamma = (-L, L) \times (M, M)^d \subseteq \mathbb{Z}^{d+1}$ with the boundary conditions of the spins s and for the sites $(x, y) \in \mathbb{Z}^d$ as follows:

$$\left. \begin{array}{l} s_x = -1, \text{ if } x \notin \Gamma; \ x_{d+1} < 0 \\ s_x = +1, \text{ if } x \notin \Gamma; \ x_{d+1} \geq 0 \end{array} \right\} \tag{3.142}$$

when a planar interface in the direction of x is considered. Consequently an infinite flat contour outside Γ separates the positive and negative spins (between the sites $x_{d+1} = 0$ and $x_{d+1} = -1$). Thus every configuration in Γ displays a connected extension of that contour which is referred to as an "interface". The above model can be completely analyzed for the low temperature range (see also Dobrushin [220]). For the higher temperature range one can consider another model known as the "solid-on-solid or (SOS) model" [210] or a discrete Gaussian model, in which at each site x of the d-dimensional lattice a certain variable $\phi_x \in X$ is attached and where the Hamiltonian in $\Gamma \subset \mathbb{Z}^d$ is then expressed by:

$$H_{\Gamma,\alpha}(\phi) = \sum_{(x,y)\cap\Gamma\neq\emptyset} |\phi_x - \phi_y|^2 \qquad (3.143)$$

where the summation extends over the nearest neighbour pair (x, y) and the coefficient α is taken to be equal to one. By choosing the boundary conditions so that $\phi_x = 0$ in (3.143), $x \notin \Gamma$, this model produces an "Ising interface" in which ϕ_x represents the variable height of the interface in the direction x_{d+1}. Thus, $\phi_x = 0$ corresponds to an interface between $x_{d+1} = -1$ and $x_{d+1} = 0$. However, the discrete Gaussian model with $\alpha = 2$ is more convenient for the interface analysis since it relates to a continuum model in which the sum over the variable $\phi \in \mathbb{Z}$ can be replaced by an integral of $\phi \in \mathbb{R}$ and the corresponding Lebesgue measure. Furthermore it can also be solved in an exact form. For both of these models the partition function can be written as:

$$Z_{\Gamma,\alpha,\beta} = \sum_{\phi_x \in \mathbb{Z}, x \in \gamma} \exp[-\beta H_{\Gamma,\alpha}(\phi)]. \qquad (3.144)$$

A rigorous analysis and discussion of results with regard to the validity of these models in the low and high temperature range with a possible "phase-transition" is given in ref. [221].

By adopting an ordered structure or lattice model the possible bond-breakage will however create two internal surfaces "$^\alpha S, ^\beta S$" or an "interface zone $z(A)$" of a certain thickness particularly, if the interface separates two media of different molecule size. This is indicated by the sketch of a discrete lattice model in Fig. 25(a). This zone will be considered further in the stochastic modelling of the interface. A "continuous interface" that may exist between the liquid-vapour states on the assumption of a planar separation is schematically shown in Fig. 25(b) that also indicates the "height variation" with respect to a reference plane representing the equilibrium between the two phases. Finally Fig. 25(c) indicates an interface model where the interface is formed by a sequence of "unit cells" or spheres the diameter of which is dictated by the range of interaction between nearest neighbour molecules. By using a phenomenological approach and a continuum model, it has been suggested in ref. [213] that the "transverse component" of the fluid pressure in the case of a planar interface can be taken as:

$$p_T = \rho\mu - \epsilon \tag{3.145}$$

where μ is the chemical potential of the homogeneous fluid at density ρ and ϵ the local free energy density. This transverse pressure component can be related to the surface tension γ_s in the interface by:

$$\gamma_s = \int\limits_{-\infty}^{\infty} (p - p_T)dz = \int\limits_{-\infty}^{\infty} (p + \epsilon - \rho\mu)dz \tag{3.146}$$

where p is the normal pressure component usually taken as constant and equal to the bulk pressure of the fluid on both sides of the planar interface. The density profile across the interface can then be determined by the profile which minimizes γ_s between the liquid and gas phases at bulk densities n_ℓ and n_g, respectively.

Since interactions may occur both at the molecular and long-range scale the "location of the interface" becomes significant. For this reason Percus and Williams [221] have more recently considered an "intrinsic profile" on the assumption of a planar interface between a liquid in equilibrium with its vapour to exist at $z = 0$ (Fig. 25(b)) and where the two phases have bulk densities $n_\ell > n_g$. They designated the location of the interface by a function:

$$z = \xi(\mathbf{x}, t) \tag{3.147}$$

in which ξ is a dynamic variable or a function of the particle phase-space and $\mathbf{x}$ the set of transverse local coordinates. The microscopic density of the system can thus be stated as:

$$\rho(\mathbf{r}, t) = \sum_i \delta(\mathbf{r} - \mathbf{r}_i(t)) \; ; \; i = 1 \dots n \tag{3.148}$$

so that the instaneous profile can be expressed by:

$$\rho_\ell(\mathbf{r}, t) = \rho[\mathbf{r} + \alpha(z)\xi(\mathbf{x}, t)\hat{z}, t] \tag{3.149}$$

in which $\alpha(z)$ is a slowly varying function satisfying: $\alpha(0) = 1$. This means, if $\alpha(z) \to 0$ as $|z| \to \infty$ the "relative fluid" can be made to heal from the bond breakage (Fig. 25(b)) on each side of the interface. For most purposes one can use $\alpha(z) = 1$. The argument of $\rho[.,.]$ in (3.149) represents qualitatively a "local coordinate" expected to equilibrate, whilst ξ may be regarded to evolve in a hydrodynamic manner. It is further assumed in this model that $\rho_\ell(\mathbf{r}, t)$ and $\xi(\mathbf{r}', t')$ are essentially independent of one another so that the phase-space average becomes:

$$n_\ell(z) = < \rho_\ell(\mathbf{r}, t) > \tag{3.150}$$

Lattice of α – molecules

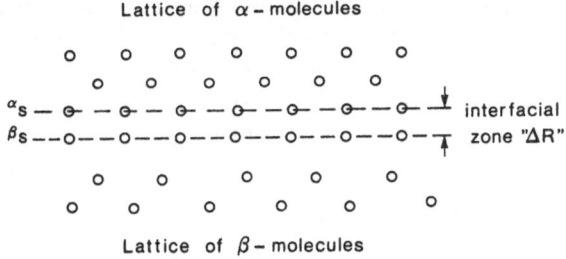

(a) Discrete lattice model (surfaces $^{\alpha}$s, $^{\beta}$s)

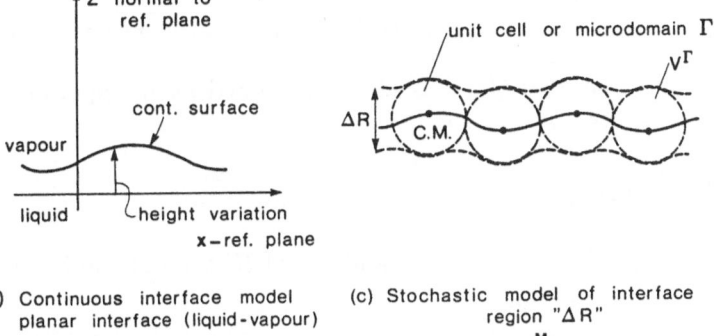

(b) Continuous interface model   (c) Stochastic model of interface
planar interface (liquid-vapour)   region "ΔR"

Fig. 25. Discrete and continuous interface models.

which is called the "intrinsic profile". By this notion and if the average value $< \rho_\ell >$ conditional on the knowledge of $\xi(\mathbf{x}, t)$ is independent of the latter, one obtains a "mean density profile" in the form of:

$$n(z) = \int < \rho(\mathbf{r}, t) | \xi(\mathbf{x}, t) > P[\xi(\mathbf{x}, t)] d\xi(\mathbf{x}, t) \qquad (3.151)$$

where $P(\xi)$ is the probability distribution of the dynamic variable ξ. The convolution of this relation leads then to values of $n(z)$, but it should be noted that ξ may be controlled by the surface tension γ_s. In this case the determination of the free energy in excess to that of the planar interface may also be involved. A rigorous analysis to this effect as well as other approaches, ie. mean field, self-consistent field, Ising models, etc. has been given in ref. [221].

It is apparent that changes in the location of the interface as well as possible changes of the local temperature will induce micro fluctuations from the thermodynamic equilibrium level. Such small fluctuations are usually formulated in terms of time-correlation functions that can describe equally the fluid response to weak external perturbations. Recalling from statistical mechanics (see also section 2.2) that the system Hamiltonian of a classical N-particle system with an interaction potential $^{\alpha\beta}\Phi$ or Φ^I is given by:

$$H = \sum_{i=1}^{N} \frac{\mathbf{p}_i^2}{2m} + \frac{1}{2} \sum_{i,j}^{N}{}' \Phi^I(\mathbf{r}_i - \mathbf{r}_j)$$

where $\mathbf{r}_i, \mathbf{p}_i$ are the position and momentum of the particle i, and where Σ' indicates that there is no summation over terms with $i = j$, the equilibrium distribution $P_e(\Gamma)$ is then:

$$P_e(\Gamma) = \exp[-\beta H]\{\exp[-\beta H]^{-1}d\Gamma\}; \quad \Gamma = \{\mathbf{r}_1 \ldots \mathbf{r}_N; \mathbf{p}_1 \ldots \mathbf{p}_N\} \quad (3.152)$$

The dynamic variable $\xi(\mathbf{r}, t)$ is also determined by its equation of motion in the "phase-space" or:

$$\frac{\partial \xi(\mathbf{r}, t)}{\partial t} = iL\xi(\mathbf{r}, t) \quad (3.153)$$

where L is the Liouville operator and $iL \equiv [\ H]$ the Poisson bracket and where the L-operator is given by:

$$L = \sum_{i=1}^{N} \frac{\mathbf{p}_i}{2m} \cdot \frac{\partial}{\partial \mathbf{r}_i} + \frac{1}{2} \sum_{i,j}{}' \frac{\partial \Phi^I}{\partial \mathbf{r}_j} \left(\frac{\partial}{\partial \mathbf{p}_j} - \frac{\partial}{\partial \mathbf{p}_i} \right) \quad (3.154)$$

The formal solution of (3.153) is:

$$\xi(\mathbf{r}, t) = e^{iLt}\xi(\mathbf{r}, 0) \quad (3.155)$$

indicating that the time dependency of $\xi(\mathbf{r}, t)$ arises from its dependence on the particle coordinates. Since P_e is an equilibrium distribution one has:

$$LP_e(\Gamma) = 0. \quad (3.156)$$

As shown above, it is often useful to consider the average of $\xi(\mathbf{r}, t)$ which depends on the thermodynamic state of the system so that:

$$< \xi(\mathbf{r}, t) >= \int_{\Gamma} P_e(\Gamma)\xi(\mathbf{r}, t)d\Gamma \quad (3.157)$$

where the explicit form of $P_e(\Gamma)$ is given by:

$$P_\ell(\Gamma) = Q_N^{-1} \exp[-\beta U] \prod_{i=1}^{N} P_0(\mathbf{p}_i) \qquad (3.158)$$

in which $U(\mathbf{r}_1 \ldots \mathbf{r}_N)$ is the potential energy of the system corresponding to the Hamiltonian H. The dynamic variable of particular interest here is the "number density" of the molecules defined by:

$$n(\mathbf{r}, t) = \frac{1}{\sqrt{N}} \sum_{i=1}^{N} \delta(\mathbf{r} - \mathbf{r}_i(t)) \qquad (3.159)$$

and another quantity known as "current density" [104], or:

$$\mathbf{j}(\mathbf{r}, t) = \frac{1}{\sqrt{N}} \sum_{i=1}^{N} \mathbf{v}_i(t) \delta(\mathbf{r} - \mathbf{r}_i(t)). \qquad (3.160)$$

For a single particle with velocity $\mathbf{v}_i(t)$:

$$n_s(\mathbf{r}, t) = \delta(\mathbf{r} - \mathbf{r}_i(t)).$$

Hence the average values are obtained as:

$$< n(\mathbf{r}, t) >= \frac{\sqrt{N}}{V} \; ; \; < n_s(\mathbf{r}, t) >= V^{-1} \; ; \; < \mathbf{v}_i(t) >=< \mathbf{j}(\mathbf{r}, t) >= 0.$$

Of further interest is the "range and strength" of the interaction Φ^I between the N-particles of the system. Thus in the simple 1-dimensional case considered by Oelschläger [222], who introduced a parameter ν as a measure of the range and strength of $\Phi^I \equiv \phi_{N,\nu}$, where:

$$\phi_{N,\nu} = N^\nu \Phi_1(N^\nu, \nu) \; ; \; \nu \in (0, 1). \qquad (3.161)$$

This parameter can be used to indicate the interaction effect in the following manner:

(i) if $\nu = 0$, any fixed particle interacts only weakly with a large number of particles of the whole system.

(ii) if $\nu \in (0, 1)$, any fixed particle interacts weakly with other particles in a small neighbourhood

(iii) if $\nu = 1$, any fixed particle interacts strongly with a finite number of other particles in a small neighbourhood.

Evidently the case (ii), ie. $\nu \in (0, 1)$ is of interest here. Returning now to the space $\mathbb{R}^n$ and denoting a dynamic variable of the N-particle system by $X_N(t)$, its time evolution may be regarded as an empirical process in that space. Considering again for simplicity the 1-dimensional case ($N \in \mathbb{Z}$) then a deterministic process can be represented by a set of ordinary differential equations or:

$$\frac{d}{dt}X_N^i(t) = -\frac{1}{N}\sum_{\substack{i=1 \\ i\neq j}}^{N} \phi'_{N,\nu}\{X_N^i(t) - X_N^j(t)\} \; ; \; i = 1\ldots N, \qquad (3.162)$$

and by the stochastic analogue in the form of:

$$dX_N^i(t) = -N\sum_{i=1,i\neq j}^{N} \phi'_{N,\nu}\{X_N^i(t) - X_N^j(t)\}dt + dW^i(t), \qquad (3.163)$$

$$(i = 1\ldots N),$$

in which the gradient interaction $\phi'_{N,\nu}(.)$ in both these cases corresponds to a "scaled potential" or:

$$\phi'_{N,\nu}(x) = N^\nu \phi_1(N^\nu, x) \; ; \; \nu \in (0,1) \qquad (3.164)$$

with $x \equiv d$, being the distance between neighbouring particles and $N^i, i = 1, 2\ldots$ independent Brownian motions or a Wiener process. Associated with the above relations is the "hydrodynamic limit". The latter has been discussed by Presutti [223]. More recently Mürmann [224] considered this limit for a one-dimensional nearest neighbour gradient system. The time evolution of a "configuration of particles" located at $x_i \in \mathbb{R}^d, i \in \mathbb{Z}$ is then expressed by:

$$\mathbf{v}_i(t)\frac{dx_i}{dt} = \sum_{i\neq j}\nabla\Phi^I = -\sum_{i\neq j}\nabla\Phi_{N,\nu} \; ; \; i \in \mathbb{Z} \qquad (3.165)$$

where an interaction force: $\mathbf{f} = -\nabla\Phi^I \equiv -\nabla\phi_{N,\nu}$ acts in the direction of $\mathbf{x}_i$.

The stochastic interface model considered here is conceptually based on Fig. 25(a,c). Thus a possible bond breakage due to external perturbations (application of a shear for instance creates the interaction zone $z(A)$, which is controlled by the range and strength of the interaction. This zone may be regarded to consist of a sequence of "unit cells or microdomain Γ"' (spheres) with a fixed radius: $a = \Delta R/2$, ΔR being the range of the interaction potential Φ^I. If x_i denotes the position vector to the C.M. of each unit cell Γ, $\mathbf{x}_i \in \mathbb{R}^n$; $i \in \mathbb{Z}$, the set $\mathbf{x} = \{x_1, x_2\ldots x_n\} \in \mathbb{R}^{n-1}$ forms a hyperplane, ie.: $\mathbb{R}^n = \mathbb{R}^{n-1} \times \mathbb{R}^+$. Hence small fluctuations $\xi(t)$ with respect to an equilibrium reference plane ($\xi = 0$) can be seen as a stochastic process or:

$$\xi(t) = \{\xi_1(t_1), \xi_2(t_2)\ldots\xi_n(t_n)\} \in \mathbb{Z}; \; t \in T \; . \qquad (3.166)$$

Evidently the most significant dynamic variable involved in this process is the number density $^\Gamma n$ within a unit cell of volume $^\Gamma V$.

For a single component fluid due to the interface there are two number densities, ie. $^{\alpha}n$ and ^{I}n (interface) so that:

$$\{^{\alpha}n\} \cup \{^{I}n\} \Rightarrow {}^{\Gamma}V \subset {}^{M}V \tag{3.167}$$

where the superscript M refers to a specific mesodomain of the macroscopic fluid body. If a liquid and vapour phase coexist in each $^{\Gamma}V$ and since generally $^{\alpha}n \neq {}^{\beta}n$, one has analogously:

$$\{^{\alpha}n\} \cup \{^{I}n\} \cup \{^{\beta}n\} \Rightarrow {}^{\Gamma}V \subset {}^{M}V. \tag{3.168}$$

It is seen that this model considers the fluid interface by the finite covering of unit cells $^{\Gamma}V \subset {}^{M}V$ (mesodomain of the fluid). Since the number densities are also functions of the molecule position coordinates "x_i", one can state the "local density" as:

$$\{^{\Gamma}n(x_i, t)\}_{i=1}^{n} \tag{3.169}$$

and for a set of non-intersecting mesodomains forming the macroscopic fluid body, ie. $^{M}V \subset \mathbb{R}^n \colon \{^{\Gamma}n(x_i, t)\}_{M=1\ldots P}$, so that the stochastic process expressing the interface dynamics is given by:

$$\frac{d}{dt}\{^{\Gamma}n(\cdot)\}_{M=1}^{P} = \{^{\Gamma}\mathbf{f}\}_1^{P} = -\sum_1^{P} \nabla\phi^I \tag{3.170}$$

which analogous to relation (3.163) leads to the Langevin type equation, ie:

$$d^{\Gamma}n_i(x_i, t_i, \Gamma_i) = \sum_{M=1}^{P} \nabla\phi^I dt + \beta^{-1}dw_i(t) \tag{3.171}$$

$$(i = 1\ldots n, M = 1\ldots P)$$

involving the symmetric, convex interaction potential Φ^I and the independent Brownian motions due to the small fluctuations $\xi(t)$. A more detailed analysis will be considered in the next chapter concerned with a more complex fluid.

3.3 Introduction to geometric probabilities:

(i) Introduction:

The concepts of mathematical probability and the use of stochastic processes in the development of the stochastic mechanics of discrete media have been emphasized in the preceding sections. It is frequently necessary however, for the verification of the analytical models as well as the evaluation of experimental results to use "geometric objects". Fundamental for the application of probabilistic concepts and for obtaining results of geometrical interest are "random sets" and their corresponding measures. Since the random elements are geometric objects such as points, lines, areas, congruent sets, etc., the definition of an appropriate measure largely depends on the geometry under consideration. It is desirable for this purpose to choose a measure, which remains "invariant" under groups of transformations. Thus the mathematical concepts involved in the theory of geometric probability are measures, groups, topology and geometry, which form the branch of mathematics called "integral geometry". For the main contributions to this field the reader is referred to Santaló [225], Blaschke [226], Hadwiger [227, 228], Kendall [229], Krickeberg [230] and Miles [231–233]. A rigorous treatment of integral geometry in terms of set theory and topology is due to Matheron [234]. More recently the mathematical analysis of geometrical objects including their shapes, texture, etc. and their representation by convex sets known as "mathematical morphology and image analysis" has been given by Serra [235].

It is apparent that the present discussion can only deal with some fundamental notions of mathematical morphology and hence for a more extensive study the reader is referred to the above references.

(ii) Random sets:

A large number of problems in geometric probability is associated with "convex sets" already mentioned in Chapt. 1. The micrographs of actual materials (Fig. 1–7) and their corresponding idealized structures given in section (2.1) of Chapt. 2, clearly show that they are composed of structural or "geometric elements". The latter can be regarded as "random sets" (convex) in Euclidean spaces and which have certain properties. Matheron [234] makes a distinction between "convex", "compact convex" and "closed convex sets" in the spaces designated by $C(X), C(\mathbb{K})$ and $C(\mathcal{F})$, respectively. He gives with respect to convexity the following formal definitions:

$$\left. \begin{array}{ll} \text{The set } X \in C \Leftrightarrow (x_1, x_2) \in X \Rightarrow [x_1, x_2] \in X & \text{(a)} \\ \text{and } X \in C(\mathcal{F}) \Leftrightarrow X = \cap_i \Pi_i \; ; \; \Pi_i \supset X & \text{(b)} \end{array} \right\} \; (3.172)$$

so that the set X of the space $\mathcal{P}(\mathbb{R}^n)$ is convex, if for every pair of points $x_1, x_2 \in X$, the whole segment $[x_1, x_2]$ belongs to X. The second definition (3.172b) is purely analytical and has no equivalent for the space of lattice points $\mathbb{R}^d$. However, one can consider the convex sets as an "affine linear subspace L" of $\mathbb{R}^n$ in which the subsets L have the characteristics that for every $x_1, x_2 \in L$ the whole line $[\alpha \cdot x_1 + (1 - \alpha)x_2 : \alpha \in \mathbb{R}]$ through x_1, x_2 also lies in L. The dimensions of such an affine linear subspace is the "smallest whole number n" such that L for some points $x_1 \ldots x_{n+1}$ in $\mathbb{R}^d$, one has:

$$L = [\alpha_1 x_1 + \ldots \alpha_{n+1} x_{n+1} \colon \alpha_1, \ldots \alpha_{n+1} \in \mathbb{R}\ ;\ \Sigma \alpha_i = 1]. \tag{3.173}$$

In general affine linear subspaces of dimensions "r" are referred to as "r-flats" or "r-planes" and if the r-flat contains the origin 0, it is called a "r-subspace". In particular the $(d-1)$ dimensional flats are called "hyperplanes" [230]. Although r-flats are closed sets they are not bounded and thus not compact. Examples of closed sets are "unit spheres", closed and bounded "hypercubes" and "closed discs" (ie. the intersections of unit spheres with 2D-flats).

Open spheres and open hypercubes are convex but not compact. It is important to note that a convex compact set $X \subset C(\mathbb{K})$, which contains the origin 0, has a "support function" s_k^0":

$$s_k^0 \colon \partial b(0,1) \to \mathbb{R} \text{ given by: } s_k^0(u) = \sup <u,x>;\ x \in \mathbb{K} \tag{3.174}$$

in which $< u,x >$ is the scalar product of $u = \{u_1, u_2 \ldots u_d\}$ and $x = \{x_1 \ldots x_d\}$. The "support function s_k^0" is continuous on the boundary of the unit sphere. Generally the sphere with its centre at a and the radius r has the boundary:

$$\left.\begin{aligned} &\partial b(a,r) = \{x \in \mathbb{R}^d \colon \| a - x \| = r\} \\ &\text{and the "unit-sphere":} \\ &\partial b(a,r) \equiv \partial b(0,1) \in \mathbb{R}^d \end{aligned}\right\} \tag{3.175}$$

If the set X is symmetric ie. $X = \check{X}$, then s_k^0 is uniquely determined by its value on one hemi-sphere of $\partial b(0,1)$. One often uses a "modified" support function, that is defined in a set X_1 of all lines passing through the origin. Thus, if a line $\ell \in X_1$ and a point $x(\ell)$ on $\ell \cap \partial b(0,1)$ (upper hemi-sphere: $x_d \geq 0$) is given, then the modified support is expressed by:

$$s_\ell = s^0(x(\ell)) \text{ for } \ell \in X_1.$$

In the 2D-case "ℓ" can be replaced by the angle formed by the line and the x_1-axis of the upper hemi-sphere so that s_ℓ is defined on $(0, \pi)$.

Apart from the set operations on subsets of the Euclidean space discussed earlier (Chapt. 1), some additional operations are required for morphological considerations. It may be recalled that the vector space operations on a Euclidean space permits addition and scalar multiplication. Thus multiplication by real numbers: $\alpha X = \{\alpha \cdot x \colon x \in X\}$, where α is real and $X \subset \mathbb{R}^d$. If $X = \check{X}$, X is said to be "symmetric" and a translation means: $X_x = X + x = \{y + x \colon y \in X\}$ for $x \in \mathbb{R}^d, X \subset \mathbb{R}^d$. Two operations are important here for "random closed" sets, namely the "Minkowski addition and subtraction".

Thus for two subsets X, Y of $\mathbb{R}^d$ one has the following:

(a) Minkowski addition:

$$X \oplus Y = \{x + y : x \in X, y \in Y \; ; \; X, Y \subset \mathbb{R}^d\} \tag{3.176}$$

This addition is both associative and commutative. It enlarges, translates and deforms a given set X. Other important characteristics are:

$$\left.\begin{array}{l} \text{(i) } X \oplus Y = \bigcup_{y \in Y} X_y = \bigcup_{x \in X} X_x \; ; \; X_x = X \oplus \{x\} \\[2mm] \text{(ii) } X \oplus Y = \{x : X \cap (\check{Y})_x \neq \emptyset\} \\[2mm] \text{(iii) } X \oplus (Y_1 \cup Y_2) = (X \oplus Y_1) \cup (X \oplus Y_2) \; ; \; X, Y \subset \mathbb{R}^d \end{array}\right\} \tag{3.177}$$

(b) Minkowski subtraction:

$$X \ominus Y = \bigcap_{y \in Y} X_y \text{ or } X \ominus Y = (X^c \oplus Y)^c \tag{3.178}$$

where in general the subtraction is not an inverse to the addition operation, but one also has:

$$(X \ominus \check{Y}) \oplus Y \subseteq X \subseteq (X \oplus \check{Y}) \ominus Y.$$

The above algebraic operations are used in mathematical morphology and image analysis and are also known as "dilation" and "erosion" of sets. Hence the former is characterized by: $X \to X \oplus \check{Y}$ and the latter by: $X \to X \ominus \check{Y}$.

The mathematical theory of morphology and image analysis as developed by Matheron and Serra is based on the Euclidean set operations. The shown Minkowski relations play a dominant role in the morphology of discrete media. The operation (3.176) will enlarge the set X. If the "structuring element Y", or in Serra's notation denoted by "B", is a ball, it will smooth the set X. Hence generally a "dilation of the set" fills out the pores or cavities and may even join separated parts of an image. If X is a realization of a random (closed) set, a dilation by rY (or rB) with the subsequent measurement of the area of $(X + rY)$ permits the estimation of a so-called "contact distribution" that has been introduced by Stoyan, Kendall and Mecke [236].

The subtraction in (3.178) which is the dual operation to (3.176) contracts the set X with the tendency to produce smaller fragments so that connected sets become separated into several subsets. This may be useful for the estimation of the "number of particles" forming an image. Generally the "erosion of the set" is an important operation for the quantitative analysis of random sets. Thus for instance, if the "structuring element" Y is a two point set: $\{x, y\}$ ($\| x - y \| = r$), the area of the "eroded set", ie. $X \ominus \check{Y}$ is an estimate for the set covariance function "$C(r)$", (see also [235] for morphological opening and closings, and structuring functions).

The "opening of a set X" is an attempt to reverse the performed "erosion" by a "dilation". It is analytically expressed in form of:

$$X \to X_y = (X \ominus \check{Y}) \oplus Y. \tag{3.179}$$

It gives X a similar appearance to the original set, but is constructed only on parts of the image that prevail in the initial erosion. It can be useful in eliminating of image defects or "noise". The dual operation is that of the "closing of a set X" by Y and the attempt to reverse the performed "dilation". It is expressed by:

$$X \to X_y = (X \oplus \check{Y}) \ominus Y. \tag{3.180}$$

Again X_y has an approximate resemblance to the original set X. Both of these actions, ie. the opening and closing of the set furnish however more distinct images. Although these algebraic operations are basic in morphology and image analysis of discrete media, they represent a "local" view point in the integral geometry of convex sets. A theorem by Hadwiger [228] shows that numerical parameters associated with "compact convex sets" and which satisfy certain properties of "invariance" are linear combinations of small subsets of such sets and are referred to as "Minkowski functionals" (Chapt. 1). It is known from integral geometry that the projections of a compact convex set $X \subset C(\mathbb{K})$ or its sections (defined by translates of subspaces with smaller dimensions) permit the determination of its Minkowski functionals. Furthermore there are relations due to Cauchy [237] and Crofton [238] with reference to the projections and sections, respectively that allow their use in "stereology".

Since a functional is by definition a "global parameter" associated with a set or the mapping from $\mathbb{R}^n$ onto $\mathbb{R}$, there exist for every $X \in C(\mathbb{K})$ in $\mathbb{R}^n$, $(n+1)$ Minkowski functionals. For the set $X \in C(\mathbb{K})$, the "connectivity number $N^{(n)}(X)$" is the functional, which is equal to one, if $X \neq \emptyset$ and zero, if not. If the boundary of the set ∂X is fairly regular, ie. to have $(n-1)$ "radii of curvature R_j" at each point, then the Minkowski functionals $W_k^{(n)}$, $(k = 1 \ldots n)$ can be represented as a surface integral of the fundamental symmetric functions of the "curvature". Thus assuming that $X \in C(\mathbb{K})$ in $\mathbb{R}^n$ has $(n-1)$ radii of curvature R_i, $(i = 1, \ldots n-1)$, then at any point $x \in \partial X$, the ith elementary functions of the principal curvature become:

$$\phi_1 = \Sigma(R_i)^{-1} \; ; \; \phi_2 = \Sigma\Sigma(R_i R_j)^{-1} \ldots..$$

and hence the Minkowski functional is given by:

$$W_k^{(n)}(X) = \frac{1}{k\binom{n}{k}} \int_{\partial X} \phi_{k-1} dS \; ; \; (k = 2 \ldots n). \tag{3.181}$$

"Structuring elements" in $\mathbb{R}^n$ or in a relevant subspace are based on points, lines, planes and directions, that can be parametrized by noting that a point x has the coordinates $(x) \in \mathbb{R}^1$, $(x_1, x_2) \in \mathbb{R}^2$ and (x_1, x_2, x_3) in $\mathbb{R}^3$, respectively. Denoting the "unit vectors in the plane and in the 3D-case by α, ϑ", and if the unit vector has its centre at the origin 0 of the plane or the "unit sphere", respectively, it denotes either the perimeter U of the unit circle or the surface S_u of the unit sphere. In the latter case the equivalent to the Lebesgue measure is then the "unit mass" uniformly distributed over S_u. Hence the corresponding densities are $d\alpha/2\pi$ in $\mathbb{R}^2$ and $d\vartheta/4\pi$ in $\mathbb{R}^3$.

It is often required to consider sectioning of the set X so that lines and planes passing through a point x can be denoted by $\Delta_\alpha(x, \alpha)$ and $\Pi_\vartheta(x, \theta)$. In the symbolism of Serra [235], where the sign "|" means the "projection of X onto Δ", the following defintion of the Minkowski functionals in $\mathbb{R}^n$ can be given:

$$\text{For } n = 1: \ W_0^{(1)} = L(X) \ ; \ W_k^{(1)} = 2$$
$$n = 2: \ W_0^{(n)} = V^{(n)}(X)$$

and generally for $1 \le k \le n$:

$$W_k^{(n)} = \frac{1}{nb_{n-1}} \int\limits_{\{\vartheta\}} W_k^{(n-1)}[X|\Pi_\vartheta^{(n-1)}]d\vartheta \tag{3.182}$$

in which $V^{(n)}(X)$ is the n-volume of the set X, b_n the volume of the unit sphere, ϑ and $\{\vartheta\}$ the directions and set of directions (unit spheres) in $\mathbb{R}^n$.

A similar definition for the lattice space $\mathbb{R}^d$ has been given by Stoyan, Kendall and Mecke [236] by using the concept of "parallel sets", ie. the set $X \oplus b(0, r)$ at a distance "r" from $X \in \mathbb{R}^d$. The operation of taking parallel sets preserves the convexity and compactness properties as well as of being a unit sphere $[b(0, 1)]$. Hence for the one, two and three dimensional case of $\mathbb{R}^d$, one has the following results:

$$\left.\begin{array}{l} \text{For} \\ d = 1: \text{ the length } L(X): \ L(X \oplus b(0, r)) = L(X) + 2r \\ d = 2: \text{ the area } A(X): \ A(X \oplus b(0, r)) = A(X) + U(X)r + \pi r^2 \\ d = 3: \text{ the volume } V(X): \ V(X \oplus b(0, r)) = V(X) + A_s(X)r + \\ \qquad\qquad\qquad\qquad\qquad\qquad + 2\pi\bar{b}(X)r^2 + \frac{4\pi}{3}r^3 \end{array}\right\} \tag{3.183}$$

in which U is the perimeter of $X^{(d-2)}$, A_s the surface area $\equiv S_u$ and the quantity $\bar{b}$ a regular real-valued functional or "mean breadth" of $b(X)$ over all lines through the origin 0 using an invariant measure. These simple relations (3.183) were obtained on the basis of "Steiner's" formula (see also Serra [235].

It is often necessary for experimental reasons to consider "thick" sections of material specimen and the corresponding sets. In this case the formulae given by Crofton apply, which also form a link to the possible "digitalization" of the sets under consideration and which are important in image analysis. The Minkowski functionals according to Hadwiger's characterization theorem [228], exhibit four basic properties, ie. that of continuity, homogeneity, invariance (translations and rotations) and the C-additivity for the $C(\mathbb{K})$ class of compact sets. However since these functionals are global parameters their application in mathematical morphology requires a "local" representation which can be obtained by the use of Steiner's relations and has been discussed by Serra. In particular this concerns the procedures of "dilation and erosion" of the sets X. For a definition of the "local form" of the Minkowski functional, one needs however a corresponding measure. Thus recalling from Chapt. 1 and the definition of a Lebesgue measure, that a family of sets generated by a countable union of "closed intervals" in $\mathbb{R}^n$ and the complement of this union defines a class of Borel sets: $\mathcal{B}(\mathbb{R}^n)$. This leads in morphology to the introduction of so-called "measuring masks or measuring fields Z" of an image. They are assumed to belong to the class of Borel sets $\mathcal{B}(\mathbb{R}^n)$. Hence a set of measures $\mu(Z)$ can be associated with each mask $Z \in \mathcal{B}(\mathbb{R}^n)$ such that $\mu(\emptyset) = 0$, $\mu(\cup_i Z_i) = \Sigma_i \mu(Z_i)$ for any countable sequence $\{Z_i\}$ where $Z_i \cap Z_j = \emptyset, i \neq j$. The latter condition is the σ-additivity. It permits the determination of the measure $\mu(Z)$ by summing "elementary disjoint masks dZ", as a partition of Z.

These measures can be extended to functions by noting, that any real valued function $f(x), x \in \mathbb{R}^n$ is measurable, if for any real number α, the set of all x for which $f(x) > \alpha$ is a "mask" in $\mathbb{R}^n$. If the function takes on only a finite number of values, it can be expressed by:

$$\left. \begin{array}{c} f(x) = \sum_{i=1}^{n} \lambda_i I_{z_i}(x) \; ; \; \lambda_i > 0, Z_i \in \mathcal{B}(\mathbb{R}^n), \\[4mm] n < \infty, \; I_z \equiv \text{indicator function.} \end{array} \right\} \quad (3.184)$$

Associated with this function is then the measure:

$$\mu(f) = \sum_{i=1}^{n} \lambda_i \mu(Z_i). \tag{3.185}$$

Considering a compact convex set X and a point $x \in X$ in $\mathbb{R}^n$, there exists a unique point x' that minimizes the distance $|x - y|$ as the point y moves over X. The point x' is the projection of x onto X, ie. $x' = x|X$. In mathematical morphology both x' and the distance $\rho = |x - x'|$ are of interest and hence the transformation of $x \to (x', \rho)$ or the mapping of $\mathbb{R}^3$ or $\mathbb{R}^2$ onto $X \times \mathbb{R}^+$. Such a mapping transforms the Lebesgue measure in $\mathbb{R}^3$ (or $\mathbb{R}^2$) into a "Minkowski measure" given by Serra in the following forms:

For the 2D-case:

$$\int_{X \oplus \rho b} f(x|X)dx = \int_{\mathbb{R}^2} f(x)[a(dx) + \rho u(dx) + \pi \rho^2 n^{(2)}(dx)]. \tag{3.186}$$

For the 3D-case:

$$\int_{X \oplus \rho b} f(x|X)dx = \int_{\mathbb{R}^3} f(x)[v(dx) + \rho s(dx) + \rho^2 m(dx) + \frac{4\pi}{3}\rho^3 n^{(n)}(dx)]. \tag{3.187}$$

where the quantities on the r.h.s. of the above relations relate to those given in (3.183) as follows:

$$\int v(dx) = V(X); \int a(x)dx = A(X); \int u(dx) = U(X);$$

$\int s(dx) = A_s(X)$ and $n^{(2)}, n^{(3)}$ to the "connectivity numbers" $N^{(2)}, N^{(3)}$ of the set X respectively. It is to be noted that the integration is over the whole space although the support of the measures is limited to X or its boundary ∂X. These local forms of the Minkowski functional are important morphological parameters and are discussed in detail in ref. [235].

It is evident that in dealing with random closed sets special attention must be given to the topological aspects. The linkage between the notion of compact spaces and the axiomatics of probability theory by regarding these sets as random variables and the underlying "event structure" are rigorously discussed by Matheron and Serra in the previously cited references.

(iii) Random point models:

Experimental observations on sections of various discrete media by means of electron microscopy, X-ray diffraction and other optical processing procedures reveal the resulting images to have irregular random patterns. Thus random closed sets may be used to construct mathematical models for the description of such patterns. The development of the modern theory of random sets is due to Choquet [239], Matheron [234], Kendall and Moran [240] among others. Some basic random set models that are useful in the modelling of microstructures of various discrete media are briefly given below. These models have to satisfy the principles of stochastic mechanics as stated in Chapt. 2, the requirements of a morphological analysis and the possibility of a quantitative evaluation. The most exclusively used class of random sets are "random point models". They are associated with elementary open sets to which the Lebesgue measure

applies. The basic types from which more complicated models can be derived
are the "Boolean models", "Poisson flats" and the "Poisson point processes".
The latter will also be considered subsequently. By neglecting possible interactions, a "random point pattern" or a point process $\varphi = \{x_1, x_2 \ldots x_n\}$ can
be regarded as a random set of points. It can also be considered as a "random
measure" applied to sets in which for each Borel set B the measure $\varphi(B)$ or
$N(B)$ is given by the number of points of φ that fall into the set B. Since the
random point patterns are all "locally finite", the set function "$\varphi(B)$" is also
locally finite and a σ-additive Borel measure $(N(B) \in \mathcal{B}(\mathbb{R}^n))$. The random
point set φ can be intersected with other sets. If B is a Borel set so is $B \cap \varphi$ a
random set of points belonging to B.

(a) The Boolean model:

The Boolean model is the most extensively used model in the description
of empirical random sets and as a good approximation to more complicated
sets encountered by experimental observation of sections of discrete media.
It is closely associated with a "stationary Poisson point process" and hence
some further remarks concerning these processes may be indicated. Poisson
point processes are often referred to simply as "random distributions of points"
on a line. Thus, if such a process allocates points on a half-open interval
$(a_k, b_k] \in \mathbb{R}^+$ in the Euclidean $\mathbb{R}^d$-space, the probability of the "number of
events" of the process falling in this interval can be expressed by:

$$P\{N(a_k, b_k] = n_k, k = 1 \ldots n\} = \prod_{k=1}^{n} \frac{\lambda[b_k - a_k]^{n_k}}{n_k!} e^{-\lambda(b_k - a_k)} \qquad (3.188)$$

in which $\lambda > 0$ is a positive constant. This process has the following characteristics:

(i) the number of points in each finite interval of $\mathbb{R}^+$
has a Poisson distribution,

(ii) the number of points in disjoint intervals are
independent variables, $\qquad\qquad (3.189)$

(iii) the distributions of the number of points are
"stationary" or independent of time and depend only
on $(b_k - a_k)$ or the length of the intervals on $\mathbb{R}^+$.

Hence the mean value and the variance of the number of points that fall into
the interval $(a, b] \in \mathbb{R}^+$ are given by:

$$E\{(a, b]\} = \lambda(b - a) \; ; \; \lambda > 0 \qquad (3.190)$$

in which the parameter λ can be regarded as the "mean rate" or "mean density"
of points of the process. In the context of experimental work, it is often required

to find the "likelihood" of a finite "realization" of the Poisson point process. It can be defined as the probability of obtaining the given number of observations for a specific observation period $(0, T]$ times the joint conditional density for the locations of these observations given their number. Thus, if N-observations are made at times $(t_1 \ldots t_N) \subset (0, T]$ then from (3.188) the probability of "single events" in $(t_k - \Delta, t_k)$ and no points for the remaining part of $(0, T]$ can be expressed by:

$$e^{-\lambda T} \prod_{i=1}^{N} \lambda \Delta$$

which upon dividing by Δ^N and taking the limit, ie. $\Delta \to 0$ gives the density. Hence the required likelihood function "L" is given by:

$$L_{(0,T]}(N; t_1 \ldots t_N) = \lambda^N e^{-\lambda T}. \tag{3.191}$$

It is sometimes necessary in the modelling of discrete media to consider "inhomogeneous point processes", which are characterized by a time-varying rate $\lambda(t)$. Such processes can still be defined in accordance with (3.188), but the quantities $\lambda(b_k - a_k)$ must be replaced by:

$$\Lambda(a_k, b_k] = \int_{a_k}^{b_k} \lambda(x) dx \; ; \; x \in \mathbb{R}^+ \tag{3.192}$$

although the joint distributions are still Poissonian and the independence property of (3.189, ii) still applies. A more general form of $L_{(0,T]}$ and the case of nonstationary Poisson processes are rigorously discussed by Daley and Vere-Jones [241]. Models based on stationary Poisson point processes are schematically shown in Fig. 26(a–c).

The realization of a stationary Poisson process φ with intensity λ is schematically shown in Fig. 26(a), where the points $\{x_1, x_2 \ldots x_n\}$ are scattered in the x, y plane. The area under consideration, ie. an "observation window (O.W.)" which may vary in its dimensions in accordance with experimental constraints under which such observations can be made in scanning electron microscopy or other optical procedures to obtain the image (see also [343, 344]). By attaching to each point of the image a disc of radius "r" yields the simplest planar Boolean model as indicated in Fig. 26(b). In image analysis the points of φ, ie. $\{x_n\}$ are called "germs" and the "discs" or sets the "primary grains X" of the model. This construction is readily generalized to give a more general stationary Boolean model by replacing the discs by "independent realizations" of a random compact set X. Thus, if $\varphi = \{x_1, x_2 \ldots\}$ is a stationary Poisson process in $\mathbb{R}^d$ and $X_1 X_2 \ldots$ a sequence of i.i. distributed "random compact sets" in $\mathbb{R}^d$, which are independent of φ and satisfy certain conditions (see

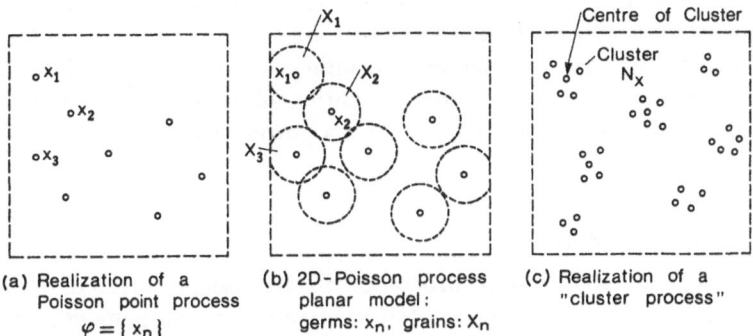

(a) Realization of a   (b) 2D-Poisson process   (c) Realization of a
    Poisson point process   planar model: "cluster process"
    $\varphi = \{x_n\}$ germs: x_n, grains: X_n

Fig. 26. Schematics of Point models.

Matheron [234]), the Boolean model can be constructed as the union of the "germs x_n" and the "primary grains X_n" as follows:

$$X = \bigcup_{n=1}^{\infty}(X_n + x_n) = (X_1 + x_1) \cup (X_2 + x_2) \cup \ldots \qquad (3.193)$$

One can also designate by $X_1 \ldots X_k$ the sequence of disjoint Borel sets, $X \in \mathcal{B}(\mathbb{R}^d)$ and consider $\varphi(X_1), \ldots \varphi(X_k)$ as independent random variables, the mean value of which is obtained in accordance with (3.190). Further by introducing the Lebesgue measure or the volume of the unit sphere, one obtains analogous to relation (3.188) the joint probability function in the following form:

$$P\{\varphi(X_1) = n_1 \ldots \varphi(X_k) = n_k\}$$

$$= \lambda^{n_1 \ldots n_k} \frac{[\nu(X_1)]^{n_1} \ldots [\nu(X_k)]^{n_k}}{n_1! \ldots n_k!} \exp(-\sum_{i=1}^{k} \lambda \nu(X_i)) \qquad (3.194)$$

A more rigorous treatment of the Boolean model in terms of the union of compact convex sets that are based on the stationary Poisson point process and are subsets of $C(\mathbb{K})$ is considered by Matheron [234].

The Boolean model is basic in stochastic geometry and stereology. In practical application, it is a natural model for discrete media, when the particles or structural elements are not too densely arranged within the microstructure. Hence "sparse discrete systems" can be readily modelled, when the value of the density λ of the Poisson process is rather low. Thus, if λ is small compared to the size of the grains (structural elements), then the primary grains will not often overlap and the set X will essentially consist of separated elements. An increase of λ causes however an increase of overlaps and an adjustment in the formulation becomes necessary. In general the grains of the Boolean model are not required to be "connected sets", they may form a set of discrete points or a point process in form of "clusters" as indicated in Fig. 25(c). Such processes are considered subsequently. The application of Boolean models in the morphological modelling of images has been rigorously discussed by Jeulin [259, 260].

(b) Other point models:

The extensive use of "point processes" in the interpretation of experimental observations and often in the simulation of the response behaviour of discrete media warrants some remarks concerning the general theory of such processes. The latter are particularly significant in the analysis of two or three-dimensional point patterns indicated in Fig. 26(a–c). In such cases both "cluster" and "doubly stochastic" models can be defined on the basis of measure theoretical arguments. Thus considering the previously mentioned "counting measure $N(x)$" defined on the Borel subsets of a complete separable metric space X and denoting the space of all finite integer-valued measures $N(x)$ by $\mathcal{N}_x$, the "simple counting measures forming $\mathcal{N}_x$ can be defined as follows:

$$\mathcal{N}(x) \equiv N(\{x\}) = 0 \text{ or } 1 \text{ for all } x \in X \ . \tag{3.195}$$

An element of $\mathcal{N}_x$ is a simple counting measure, iff it can be expressed by:

$$N(x) = \sum_i k_i \delta_{x_i} \tag{a}$$

and the support counting measure by:

$$N^* = \sum_i \delta_{x_i} \tag{b}$$

$$\left.\begin{array}{c} \\ \\ \\ \\ \end{array}\right\} \tag{3.196}$$

where each k_i is a positive integer and the $\{x_i\}$ are distinct points indexing the "atoms of the measure" or equivalently the "Dirac measure δ" forming a countable set with finitely many x_i's in a bounded Borel set $\mathcal{B}(X)$. It is to be noted that $N(x)$ is a simple counting measure only, iff $k_i = 1$ for all i's in (3.196a) or if it coincides with its "support counting measure" (see also Daley and Vere-Jones [241]). From the characteristics of these counting measures two important propositions can be given, ie.:

(i) The Borel sets X of $\mathcal{N}_x$ (as a metric space) coincide with the Borel sets of $\mathcal{N}_x \subset \mathcal{M}_x$ (or a more general measure space).

(ii) $\mathcal{B}(\mathcal{N}_x) \equiv \mathcal{F}_x$ is the smallest σ-algebra with respect to the mappings $N \to N(X)$ are measurable for each set $X \in \mathcal{B}(X)$.

Hence it is possible to define formally a "point process $N(x)$" as a measurable mapping from the probability space $[X, \mathcal{F}, \mathcal{P}]$ into $[\mathcal{N}_x, \mathcal{F}_x]$. A point process is "simple", when $\mathcal{P}\{N \in \mathcal{N}_x\} = 1$.

In terms of the earlier notation, a "point process φ" is a measurable mapping of the probability space $X \to [\mathcal{N}_x, \mathcal{F}_x]$ generating a distribution "P_φ" on this space. Thus one may consider a point process either in terms of "random sets of discrete points" or as "random measures" counting the number of points lying in a specific spatial region, so that:

(i) $x \in \varphi$ asserts that a point x belongs to the
random sequence φ,

(ii) $\varphi(X) = N$ asserts the set X contains N
points of φ.
$$\left.\right\} \quad (3.197)$$

Since the point process can be considered in terms of random closed sets, the theory of such sets will apply and the distribution P_φ of the process is determined by the probabilities:

$$P_\varphi(Y) = P(\varphi \in Y) = P\{(x \in X : \varphi(x) \in Y)\} \text{ for } Y \subset \mathcal{N}_x. \quad (3.198)$$

Of particular interest here are the finite dimensional distributions expressed by:

$$P\{\varphi(X_1) = N_1 \ldots \varphi(X_k) = N_k\}; \text{ where } X_1 \ldots X_k \text{ are bounded Borel sets and}$$

$$N_1 \ldots N_k \geq 0 . \quad (3.199)$$

Hence the distribution of φ on $[\mathcal{N}_x, \mathcal{F}_x]$ is uniquely determined by the system of all values for $k = 1, 2 \ldots$ and even for subsystems for which the "constituents of X_k" are pair-wise disjoint. For still smaller subsystems Stoyan, Kendall and Mecke [236] introduced a "void probability" for the sets X, where:

$$V_x = P\{(\varphi \in \mathcal{N}_x : \varphi(X) = 0\} = P\{\varphi(X) = 0\} = P \{\emptyset \cap X \text{ is empty}\} \quad (3.200)$$

for all Borel sets $X \subset X$. Thus for a simple point process the distribution is determined by the values V_k as k goes through the range of the compact sets $X \in C(\mathbb{K})$ (see also Ripley [242], Kallenberg [243, 244]). In this representation the probabilities are specified by the "non-occurence" of points in a specific region. From the interpretation of φ in terms of random sets, it follows that for any measurable function $f(x)$ on $X = \mathbb{R}^d$, the sum of $f(x)$ over x in φ can be expressed by:

$$f(x_1) + f(x_2) + \ldots = \sum_{x \in \varphi} f(x)$$

$$\text{or equivalently as } \int f(x)p(x)dx \; ; \; p(x) = \sum_{y \in \varphi} \delta(x - y)$$
$$\left.\right\} \quad (3.201)$$

where δ is the Dirac-delta function. Evidently the expected value of the sum of $f(x)$ will be:

$$E\{\sum_{x \in \varphi} f(x)\} \text{ or } \int \sum_{x \in \varphi} f(x) P(d\varphi)$$

so that the number of points in the Borel set X and its mean value become:

$$\varphi(X) = N(X) = \sum_{x \in \varphi(X)} I_X(x) \; ; \; E\{\varphi(X)\} = E\{\sum_{x \in \varphi} I_X(x)\}$$

$$= \int \sum_{x \in \varphi} I_X(x) P(d\varphi) \qquad (3.202)$$

where I_X is the indicator function of the set X. A point process is "stationary", if its statistics are invariant under translation, ie. $\varphi\{x_n\}$ and $\varphi = \{x_n + x\}$ have the same distribution for all x in $X = \mathbb{R}^d$. This means that:

$$P\{\varphi \in Y\} = \{\varphi_x \in Y\} \text{ for all } Y \text{ in } \mathcal{N} \text{ and all } x \text{ in } X \equiv \mathbb{R}^d. \qquad (3.203)$$

Similarly the notion of "isotropy" here means that the statistics of φ are also invariant under rotations r, ie. φ and $r\varphi$ have the same distribution for every rotation around the origin: $P(Y) = P(rY)$. If stationarity and isotropy together apply, one has a "motion invariant" process, ie. φ has the same distribution as "$m\varphi$" for all Euclidean motions m in $\mathbb{R}^d$ (see also Lipster and Shiryayev [138]).

A more general definition of a finite point process using the concept of "unordered sets" that is often employed for the simulation of point processes is given by Daley and Vere-Jones [241]. These authors state the following conditions for such a process:

(i) the points of the process φ are located in a complete separable metric space $X(X = \mathbb{R}^d$ for example).

(ii) A distribution: $P_N = \{P_N\}$; $(n = 0, 1 \ldots)$ is given determining the total number of points $N(x)$ in the population with $\sum_{n=0}^{\infty} P_N = 1$.

(iii) for each integer $n \geq 1$ a probability distribution P_N is given on the Borel sets X of $X^{(n)} = X \times \ldots \times X$ and it determines the joint distribution of the positions of the points of the process given that their total number is $N(x)$.

It is further stipulated that the distribution P_N should give equal weight to all $n!$ permutations of the position coordinates $(x_1 \ldots x_n)$, ie. that P_N should be symmetric. If this is not the case the latter can be symmetrized for any partition of X by:

$$P_n^s(X_1 \times \ldots \times X_n) = (n!)^{-1} \sum_P P_N(X_{i_1} \times \ldots \times X_{i_n}) \qquad (3.204)$$

in which P_n^s refers to the symmetric form of P_N, $\sum\limits_{P}$ is taken over all permutations $(i_1 \ldots i_n)$ of the integers $(1 \ldots n)$ and $(n!)^{-1}$ is a normalizing factor. For the simplification of the combinatorial expression, it is advantageous to use non-probability measures of the form:

$$\left. \begin{aligned} J_n(X_1 \times \ldots \times X_n) &= P_N \sum_p P_N(X_{i_1} \times \ldots \times X_{i_n}) \\ &= n! P_N P_n^s(X_1 \times \ldots \times X_n) \end{aligned} \right\} \quad (3.205)$$

This type of measures has been referred to by Srinivasan [245] as the Janossy measures, first introduced by Janossy [246] (see also Chapt. 1). The quantity P_n^s is also referred to as the "exclusion probability" (see for instance Macchi [247]). In the present context the Janossy measures may be interpreted in a simple manner. Assuming that derivatives exist for $X = \mathbb{R}^d$, and by denoting the density of J_n with respect to the Lebesgue measure on $(\mathbb{R}^d)^{(n)}$ with $x_i \neq x_j$ for $i \neq j$, by $j_n(x_1 \ldots x_n)$, then the probability that there are exactly n points in φ, one in each of the n-distinct neighbourhoods $(x_i, x_i dx_i)$ is given by:

$$P_N = j_n(x_1 \ldots x_n) dx_1 \ldots dx_n. \quad (3.206)$$

In the context of the previously mentioned cluster processes and on the assumption that the clusters are independently and identically distributed in $X = \mathbb{R}^d$ and that:

$$P(X) = \int_X p(x) dx \quad \text{for some density function } p(x),$$

the joint density functions of an "ordered sequence" of n points $x_1 \ldots x_n$ is given by:

$$j_n(x_1 \ldots x_n) = P_N n! p(x_1) \ldots p(x_n) \quad (3.207)$$

in which j_n and not P_N is the probability density of finding a microelement (particle) at each of the n-points $(x_1 \ldots x_n)$. The factorial characterizes here the number of ways in which the particle can be allocated to these positions.

An important characteristic of point processes is the "generating functional", that can be defined analogously to the "generating function" of a non-negative integer valued random variable x. Thus letting $P(x = k) = p_k$; $(k = 0, 1, 2 \ldots)$, then if the mean value and standard deviation exist, one has:

$$E\{x\} = \sum_{k=1}^{\infty} k p_k \; ; \; D^2\{x\} = \sum_{k=1}^{\infty} k^2 p_k - (\sum_{k=1}^{\infty} k p_k)^2 \quad (3.208)$$

Denoting by $P(z)$ the "generating function" of the sequence $\{p_k\}$, then for any complex number z such that $|z| \leq 1$ the probability generating function is given by:

$$P(z) = \sum_{k=0}^{\infty} p_k z^k. \tag{3.209}$$

In the context of point processes an appropriate generalization is the "probability generating functional" (p.g.f.). Thus by letting f be any bounded complex valued Borel measurable function, then for a realization $\{x_i; i = 1 \ldots N\}$ of a finite point process the product $\prod_{i=1}^{N} f(x_i)$ is well defined with $|f(x)| \leq 1$ for all $x \in X$ and the expectation will exist and will be finite. A somewhat more restricted definition is given by Westcott [248] in which by designating the class of real-valued Borel functions h on X by $\mathcal{V}(X)$ with $(1 - h)$ vanishing outside some bounded Borel set satisfying $0 \leq h(x) \leq 1$ for all $x \in X$ the p.g.f. of the point process $N(x)$ or $\varphi(x)$ can be specified by:

$$G_N(h) = E\{\exp[\int_X \log h(x) N(dx)]\}. \tag{3.210}$$

Since the process is finite on the bounded set and $(1 - h)$ does not vanish, one can replace the integrand above by a product, so that:

$$G_n(h) = E\{\prod_i h(x_i)\}, \tag{3.211}$$

where the product is taken over the points x_i of each realization of $N(x)$ with the condition that it is zero if $h(x_i) = 0$ and equal to unity for any x_i, if there are no points of $N(x)$ within the support of $(1 - h)$.

Another characteristic of point processes are the factorial moments and cumulant measures. The former are based on the corresponding concepts of non-negative integer-valued random variables that have not been mentioned earlier in Chapt. 1. Thus for any integers n, α the "factorial powers $n^{(n)}$" are given by:

$$n^{(n)} = \begin{cases} n(n-1)\ldots(n-\alpha+1); & \alpha = 0 \ldots n \\ 0 & \alpha > n \end{cases} \right\} \tag{3.212}$$

and the factorial moments $m^{(\alpha)}$ of a random variable N, by:

$$m^{(\alpha)} = E\{N^{(\alpha)}\}. \tag{3.213}$$

Hence, if the variable "N" has the probability distribution $\{P_N\} = P\{N = n\}$, the factorial moment is simply:

$$m^{(\alpha)} = \sum_{n=0}^{\infty} n^{(\alpha)} P_n.$$

These factorial moments are related to the Taylor series expansion of the p.g.f. about $z = 1$ in (3.209) so that:

$$P(z) = E\{z^N\} \; ; \; (|z| \le 1). \tag{3.214}$$

Hence for the point process N with the kth functional moment which is finite, the p.g.f. in accordance with Daley and Vere-Jones (Prop. 5.2III) becomes:

$$P(1 + \eta) = 1 + \sum_{\alpha=1}^{k} \frac{m^{(\alpha)} \eta^{\alpha}}{\alpha!} + o(\eta^{\alpha}) \tag{3.215}$$

for all η such that $|1 + \eta| \le 1$ and the Taylor series expansion then yields the p.g.f. of the form:

$$P(1 + \eta) = 1 + \sum_{\alpha=1}^{\infty} \frac{m^{(\alpha)} \eta^{\alpha}}{\alpha!} \tag{3.216}$$

which is valid for some $\eta > 0$, iff all moments exist and the series has non-zero radius of convergence in η. For the point process $N(x)$ by assuming that the kth moment measure exist for some positive integer k, then for the family of functions $(1 - \eta) \in \mathcal{V}(X)$ and an arbitrary $\epsilon : (0 < \epsilon < 1)$, the p.g.f. is obtained in the form of:

$$G(1 - \eta) = 1 + \sum_{i=1}^{k} \left(\frac{(-\epsilon)^k}{k!}\right) \int_X \cdots \int_X \eta(x_1)..\eta(x_i) m^{(i)}(dx_1 \times .. \times dx_i) \tag{3.217}$$

(see also [242]). In order to establish suitable point models for a given point pattern and the required statistical properties of them, the theory of statistics of point processes is usually employed (see Lipster and Shiryayev [138], Brémaud [47], Daley and Vere-Jones [241]). This theory is essentially concerned with the spatial statistics and deals in practical applications with observations of one material sample only. Such observations are made by means of a "compact sampling window" or "observation window (O.W.)", mentioned earlier and hence is restricted to a two-dimensional statistics. It is assumed that these observations are truly unique samples of the stochastic phenomena imparted on the given microstructure. It also implies that the "observed patterns" are samples of "stationary ergodic point process". A rigorous analytical treatment and the conditions for such processes is discussed in ref. [241]. Another inherent problem in the spatial statistics is the occurrence of "edge effects" caused by the dimensions of the "observation windows". These and other factors pertaining to the chosen parameters in the statistical analysis in terms of the notion of "estimators" is discussed by Daley and Vere-Jones [241], Ripley [249, 250], Stoyan et al. [236] among others. Certain operations on point processes are important in modern image analysis. They permit to produce new point processes from the original ones. Several important models for the morphology of discrete media can be derived from simpler ones by the operations known as "thinning", "clustering" and "superposition". The thinning operation is based on some formula to delete points of a point process resulting in a "thinned process". By using the earlier notation of a point process, and regarding $\varphi(x)$ as a closed random set, the thinned process φ_{th} is then a subset of $\varphi(x)$ or $\varphi_{th} \subset \varphi(x)$. In the simplest case of such a random "deletion" each point of $\varphi(x)$ has the probability $(1 - p)$ of undergoing deletion, which is independent of both the positions and possible deletion of any other point of the process. Generally the thinning probability depends on the locations of $x \in X$ and is a measurable function $p(x)$ of X; $(0 \leq p(x) \leq 1)$ such that $[1 - p(x)]$ designates the probability of deleting a point located at $x \in X$. It is also assumed that the points are deleted from the original process or its corresponding point pattern independently from each other. This type of thinning is known as "independent thinning" and excludes any interactions between points of the process. The generating functional G_N in this case, by assuming a stationary process can be expressed by:

$$G_N[h|x] = p(x)h(x) + [1 - p(x)] \tag{3.218}$$

which is related to that of the "thinned process G_{th}" by:

$$G_{th}[h] = G_N[1 - p + ph] \; ; \; h \in \mathcal{V}(X) \tag{3.219}$$

Similarly, if the original process has a factorial measure of any given order k, the thinned process has a similar measure, which in the case of a simple point process are related by:

$$M^{(k)}(dx_1 \times \ldots \times dx_k) = M_N^{(k)}(dx_1 \times \ldots \times dx_k) \prod_{i=1}^{k} p(x_i). \tag{3.220}$$

A generalization of the above "thinned process" is possible so as to introduce the so-called "dependence" on the configurations of $\varphi(x)$. This leads then to the class of "dependent thinning processes". An example of this class of thinning is the Matern or "hard-core" process, when the dependent thinning is applied to $\varphi(x)$, being a stationary Poisson process with intensity λ (see also Daley and Vere-Jones [241], Stoyan, Kendall and Mecke [236]). In the "clustering" operation every point of the process $\varphi(x)$ or its point pattern is replaced by an $N_{cl}(x)$ of points. These clusters are themselves point processes, which in most cases have a finite number of points each. The union of the clusters is then a "cluster points process" or $\varphi_{cl} = \underset{x \in \varphi}{\cup} N_{cl}(x)$. It is however assumed that with probability one, no cluster intersects another, ie. that $N_{cl}(x) \cap N_{cl}(y)$ is empty, if $x \neq y$ and that φ_{cl} is locally finite. In the description of cluster models a typical interpretation of the process φ_{cl} or N_{cl} is that it is regarded as an ensemble of "original or parent" points and the points of the cluster as "daughters or satellite" points of the parent points. In general, if the original process $\varphi_N(x) = \{x_1, x_2 \ldots\}$ is stationary and the clusters denoted by $N_{cl}(x_i)$, then:

$$\varphi_N(x_i) = N_{cl}(x_i) + x_i \text{ for each } x_i \text{ in } \varphi_N(x_i) \tag{3.221}$$

Thus the $N_{cl}(x_i)$ form a family of independent identically distributed finite point sets with the distribution P_{cl}, which is independent of the original or parent process. This is normally referred to as "homogeneous independent thinning". An important example in this context is the Neyman-Scott process [251], characterized by the homogeneous independent clustering applied to a stationary Poisson process. Here the "parent points" are the points of the Poisson process with intensity λ and the "daughter or marginal" points are the points of a representative cluster N_{cl}^0, which are random in number and scattered independently about 0 with identical distributions. This process can also be regarded as a particular type of the Boolean model. Generally the intensity of the cluster process φ_{cl} can be given by:

$$\lambda_{cl} = \lambda \bar{c}_m \tag{3.222}$$

where $\bar{c}_m$ is the mean value of the marginal points/parent points. Hence the following probability generating functional can be given:

$$G_{cl}[h] = \exp\{-\lambda \int_{X=\mathbb{R}^d} [1 - g \int_{\mathbb{R}^d} h(x+y)p(y)dy]dx\} \tag{3.223}$$

in which g is the generating function of the random number of points of N_{cl}^0 and $p(\cdot)$ the probability density of the location of a typical marginal point of the N_{cl}^0. This definition is based on the use of the so-called "Palm measure" as treated by Stoyan, Kendall and Mecke [236]. Another definition is given by Daley and Vere-Jones [241] who consider the Neyman-Scott process as a "centre-satellite" cluster process, where the probability distribution of the cluster members centered at a point x is $P(dy|x)$ and $Q(z|x)$ the p.g.f. of the total finite cluster size. Hence the probability generating functional in this case has the form:

$$G_m[h|x] = Q\{\int_{x=\mathbb{R}^d} h(y)P(dy|x)|x\} \tag{3.224}$$

and the corresponding factorial measures are expressed by:

$$M^{(k)}(dy_1 \times \ldots dy_k|x) = m^{(k)}(x)\prod_{i=1}^{k} P(dy_i|x). \tag{3.225}$$

In conclusion the superposition operation in the simple case of two point processes $\varphi_1(x), \varphi_2(x)$ can be expressed by the union of two point sets, ie. $\Phi_s = \varphi_1(x) \cup \varphi_2(x)$, where the latter have the corresponding distributions p_1, p_2, respectively and where it is assumed that the points of the sets to coincide have zero probability. A more general superposition is possible, but involves then the application of convergence concepts and limit theorems (see for instance Daley and Vere-Jones [241]).

If the point process is a stochastic Poisson process with a randomized intensity measure such a generalization leads to the class of "doubly stochastic processes" or Cox processes. They are discussed by Snyder [252], Kallenberg [243], Krickeberg [253], Karr [254], Serra [235] and others.

3.4 Some fundamental concepts of stereology:

The evaluation of test results of various discrete media obtained in form of micrographs or similarly produced images of microstructures involves the statistics of random variables or the values of the stochastic processes that were adopted in the modelling of the material behaviour. In both of these cases it becomes necessary to deal with sets of points, lines, areas, etc. in the characterization of the microstructure. The calculations are usually carried out on the basis of "stereological" formulae, which are an application of the stochastic geometry concerning a particular microstructure. Stereology can be regarded as the synthesis of integral theory, the theory of point processes and random measures. Its main aim is to establish the three-dimensional properties of a discrete medium from the two-dimensional information experimentally obtained from planar sections of test samples. It is to be noted that the stereological formulae are based on the same assumption as in the analytical modelling, ie. that the microstructure exhibits a certain randomness. The analytical basis of modern stereology is due to several researchers, the work of which is cited in the references of this chapter.

The previously given definition of "convex sets" (section 3.3) is now considered more specifically for "planar sets" due to the above given reason. It is known from integral geometry (Santaló [225]) that a "set of points X" in the plane is called "convex", if for each pair of points $A, B \in X$, $AB \subset X$ is true. A curve connecting points P, Q is convex, if its point set together with the segment PQ bounds the convex set. If the convex set is bounded and has interior points, the boundary ∂X of X is a "closed convex curve". A line in the plane (x, y) is determined by its distance "d" from 0 and the angle ϑ of the normal with the x-axis so that:

$$d = x \cos \vartheta + y \sin \vartheta \ .$$

If the envelope of areas is the boundary of the set X and 0 an interior point, then $d = d(\vartheta)$ is the support function of X or the support function of the convex curve ∂X with respect to 0. Since this characterizes support lines of X, the length "L" of a closed curve will be: $L = \int\limits_{0}^{2\pi} dd\vartheta$ and the area A of X will be obtained from the support function by decomposing it in form of elementary triangles (with a base ds and a height d) so that:

$$A = \frac{1}{2} \int_{\partial X} d \cdot ds = \frac{1}{2} \int_{0}^{2\pi} (d^2 - d'^2) d\vartheta \qquad (3.226)$$

in which the prime sign indicates here differentiation. Thus using L or the perimeter of X the "mean breadth of X" in the direction of ϑ is given by:

$$E\{b(\vartheta)\} = L/\pi. \qquad (3.227)$$

It can also be defined by the length of the orthogonal projection of X on a line parallel to the ϑ-direction. The "least breadth" of X is called the width of the set and the "diameter of X" is the greatest distance between two points of X.

Of special interest are the set of points in a plane or those derived from a "planar Poisson process". It is sometimes useful to consider measures of such sets. From the theory of geometric probabilities a measure $m(X)$ of the set is defined by the integral over the set of a differential form, which by using the notation of exterior calculus (Cartan [256]) is written as:

$$m(X) = \int_X f(x,y) dx \wedge dy, \qquad (3.228)$$

in which the function $f(x,y)$ has to be chosen so that $m(X)$ remains invariant under the groups of motion in the (x,y)-plane. The density of points can be given as:

$$dP = dx \wedge dy$$

which in the case of n-independent points $P_1, P_2 \ldots P_N$; $P_i(x_i, y_i)$, becomes:

$$dP_1 \wedge dP_2 \wedge \ldots \wedge dP_N \; ; \; dP_i = dx_i \wedge dy_i \qquad (3.229)$$

and is unique up to a constant factor. Hence the probability of a random element "α" considered either as a point or as a finite number of points in a set or domain $\mathcal{D}$, when it is known to be in the set X is given in form of a so-called "quotient measure", ie.:

$$\mathcal{P}_\alpha = \frac{m(\mathcal{D})}{m(X)} \; ; \; \alpha \subset X. \qquad (3.230)$$

This becomes significant for the determination of "areal distributions" of microelements of the structure belonging to a specific domain $\mathcal{D}$ contained in M (meso-domain) of a plane section of the material sample. Considering next two such domains $\mathcal{D}_0, \mathcal{D}$ with $\mathcal{D}_0 \supset \mathcal{D}$ and corresponding areas $A_0 \supset A$ respectively, then by choosing n-random points in $\mathcal{D}_0$, the probability that exactly m of them lie in $\mathcal{D}$ is given by a binomial distribution, ie.:

$$P_m = \binom{n}{m} (\frac{A}{A_0})^m [1 - \frac{A}{A_0}]^{(n-m)} \tag{3.231}$$

On the assumption that the domain $\mathcal{D}_0$ extends over the whole plane and that $A_0 \to \infty$ when $n/A_0 \to \lambda$ or a positive constant, Santaló [225], gives the following limit:

$$\lim_n P_n = \frac{(\lambda A)^m}{m!} e^{-(\lambda A)} \tag{3.232}$$

indicating that the limit of P_n is a Poisson distribution that depends on the coefficient (λA). Hence this model represents points in the (x, y)-plane derived from a homogeneous planar Poisson point process with intensity λ. This distribution also signifies that the "number of points" in a certain region $\mathcal{D}$ of the plane is a subset of M (meso-domain) and is a random variable depending on A only and not on the shape or position of $\mathcal{D}$ within M. Such processes have been considered more generally by Miles [258] and in the context of "random tessellations" by Stoyan, Kendall, Mecke [236], Weibel [257], and Serra [235] among others.

In the stereological analysis of discrete media, by regarding their structural elements as geometrical objects, not only their dimensions (mostly volumes), surfaces or boundaries with adjacent spaces are required for their definition, but also other parameters such as shape, connectivity and topological characteristics. To obtain the necessary stereological quantities usually sectioning by means of "microtoming" is employed. This experimental procedure creates a surface by pushing a knife edge through the material samples as diagramatically shown in Fig. 27(c). The surface (cross-section) A, then reveals the 2D structural arrangement of the elements α, whereby the three-dimensional specimen is cut into two spatial regions (upper and lower part) of the sample and the surface separating them (Fig. 27(d)).

The carefully produced surface can then be subjected to observations in a transmission or scanning electron microscope which provide "micrographs". The latter are then evaluated by taking into account the statistics of the involved random variables by means of a fully automated procedure, ie. including scanners, digitizer and microcomputers. The sectioning of solid microstructures indicated in Fig. 27(a,b) can be performed by producing either one surface (cross-section) of a material specimen (often used in metallurgical investigations) and electron-microscopy, or two surfaces forming a thin slice or "probe". The latter is frequently required in biological stereology and the morphological analysis of tissues. The evaluation of observations is in general based on stereological formulae that relate the 2D-measurements to the 3D-geometrical

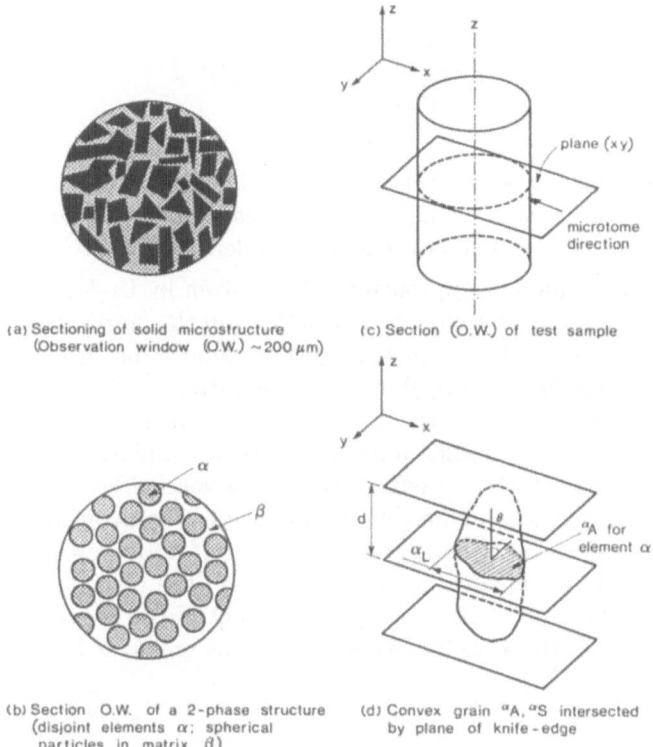

(a) Sectioning of solid microstructure
(Observation window (O.W.) ~ 200 μm)

(c) Section (O.W.) of test sample

(b) Section O.W. of a 2-phase structure
(disjoint elements α; spherical
particles in matrix β)

(d) Convex grain "A, "S intersected
by plane of knife-edge

Fig. 27. Sectioning of solid microstructures.

properties of the structures. From an experimental view point, it is often useful to carry out the "microtoming" not only in the direction of the (x, y)-plane but also perpendicular to it in order to obtain a sufficient number of the characteristics of an "element or cell α" of the structure such as the sectional area $^\alpha A$ Fig. 27(c,d), the enclosing surface $^\alpha S$, the perimeter $^\alpha L$, an intersect length $^\alpha \ell$, etc. according to the requirements of statistical analysis.

Conceptually it may be visualized that a 3D-lattice formed by such cutting planes having a spacing "d" (Fig. 27(d)) which contain a finite number of the randomly arranged convex grains α in a polycrystalline solid for example. Considering the random 2D-section $^\alpha A$ of an individual grain which has an enclosing surface $^\alpha S$, then the latter will be specified by the coordinate $z(0, d)$ of any fixed point on it and the orientation in terms of three angles briefly designated here by Ω. Hence an elemental volume of the grain will be given by:

$$d^\alpha V = {}^\alpha A(\Omega, z)dz$$

which by integration over these volumes gives the mean value of $^\alpha A$ as:

$$E\{{}^{\alpha}A\} =< {}^{\alpha}A >= \frac{\int\limits_{0}^{d} {}^{\alpha}A(\Omega, z)dz \int d\Omega}{\int d\Omega \int\limits_{0}^{d} dz} = {}^{\alpha}V/d \qquad (3.233)$$

which is valid for z varying between $(0, d)$. Similar consideration can be given to the intersect lengths ${}^{\alpha}\ell$ and sectional areas ${}^{\alpha}A$ so that an average of their ratio can be obtained. The latter is independent of the lattice spacing d.

A somewhat different approach has been given by Underwood, who uses an array of "test planes" in a cube of length "ℓ" and the "projections of the convex element" to one of the test planes. As a consequence the "mass projected area" of a microelement "α" averaged over all orientations is the same as the mean of the projected areas of the system of "identical randomly oriented" elements and further the fraction of parallel uniformly distributed planes or lines intersecting a convex microelement "α" in the test cube of volume $V_T = \ell^3$, is equal to the ratio of projected areas or heights of the element to projected area or height of V_T. Hence, a parallel array of test planes $((x, y)$-planes$)$ with total area $A_T = n_T \ell^2$, will penetrate the cube uniformly (the test planes being parallel to one side of the cube). If the projected height of the element is h_α and ℓ for the test cube and the two-dimensional fractions are $\phi_z = h_\alpha/\ell$ one has for the element α:

$$n_\alpha = \phi_z \cdot n_T$$

and the ratio is therefore given by:

$$\frac{n_\alpha}{A_T} = \frac{n_T - h_\alpha}{A_T \ell} = \frac{h_\alpha}{\ell^3} . \qquad (3.234)$$

The mean value of $< h_\alpha >$ can be determined for grains or microelements of simple shapes as:

$$< h_\alpha >= \int\limits_{(h_\alpha)_{min}}^{(h_\alpha)_{max}} h_\alpha P(h_\alpha)dh_\alpha \qquad (3.235)$$

in which $P(\cdot)dh_\alpha$ is the probability that a given microelement has a projected height h_α within $h_\alpha, h_\alpha + dh_\alpha$ on the projection plane.

Underwood has provided a valuable table of geometric properties of convex elements with known surfaces of revolution, that is very useful for an approximation of more complex shapes. Perhaps the oldest formula concerning the above introduced fractions goes back to Delesse [262], who found that the following ratios are equivalent:

$$\phi = \frac{{}^{\alpha}v}{V_T} = \frac{{}^{\alpha}A}{A_T} = \frac{{}^{\alpha}L}{L_T} = \frac{{}^{\alpha}N}{N_T} . \qquad (3.236)$$

where the suffices α refer to a convex element and T is the total volumen, area, lineal intersections etc. of the test sample. The two dimensional measurements or often carried out by using observation areas of an "observation window (O.W.)" as mentioned earlier, which has a specified size due to an achievable resolution in the electron-microscope and by neglecting edge effects (see also [302, 303]). The latter are discussed by Serra [235], Stoyan, Kendall and Mecke [236] and Underwood [261] among others. If $x_1, x_2 \ldots x_n$ observations are made the mean value is given by:

$$< x >= \frac{1}{n} \sum_{i=1}^{n} x_i \tag{3.237}$$

which is often replaced by the use of frequencies "f_i" or the number of observations having the same value, ie.:

$$< x >= \sum_{i=1}^{k} f_i x_i / \sum_{i=1}^{k} f_i. \tag{3.238}$$

Denoting the nubmer of the observations x_i by N, the variance is given by:

$$\sigma^2(x) = (\frac{1}{N-1})\{(\sum_{i=1}^{N} x_i^2) - N < x >^2\}. \tag{3.239}$$

Since variances have the additivity property any measurement quantity X is a function of n independent observations and thus:

$$X = \sum_{i=1}^{n} c_i x_i \; ; \; c_i = \text{const.}$$

Hence the variance of X is related to the x_i observations by:

$$\sigma^2(X) = D(X) = \sum_{i=1}^{n} c_i^2 \sigma^2(x_i). \tag{3.240}$$

It is to be noted that different experimental procedures may be adopted for the determination of the above mentioned ratios (see also [301, 303, 344]). It is therefore necessary to use rather a "relative value" of the error involved than the absolute one. Hence one can employ a quantity known as the "coefficient of variation" defined by:

$$(C.V.) = \sigma(x)/ < x > . \tag{3.241}$$

In the terminology of section (3.3), a planar closed set resulting from the intersection of a compact convex set X_V, with a test-plane (or plane of the knife-edge) will have a mean volume fraction ϕ_V given by:

$$< \phi_V >= E\{X_V \cap {}^\alpha v\} \to \int_{X_V} P\{{}^\alpha v \in X_V\} dv_\alpha \tag{3.242}$$

where v_α is the volume of a microelement α and similarly a mean area fraction ϕ_A pertaining to a cross section A_T of the specimen:

$$< \phi_A >= E\{X_{A_T} \cap {}^\alpha A\} = \int_{X_{A_T}} P\{{}^\alpha A \in X_{A_T}\} d^\alpha A . \tag{3.243}$$

In general the determination of the expected values of stereological quantities does not depend on the type of microstructre per se, ie. whether it is random, partially ordered or completely ordered (oriented). However, if the structure does not satisfy the requirement of randomness one has to introduce randomly placed and or randomly oriented "section or test planes" (hypothetical), with properties equal to the average or expected value of all possible sections (or lines) through the microstructure of a material sample. In the stereological analysis of discrete media it is assumed, that a sufficient number of specimen can be tested so that the sampling of specimen and the number of observations made on these samples are statistically meaningful.

In practice one often uses only one test-section, which may be sufficient for the evaluation of relevant parameters of many commonly occurring microstructures, if they exhibit a randomness in their characteristics and enough of these features appear on an observation plane consisting of a finite number of observation windows (O.W.) in the microscope observations. A more detailed stereological analysis is given in the references listed above. An application of geometric probabilities concerning the class of "fibrous structures" which are of considerable importance in engineering practice is given in Chapt. 4.

4 Applications of the Stochastic Analysis

4.1 The Response Behaviour of Discrete Solids:

The preceding chapters dealt with the basic concepts and principles of the stochastic mechanics of discrete media. They include the four postulates (P.1–4) of the theory and the axiomatic definitions of section 2.1 in Chapt. 2. The stochastic mechanics approach recognizes the involved field quantities as random variables or functions of such variables. It has been shown that the theory of Markov processes can be used to construct simple models for the representation of the behaviour of discrete media. The main objective of this chapter is to present the application of the stochastic analysis in the description of the response of certain classes of discrete media to external influences. Since most of the commonly used solids are polycrystalline, amorphous or composite, their microstructures are generally not in an equilibrium state, but evolve during a "deformation process". Hence the modelling by simple Markov processes may not be sufficient and a wider class of Markov process may be necessary for the description of the material behaviour. It is evident, that in general the description of the material behaviour at the micro level (atomic, molecular) must lead to the macroscopic properties of the discrete medium, if a realistic representation is to be achieved. Such a link in the stochastic mechanics theory is provided by the concept of a "meso-domain", a finite number of which forms the macroscopic material body and thus permits the use of an "intermediate length scale".

Another important notion is that of a material functional or operator M (Chapt. 2) that allows for a mapping between the space of "observable" to that of the "unobservable" quantities involved in the representation of the material behaviour. The former are usually "microdeformations u" or "microstrains ϵ" as elements of the deformation space $\mathcal{U}$ and strain space $\mathcal{E}$, respectively, where both of these spaces are subspaces of the probabilistic function space X. The unobserved quantities are the induced "microstresses $\sigma \in \Sigma \subset X$". All these quantities are understood in the generalized sense outlined in Chapt. 2. Due to more recent developments in the analysis of binary and multi-component structures, it may be beneficial to briefly review the stochastic deformation theory of previous work (see also refs. [23], [263], [264]).

(i) A general stochastic deformation theory:

Due to the inherent nature of the physical and configurational characteristics of a large class of discrete solids, the application of statistical mechanics and the principles of continuum mechanics still requires the use of some lattice construct so that the ensuing analysis of the material behaviour is carried out with reference to a network of elements (particles) that strongly idealizes a given microstructure. In contrast the stochastic mechanics approach adopts from the onset the concepts of "sets of microelements" and of an "abstract dynamical system" in representing the material for considerations of the deformation or flow of "discrete media". The notion of an abstract dynamical system has been conventionally used in the past in the analysis of "elastic stability" of continuous and discrete systems (see for instance Knops and Wilkes [265]). It will acquire in the present analysis a somewhat different meaning. For the development of a stochastic deformation theory of discrete solids the basic kinematic relations discussed earlier are important. Although these relations referred mainly to polycrystalline solids, they are quite general and can be used for other discrete structures. It has been shown, that the overall deformation of a microelement of the structure is a function of its "internal" and the "surface" deformations, which is regarded as a generalized quantity ie. $^{\alpha}\hat{u}(\mathbf{X}, t)$ (eqns. 3.68, 3.69). To simplify the symbolism the " $^{\wedge}$ " sign will be omitted in the subsequent discussion and hence the overall deformation of an element is expressed by:

$$^{\alpha}\mathbf{u}(\mathbf{X}, t) = f[^{\alpha}\mathbf{u}^i(\mathbf{X}, t), {}^{\alpha}\mathbf{u}^s(\mathbf{X}, t)] \equiv \text{ eqn. (3.68)}$$

in which $^{\alpha}\mathbf{u}^i, {}^{\alpha}\mathbf{u}^s$ in general are not continuous functions in the strict sense, but may be regarded to be at least piece-wise continuous and that derivatives of all orders within some compact support exist. In this sense $^{\alpha}\mathbf{u}(\mathbf{X}, t)$ can be generalized so that it will be a continous function within these supports. From a system theory point of view these deformations (see also ref. [23]) can be given as follows:

$$^{\alpha}\mathbf{u}(\mathbf{X}, t) = < L_1, \mathbf{u}^i(\mathbf{X}, t) > + < L_2, \ \mathbf{u}^s(\mathbf{X}, t) > \tag{4.1}$$

where L_1, L_2 are continuous operators within the compact support in which the microdeformations $\mathbf{u}^i, \mathbf{u}^s$ are defined and where $< ., . >$ denotes the inner product of the quantities in the brackets.

Thus the overall microdeformation $^{\alpha}\mathbf{u}$ can be interpreted as a function, the value of which is given by the sum of the inner products in eqn. (4.1). It is considered in the present analysis that the above form can always be established, although the operators L_1, L_2 may not be simple ones. Some explicit forms of the operators were given earlier in refs. [22, 23]. By considering deformations only, one has the following subspaces:

$$^{\alpha}\mathbf{u}^i \in U^i \; ; \; ^{\alpha}\mathbf{u}^s \in U^s \; ; \; U^i \oplus U^s = \mathcal{U} \subset X. \tag{4.2}$$

Since X has been shown earlier to be a measurable space, the deformation space $\mathcal{U}$ as well as the subspaces U^i, U^s, are also measurable. Hence a measure "$\mathcal{P}^u$" on $\mathcal{U}$ will have the following properties:

(i) $\;\; 0 \leq \mathcal{P}^u\{E_n\} \leq 1$ for all events $E_n \in \mathcal{F}^u, n = 1, 2 \ldots$

(ii) $\;\; \mathcal{P}^u\{E_1 \cup E_2\} = \mathcal{P}^u\{E_1\} + \mathcal{P}^u\{E_2\} - \mathcal{P}^u\{E_1 \cap E_2\},$

$\qquad$ if $E_1, E_2 \in \mathcal{F}^u$ and $E_1 \cap E_2 \neq \emptyset,$

(iii) $\;\; \mathcal{P}^u\{\cup_n E\} = \displaystyle\sum_n \mathcal{P}^u\{E_n\}$, if $E_n \cap E_m = \emptyset, n \neq m,$

$\qquad\qquad\qquad\qquad\qquad\qquad\qquad n, m = 1, 2 \ldots$

$$\left. \begin{array}{} \\ \\ \\ \\ \\ \\ \end{array} \right\} \tag{4.3}$$

(iv) $\mathcal{P}^u\{\mathcal{U}\} = 1.$

This measure is closely related to the distribution function of probability theory and thus is called the distribution function of the microdeformations $^{\alpha}\mathbf{u}$. The triplet $[\mathcal{U}, \mathcal{F}^u, \mathcal{P}^u]$ can be used as the probability space of all $\mathcal{P}^u$-regular measurable functions of $^{\alpha}\mathbf{u}$ and thus forms the basis of the mathematical structure of the general stochastic deformation theory. Moreover it leads directly by recognizing the duality with the stress-space $\sum$ (section (2.3), Chapt. 2) to the establishment of constitutive relations for discrete solids in an operational form. In this context the general state-space X may be thought of as being composed of the deformation space $\mathcal{U}$ and the stress-space $\sum$. In terms of system theory, the "input" to a microelement α could be regarded as an element of $\mathcal{U}$ and the "output" as an element of $\sum$. For a given physical system a mapping between $\mathcal{U}$ and $\sum$ will always exist, whereby the topological structure of one of the spaces, say $\sum$, may be given in terms of a non-degenerate bilinear form with respect to this mapping (see Moreau [135], Tonti [266]). The function space approach requires the use of a set of norms or semi-norms in the associated probability space. By considering $\mathcal{U}$ as the space of all $\mathcal{P}^u$-regular measurable and bounded functions of $^{\alpha}\mathbf{u}$, which are discrete, one can define the expected value of $^{\alpha}\mathbf{u}$ by:

$$E\{^{\alpha}\mathbf{u}\} = \sum {}^{\alpha}\mathbf{u} p\{E(^{\alpha}\mathbf{u})\} = \sum {}^{\alpha}\mathbf{u} \Delta \mathcal{P}^u \; ; \; \alpha = 1 \ldots N \; . \tag{4.4}$$

which is also the mean value of $^{\alpha}\mathbf{u}$ or $< {}^{\alpha}\mathbf{u} >$. If in an experiment the expected value of $\mathbf{u}$, which is representative of a "macrodeformation" in the conventional sense, ie. $< {}^{\alpha}\mathbf{u} >$ is equal to zero, although "microdeformations" may still exist, it is seen that $| < {}^{\alpha}\mathbf{u} > |$ or $|E\{^{\alpha}\mathbf{u}\}|$ is in fact a semi-norm and $\mathcal{U}$ a Fréchet space. If $\mathcal{U}$ is a linear topological vector space, it can be suitably topologized by a family of semi-norms and is then a locally convex space. On the other hand, if in an experiment one is concerned with finding the average value of the magnitude of $^{\alpha}\mathbf{u}$ a similar definition to (4.4) will satisfy the properties of a norm and $\mathcal{U}$ will be a "Banach space". The standard deviation of the discrete microdeformations can be expressed in terms of the measure $\mathcal{P}^u$ by:

$$\mathcal{D}^u = \{\sum |{}^\alpha\mathbf{u} - <{}^\alpha\mathbf{u}>|^2 \Delta\mathcal{P}^u\}^{1/2} \tag{4.5}$$

where $|\mathcal{D}^u|$ satisfies the properties of a semi-norm and $\mathcal{U}$ is a Fréchet space. Evidently, if higher order statistics of a random variable or random function are used other definitions of norms or semi-norms can be given. The deformation kinematics have already been considered in [22, 23] and will be extended somewhat in the section concerned with the elastic and relaxation behaviour of polycrystalline solids. The evolution of the occurring deformations is, however, of interest in a general deformation process. Thus one can consider a sequence of deformations $\mathbf{u}(\mathbf{X})$ indexed by the time t and denoted by $\mathbf{u}_t(X)$ or $\mathbf{u}(\mathbf{X}, t)$ as a "random process". The function $\mathbf{u}(\mathbf{X}, t)$ is a random function that for a fixed time $t \in \mathbb{R}^+$ is a random vector $\mathbf{u}(\mathbf{X})$ in $[\mathcal{U}, \mathcal{F}^u, \mathcal{P}^u]$.

Considering the probability space $[\mathcal{U}, \mathcal{F}^u, \mathcal{P}^u]$ for any particular time t, the whole deformation process will then be represented by a set of function spaces or a "product space" (see also Halmos [53], Hille and Phillips [149]). In particular, if $\mathbb{R}^+ = [0, \infty)$, where for each $t_r \in \mathbb{R}^+, r = 1, 2 \ldots N$ there corresponds a triplet $[\mathcal{U}, \mathcal{F}^u, \mathcal{P}^u]$, an N-fold product of these spaces leads to a "product space" in which $\mathbf{u}(\mathbf{X}, t)$ becomes a measurable function. For convenience this N-fold product space can be extended to infinity so that $\mathbf{u}(\mathbf{X}, t)$ may be regarded as a t-continuous random function in $[\mathcal{U}_\infty, \mathcal{F}^u_\infty, \mathcal{P}^u]$. Alternatively, one can consider $[\mathcal{U}, \mathcal{F}^u, \mathcal{P}^u]$ and a "one-parameter" family $\{T_t\}$ of transformations T_t such that:

$$T_t : \mathcal{U} \to \mathcal{U} \text{ for all } t \in \mathbb{R}^+ = [0, \infty); \ \mathbf{u}_t(\mathbf{X}) \in \mathcal{U}. \tag{4.6}$$

In this sense the deformation process is defined as a measurable function, ie. a random function which is generated by the random endomorphism $\{T_t\}$ for all $t \in \mathbb{R}^+$. The σ-algebra $\mathcal{F}^u_\infty$ of the product space has a countably finite number of Borel sets $E_1(\mathbf{u}), E_2(\mathbf{u}), \ldots E_N(\mathbf{u})$ corresponding to the time sequence $t_1, t_2 \ldots t_N \in \mathbb{R}^+$ and where each Borel set is obtained from the other by the automorphism or its inverse. If $t_1 < t_2 < \ldots t_N$ one has:

$$T_{\Delta t_r} E_r(\mathbf{u}) = E_{r+1}(\mathbf{u}); \ r = 1, 2 \ldots N - 1; \ \Delta t_r = t_{r+1} - t_r \tag{4.7}$$

It can be shown that, if $\mathcal{P}^u$ is a regular measure on E_r and T_t is defined so as to satisfy (4.7), E_{r+1} is also $\mathcal{P}^u$-regular measurable. One can generalize this statement to any Borel sets $E_r(\mathbf{u}); r = 1, 2 \ldots N$ in $\mathcal{F}^u_\infty$. Hence, it may be concluded that at any time during a deformation process the random deformations are $\mathcal{P}^u$-regular measurable. If the probability measure is independent of time as in the case of a purely elastic or a "steady-state" deformation process, which will be discussed subsequently, one can write that:

$$\mathcal{P}^u\{E_{r+1}(\mathbf{u})\} = \mathcal{P}^u\{E_r(\mathbf{u})\} \ ; \ r = 1, 2 \ldots N - 1 \tag{4.8}$$

which means that the deformation process is a strictly stationary random process (see for instance, Doob [45] and others).

From a measure theoretical point of view, if the Borel set $E_r \in \mathcal{F}_\infty^u$ at time $t_r \in \mathbb{R}^+$ is given, the probability measure on the set $E_s \in \mathcal{F}_\infty^u$ at time t_s is usually required. This can be accomplished by using the concept of a "conditional probability measure" introduced by Kolmogorov [10] and Rényi [41]. It is defined as an extended real-valued set function on $\mathcal{F}_\infty^u$ with properties analogous to those specified by relations (4.3). On this basis, it is evident that to each automorphism T_t there corresponds a conditional probability measure $\mathcal{P}\{E_{r+1}|E_r\}$ such that, whenever:

$$T_t E_r(\mathbf{u}) = E_{r+1}(\mathbf{u})$$
$$\mathcal{P}^u\{E_{r+1}\} = E_{r+1}|E_r\}\mathcal{P}^u\{E_r\}. \tag{4.9}$$

The above relation between the measures $\mathcal{P}^u\{E_{r+1}\}$ and $\mathcal{P}^u\{E_r\}$ can also be obtained by considering two Borel sets E_n, E_s for which:

$$E_n \cap E_s = E_s, \text{ if } E_n \supset E_s$$

and

$$\mathcal{P}^u\{E_n \cap E_s\} = \mathcal{P}^u\{E_s\}. \tag{4.10}$$

By inserting $(r+1)$ for s, it is seen that:

$$\mathcal{P}^u\{E_r \cap E_{r+1}\} = \mathcal{P}^u\{E_{r+1}\}.$$

Using Kolmogorov's definition of the conditional probability shows that:

$$\left. \begin{aligned} \mathcal{P}^u\{E_{r+1}|E_r\} &= \frac{\mathcal{P}^u\{E_{r+1} \cap E_r\}}{\mathcal{P}^u\{E_r\}} \\ \text{and} \\ \mathcal{P}^u\{E_{r+1}\} &= \mathcal{P}^u\{E_{r+1}|E_r\}\mathcal{P}^u\{E_r\} \equiv \text{eqn. (4.9)}. \end{aligned} \right\} \tag{4.11}$$

This relation is easily generalized to read:

$$\mathcal{P}^u\{E_n\} = \mathcal{P}^u\{E_1\} \prod_{r=1}^{n-1} \mathcal{P}^u\{E_{r+1}|E_r\} \tag{4.12}$$

which is valid for any sequence of events $E_1, E_2 \ldots E_N$ corresponding to the time sequence $t_1 < t_2 \ldots t_r < \ldots t_N$ and a set of conditional probabilities $\mathcal{P}^u\{E_{r+1}|E_r\}; r = 1, 2 \ldots N - 1$. The relations (4.11, 4.12) are very significant in the analysis of a deformation process since they permit the determination of the probability distribution of the microdeformations at any time, given the distribution at any other time or the initial one, is known. This can also be recognized by writing eqn. 4.10 in a more explicit form, ie:

$$\mathcal{P}^u\{E_{r+1}, t_{r+1}\} = \mathcal{P}^u\{E_{r+1}, t_{r+1}|E_r, t_r\}\mathcal{P}^u\{E_r, t_r\} \tag{4.13}$$

where E_r corresponds to $t_r \in \mathbb{R}^+$ and E_{r+1} to $t_{r+1} > t_r$. It is readily seen that for a deformation process in which $\mathcal{P}^u\{\cdot\}$ depends only on the time difference

Δt_r (eqn. 4.7), the sequence of deformations $\mathbf{u}_t(\mathbf{X}) \in [\mathcal{U}, \mathcal{F}^u, \mathcal{P}^u]$ together with the conditional probability measure describes a homogeneous stochastic process. Letting for simplicity $t_{r+1} = t$ and $t_r = 0$, eqn. (4.13) reduces to:

$$\mathcal{P}^u(t) = \mathcal{P}^u\{\mathbf{u}(t)\} = \mathcal{P}(t)\mathcal{P}^u(0) = \mathcal{P}(t)\mathcal{P}^u\{\mathbf{u}(0)\} \tag{4.14}$$

indicating the possibility of evaluating the distribution of $\mathbf{u}$ at time t from the knowledge of the initial distribution $\mathcal{P}^u(0)$.

In representing the deformational behaviour of a structured solid by a Markov process in the Banach space $X(\mathcal{U})$ of all $\mathcal{P}^u$-measures, the conditional probability $\mathcal{P}^u\{t_{r+1}, t_r\}$ can be shown to be a contraction operator on $X(\mathcal{U})$ tó $X(\mathcal{U})$, ie. $\| P(E) \| \leq 1$. In the theory of Markov processes this operator is known generally as a "transition probability". For instance considering a closed time interval $[t, s] \in \mathbb{R}^+$, subdividing it into smaller intervals and selecting a point $\tau > t, \tau \in [t, s]$, the transition probability satisfies the Chapman-Kolmogorov functional relation (see also eqn. 1.122, Chapt. 1), ie:

$$P\{t, s\} = \int_{\mathcal{U}} P\{t, \tau\} dP\{\tau, s\} \tag{4.15}$$

which for the homogeneous process due to the dependence of P on $\{t - s\}$ only, reduces to:

$$P\{t - s\} = \int_{\mathcal{U}} P\{t - \tau\} dP\{\tau - s\} \tag{4.16}$$

A discrete solid can pass through a countable number of "states" during a general deformation process as schematically indicated in Fig. 28.

Hence a Borel set related to the deformation space $\mathcal{U} \subset X$ at time t_r corresponds then to a state "i" and the Borel set at time t_s to another state "j". Evidently for fixed i, j the Chapman-Kolmogorov relation (4.15) can be written for all possible states i, j, as follows:

$$P_{ij}(t_r + t_s) = \sum_k P_{ik}(t_r)P_{kj}(t_s) \tag{4.17}$$

Since $P_{ij}(\cdot)$ is an element of the matrix $\underline{P}(t_r + t_s)$, eqn. (4.15) can be written in matrix form as:

$$\underline{P}(t_r + t_s) = \underline{P}(t)\underline{P}(s) \tag{4.18}$$

showing the semi-group property. The latter can be used to establish certain criteria for the miocrostructural stability during a general deformation process as discussed subsequently. It can be shown that the elements $P_{ij}(t)$ of the

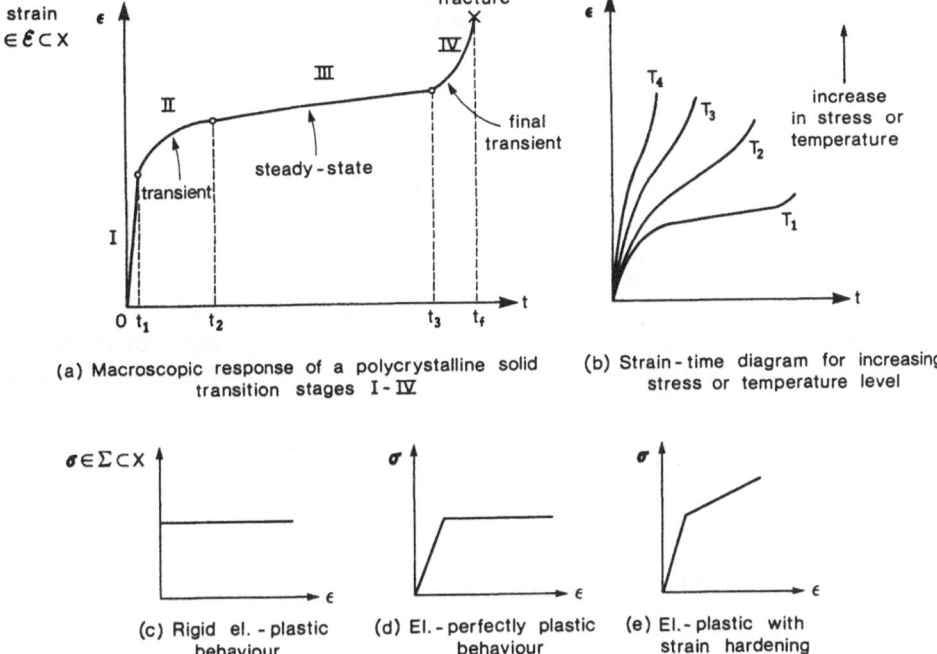

(a) Macroscopic response of a polycrystalline solid
transition stages I - IV

(b) Strain - time diagram for increasing
stress or temperature level

(c) Rigid el. - plastic
behaviour

(d) El. - perfectly plastic
behaviour

(e) El. - plastic with
strain hardening

Fig. 28. Phenomenological models of the elastic-plastic response and time-dependent behaviour of polycrystalline solids.

transition matrix in a time homogeneous process satisfy the limiting conditions given below. Thus for $(t, s) \in \mathbb{R}^+$, $P_{ij}(t)$ of such a process is a uniformly continuous function of time (see also Bharucha-Reid [44]) so that:

$$|P_{ij}(t + s) - P_{ij}(s)| \leq 1 - P_{ij}(t) \tag{4.19}$$

where the limits

$$\left.\begin{array}{l} \displaystyle\lim_{t\to 0^+} \frac{1 - P_{ii}(t)}{t} = -q_{ii} < \infty \\[3mm] \displaystyle\lim_{t\to 0^+} \frac{P_{ij}(t)}{t} = q_{ij} \end{array}\right\} \tag{4.20}$$

exist and are measurable for all $t \in \mathbb{R}^+$ so that

$$1 - P_{ij}(t) \leq 1 - e^{-q_{ij} \cdot t}. \tag{4.21}$$

In the above expressions the quantity q_{ij} is a measure of the relative transition or the intensity with which the deformation process $\mathbf{u}_t(\mathbf{X})$ progresses through time and is an element of the "transition intensity" matrix, ie. $q_{ij} \in \underline{Q}$. Using (4.21) one can express the inequality (4.19) as follows:

$$|P_{ij}(t+s) - P_{ij}(s)| \leq 1 - e^{-q_{ij} \cdot t}$$

so that for any $P(t)$:

$$\lim_{t \to 0^+} |\{\underline{P}(t+s) - \underline{P}(s)\}P(t)| = 0 \text{ for all } s \geq 0 \qquad \text{(a)}$$

or equivalently:

$$\lim_{t \to s}[\underline{P}(t) - \underline{P}(s)]P(t) = 0 \text{ for all } s \geq 0. \qquad \text{(b)}$$

(4.22)

In view of these properties of the transition probability, one can define the time-homogeneous process or correspondingly the "steady-state" deformation process, rigorously in the following manner:

(i) $\underline{P}(t)$ is a contraction on the space $X(\mathcal{U}) \to X(\mathcal{U})$,

$$\| \underline{P}(t) \| \leq 1$$

(ii) for all $(s,t) \in \mathbb{R}^+, s, t \geq 0$ and $E \in \mathcal{F}^u$ of $\mathcal{U}$ corresponds to t,

$$\lim_{t \to s}[\underline{P}(t) - \underline{P}(s)]\mathcal{P}\{E\} = 0$$

(iii) for all $(s,t) \in \mathbb{R}^+$

$$\underline{P}(t+s) = \underline{P}(t)\underline{P}(s) \text{ with } \underline{P}(0) = \underline{I} \text{ (identity matrix)}.$$

(4.23)

In general for each stage of the deformation as indicated in Fig. 28(a) the Chapman-Kolmogorov relation will hold, but the semi-group property for the transition probabilities may not be satisfied. Writing eqn. (4.15) in matrix form:

$$\frac{d\underline{P}(t,s)}{dt} = \underline{Q}(t)\underline{P}(t,s) \; ; \; \underline{P}(0) = \underline{I} \qquad (4.24)$$

yields the key equation in the functional analytic formulation of the deformation of discrete solids. It represents an "evolution equation", the solution of which will be discussed further in this chapter. In closing this section, it should be noted that the stochastic and topological structure of deformations have been considered without reference to the topology and measures on the associated "stress-space" $\sum \subset X$. However, since there exists an intrinsic relation between these spaces, measures chosen a priori on the deformation space will restrict the choice of measures on the stress-space. This is due to the required mapping between the spaces for which a measurable transformation and the property of invertibility of the material operator (see P.4 of Chapt. 2) is needed.

(ii) Deformational stability of structured solids:

In continuation of the stochastic deformation theory of discrete solids, discussed above, one can distinguish various stages during a general deformation process. Such a process is schematically shown for a polycrystalline solid, when subjected to a constant load in Fig. 28(a). This figure indicates further the phenomenological models of the "elastic-plastic" response as well as the time-dependent behaviour of polycrystalline solids. In particular the significance of the involved transition probabilities with regard to the stages of deformation shown in Fig. 28(a) will be investigated. In continuum physics the behaviour of "conservative systems" is usually represented by hyperbolic type differential equations which are associated with group transformations T_t on X with a domain $-\infty < t < \infty$. Dissipative systems however, correspond to parabolic type differential equations that only induce semi-groups $\{T_t\}$ defined for $t > 0$.

The elastic stability of dynamical systems both for the continuous and discrete case have been investigated as mentioned earlier by Knops and Wilkes [265]. It has been shown by Kolmogorov [10] that under fairly broad assumptions, the transition probabilities defined earlier can be obtained from certain parabolic differential equations that are concerned with the evolution of field variables of a physical system. The problem of integration of such equations has been widely investigated, mostly in terms of an operator formalism (see also [129]). Yosida's treatment [24] considers a field variable as an element of the Banach space X depending on the real parameter t. The solution of the corresponding evolution equation is then given by an operator $A(t)$, that in general is unbounded and has a domain $\mathcal{D}(A(t))$ and a range $\mathcal{R}(A(t))$ both in X. Similar considerations of evolution equations and their solutions are due to Mizohata [267], when field variables may have arbitrary initial values, but in which the operator $A(t)$ is bounded. From the properties of the transition probabilities $P(t)$ and the matrix differential equation (4.24), it follows that in the case of purely elastic deformations, the stochastic deformation process is time-independent and thus corresponds to a stationary random process. Under the application of a constant load (Fig. 28(a)) the elastic deformations belong to a subset $U(0^+)$ of the more general product space $[\mathcal{U}_\infty, \mathcal{F}^u_\infty, \mathcal{P}^u]$ on which the measures generated by the transition probabilities (4.9) are such that

$$|\mathcal{P}^u\{t + s\} - \mathcal{P}^u\{t\}| \to 0 \text{ for } \forall s, t \in \mathbb{R}^+ \tag{4.25}$$

which implies

$$\mathcal{P}^u\{t + s\} \stackrel{\text{a.e.}}{=} \mathcal{P}^u\{t\} \tag{4.26}$$

and where $\mathcal{P}^u\{E\}$ is time-independent. Hence the following properties of the transition probabilities will hold:

(i) the transition probability $\mathcal{P}^u(t, t + \Delta t)$ of the process $\mathbf{u}(t)$ changing from state i to another state j in the interval Δt is zero.

(ii) $\mathcal{P}^u(t, t + \Delta t)$ of no change in the interval Δt is equal to 1.

In view of these statements the Kolmogorov relation (4.24) reduces to:

$$\frac{dP^u}{dt} = 0, \text{ since } \underline{P}^u = \text{ constant matrix; } \underline{Q} = 0 \qquad (4.27)$$

indicating that the elements $q_{ij}(t) \in \underline{Q}(t)$ are time-independent infinitesimal constants, ie.:

$$q_{ij} = 0 \text{ for } i = 0, 1 \dots ; \quad q_{ij} = 0 \text{ for } j = i + 1 \qquad (4.28)$$

Equation (4.27) can be solved for an initial value of $\underline{P}(0) = \underline{I}$ resulting in:

$$\underline{P}^u(t) = \underline{I} \text{ or } P_{ij}^u(t) = \delta_{ij} ; \quad \text{for } 0 < t < t_1 \qquad (4.29)$$

where t_1 corresponds to the time-instant (Fig. 28a), at which the purely elastic response of the solid ends and a first transient with the associated microstructural changes occurs. The above result suggests, that the probability distribution of the random microdeformations occuring within the reversible or elastic response of the discrete solid under a constant load application at $(0, t_1)$ remains constant. This result is of considerable interest for a large class of discrete solids the initial distribution of which can be experimentally determined by observation of the material morphology and subsequent stereological evaluation. It is consistent with the response characteristics concerning the next stages of the deformational behaviour (Fig. 28(a)) to be discussed in the following. It may also be noticed from this figure that an increase of the stress or temperature level causes a shortening of the elastic response and a prolonged "transient" and "steady-state" deformation stage (Fig. 28b). The transient stage beginning at approximately time t_1 of the "strain-time" diagram has already been briefly discussed in the section on the "random evolution" of discrete solids (Chapt. 3) (see also [166]), where it was shown that in general the simple Markov process is inadequate for the description of the material response and the associated changes of the given microstructure. It will be further discussed in paragraph (iii) of this section. Considering next the "steady-state" deformations, the properties of the transition probabilities have been rigorously defined in (4.23). It is of interest to employ the transition probabilities and the corresponding transition intensity matrix to represent the evolution of the distributin $\mathcal{P}^u\{\mathbf{u}(t)\}$. During the steady-state deformation process the random vector $\mathbf{u}(t)$ changes from one state to another with a certain "intensity λ" which is directly related to the elements of the transition matrix. Its significance with respect to the "microstructural stability" during deformations will be shown subsequently. The "states" a discrete solid can pass through are again identified by the index of the event set or the Borel set to which $\mathbf{u}(t)$ belongs at the corresponding instant of time. Thus, if within a small time interval Δt, $\mathbf{u}(t)$ changes from an initial state 0 to a state 3 for instance, the intensity of this change will be given precisely by the element q_{03} of the matrix $\underline{Q}$. It is assumed however that "Δt" is small enough so that a change of $\mathbf{u}(t)$ is permitted from state $i \to j$ only, when $j = i + 1$ and $i = 0, 1, 2 \ldots$ Hence, one can state the following characteristics for a "steady-state deformation process:

(i) the intensity with which the random deformation $\mathbf{u}(t)$ changes from a state i to its adjacent state j in the interval Δt is given by: $\lambda \Delta t + O(\Delta t)$, where λ is a positive constant and $O(\Delta t)$ is the order of magnitude of Δt,

(ii) the intensity with which $\mathbf{u}(t)$ does **not** change from a state i to a state j in Δt is: $(1 - \lambda \Delta t - O(\Delta t))$,

(iii) the intensity with which $\mathbf{u}(t)$ changes from a state i to any other state except j, $(j = i + 1)$ in the interval Δt is $O(\Delta t)$.

It is apparent from the above criteria that the elements of the intensity matrix $\underline{Q}$ for the "steady-state" deformation process can be expressed by:

$$\left. \begin{array}{l} q_{ii} = -\lambda \text{ for } i = 0, 1, 2 \ldots \\ q_{ij} = \begin{cases} \lambda & \text{for } j = i + 1 \\ 0 & \text{otherwise.} \end{cases} \end{array} \right\} \quad (4.30)$$

The above form of the intensity matrix $\underline{Q}$ suggests that the "steady-state" deformation process by allowing the transition from a state $i \to j$ in one step only ($\lambda =$const.), can be represented by a Poisson process. Thus, if from experimental observations the macroscopic "strain-time" relations (Fig. 28a) are available, where $\epsilon \in \mathcal{E}$ (strain-space contained in X), the evolution of the strain field with time in terms of a Poisson process is expressed by the following differential equations:

$$\frac{dP_{ij}^\epsilon(t)}{dt} = -\lambda P_{ij}^\epsilon(t) + \lambda P_{i,j-1}^\epsilon(t) ; \quad t_1 \le t < \infty. \tag{4.31}$$

Disregarding any interaction effects, thermal noise etc. between the structural elements of the discrete solid and on the assumption of a "stable" transition mechanism, the general solution of (4.31) can be obtained by means of an iteration scheme so that:

$$P_{ij}^\epsilon(t) = \begin{cases} \frac{[\lambda(t-t_1)]^{(j-i)}}{(j-i)!} e^{-\lambda(t-t_1)} & \text{for } j \ge i, \ t_1 \le t < \infty \\ 0 & \text{for } j < i \end{cases} \left. \right\} \tag{4.32}$$

and for $j > i$ one also obtains:

$$\left. \begin{array}{l} 0 \le e^{-\lambda(t-t_1)} \le 1 ; \ t_1 \le t < \infty \\ \text{and } \dfrac{[\lambda(t - t_1)]^{(j-i)}}{(j - i)!} \ge 0 ; \ t_1 \le t < \infty \end{array} \right\} \tag{4.33}$$

showing that $P_{ij}^\epsilon(t) \ge 0$ for $j \ge i$ and $P_{ij}^\epsilon(t) = 0, j \le i$. Hence $P_{ij}^\epsilon(t)$ is the contraction operator $\| \underline{P}^\epsilon(t) \| \le 1$. The notion of the intensity factor λ

introduced above is rather important in stochastic mechanics since it indicates the degree of transition from one state of the material to another and hence, whether the microstructure remains "stable" or not during a particular stage of deformation. Some further remarks on this parameter should be made here. In the theory of Markov processes, if the initial state in a time homogeneous process (steady-state and isothermal conditions) is known, the change from this state to another may occur at any random instant of time. The time elapsed until $u(t)$ enters a new state is called the "waiting time τ". It is easily shown that the random variable τ has an exponential distribution with the parameter λ as a non-negative constant, ie. $F(\tau) = e^{-\lambda t}$.

The value of $\lambda = \infty$ is not excluded. If the intensity $\lambda = 0$, the process $u(t)$ remains in the same state. However, if $\lambda = \infty$, the process will leave the state instantaneously (absorbing state). For any intermediate value of λ, the transition intensity q_{ij} will be therefore:

$$\left.\begin{aligned} 0 \leq q_{ij} < \infty \quad \text{for } i \neq j \\ q_{ij} = q_{ii} = -q_i \quad \text{for } i = j \end{aligned}\right\} \quad (4.34)$$

which implies in accordance with (4.30) that for $0 \leq \lambda < \infty$ any state i will be a stable one. Once the system is in a stable state, it remains there with probability one for a positive period of time. If the system has no instantaneous states, ie. all transition intensities are finite, a necessary and sufficient condition for a stable transition mechanism is given by [42]:

$$\sum_{n=1}^{\infty} \frac{1}{\lambda_{u(\tau_n)}} = \infty \text{ with probability 1} \qquad (4.35)$$

where "τ_n" is that instant of time of leaving the state: $i_n = u_n(\tau_{n-1})$. By the use of these criteria for the intensity, one can state that the steady-state deformation process will be represented by the Kolmogorov differential equation in the form of:

$$\frac{dP_{ij}^u(t)}{dt} = -\lambda P_{ij}^u(t) + \lambda P_{i,j-1}^u(t) \; ; \; t_2 \leq t < \infty \qquad (4.36)$$

where "t_2" is the instant of time when the steady-state begins. In general this time will coincide with the time instant of the "upper limit" for the purely elastic material behaviour. Under circumstances such as "creep deformations", for instance, t_2 refers however to the upper limit of the transient stage occurring within a fixed time interval (t_1, t_2) (Fig. 28a). The combination of the reversible and steady-stages in terms of the intensities λ in (4.28) and (4.30) together with equation (4.24) forms an equivalent representation to the simple phenomenological model of the behaviour known as the elastic-plastic response under a constant load as indicated in Fig. 28(c–e). It is seen that Fig. 28(c) corresponds to the perfectly plastic response, whilst (d) represents an elastic-perfectly plastic solid and (e) the case of an elastic-plastic material with "strain-hardening". In these phenomenological models the change of the material from a reversible to an irreversible state is represented only by a single point. In actual discrete solids however this change can only occur within a finite period of time allowing for a "transient" period to take place. This period of the transient material behaviour will be discussed further in the section (iii).

For the condition of creep deformations (Fig. 28a) as mentioned before the "transient state" will be generally of a longer duration and can be represented by a Kolmogorov type differential equation. By the use of an iteration scheme (see also [268]), the general solution of (4.36) has been obtained as follows:

$$\left.\begin{array}{ll} P_{ij}(t) = \dfrac{[\lambda(t - t_2)]^{(j-i+1)}}{(j - i + 1)!} e^{-\lambda(t-t_2)} & \text{for } j \geq i, t_2 \leq t < \infty \\[2mm] \qquad\quad = 0, & \text{for } j < i. \end{array}\right\} \quad (4.37)$$

For $j \geq 1$ one obtains the solution shown earlier concerning the strain-distribution, (4.32, 4.33), ie. for the upper limit t_2 as follows:

$$\left.\begin{array}{ll} 0 \leq e^{-\lambda(t-t_2)} \leq 1 \; ; \; t_2 \leq t < \infty & \text{(a)} \\[2mm] \text{and similarly} & \\[2mm] P^u_{ij}(t) = \dfrac{[\lambda(t - t_2)]^{)j-i)}}{(j - i)!} \geq 0 \; ; \; t_2 \leq t < \infty. & \text{(b)} \end{array}\right\} \quad (4.38)$$

Using the above relations, the one-step transition probability can be expressed by:

$$P_{i,i+1}(t) = e^{-\lambda(t-t_2)} \; ; \; j = i + 1, \; t_2 \leq t < \infty. \qquad (4.39)$$

Employing this transition probability permits the assessment of the distribution of microdeformations during the steady-state from successive experimental observations, since

$$\mathcal{P}^u(t) = e^{-\lambda(t-t_2)} \mathcal{P}^u(t_2) \qquad (4.40)$$

as a consequence of the relations between $\mathcal{P}^u(t)$ and $\mathcal{P}^u(t_2)$ given earlier (4.14). During the steady-state stage the deformations are stable in the sense that λ satisfies the condition in (4.38a). This stage remains stable so long as the intensity $\lambda \equiv \lambda_s$ (steady-state) is smaller than unity.

From the point of view of "microstructural" stability, the transient stages indicated in Fig. (28a) are more significant. The first transient is characteristically associated with a rearrangement of the microstructure, whereby interaction effects between elements may reach extreme magnitudes. This behaviour is indicative of "yielding" in a ductile metallic structure. In a phenomenological theory, it is usually represented in terms of a yield function or a yield point on the corresponding stress-strain diagram (Fig. 28(c–e)). For a somewhat longer duration of this transient stage as, for instance, during creep deformations, the microstructure tends to stabilize towards the steady-state deformation. Hence according to stochastic mechanics theory, one may regard this type of behaviour as a "limiting process" connecting the end of the reversible with the beginning of the "steady-state" deformation process. The requirements for such a limiting process in the representation of the transient stage, ie. for the transition from the elastic to the inelastic or irreversible deformations, can be stated as follows:

(i) the intensity with which the deformation or strain process changes from state $i \to j$ in a small time interval Δt is:

$$\frac{\lambda t_2(t - t_1)}{t(t_2 - t_1)} \Delta t + O(\Delta t) ; \quad t_1 \leq t \leq t_2$$

(ii) the intensity with which the process u_t or ϵ_t changes from $i \to j$ except $j, j = i + 1$ in the interval Δt is $O(\Delta t)$.

(iii) the intensity with which u_t or ϵ_t does not change from the state i to a state j in Δt is:

$$1 - \frac{\lambda t_2(t - t_1)}{t(t_2 - t_1)} \Delta t - O(\Delta t) ; \quad t_1 \leq t \leq t_2.$$

It is seen from the above criteria, that the elements of the transition intensity matrix $\underline{Q}$ can be given as follows:

$$\left.\begin{aligned}
q_{ij}(t) &= -\frac{\lambda t_2(t - t_1)}{t(t_2 - t_1)} \quad \text{for } i = 0, 1 \ldots \\
q_{ij}(t) &= \frac{\lambda t_2(t - t_1)}{t(t_2 - t_1)} \quad \text{for } j = i + 1 \\
&= 0 \qquad\qquad\qquad \text{otherwise.}
\end{aligned}\right\} \quad (4.41)$$

Using two new parameters as follows:

$$a = \frac{\lambda t_1 t_2}{t_2 - t_1} \; ; \; b = \frac{\lambda t_2}{t_2 - t_1}$$

yields upon substituting them into (4.41):

$$\left.\begin{aligned}
q_{ii}(t) &= -b + a/t \text{ for } i = 0, 1 \ldots \\
q_{ij}(t) &= b - a/t \text{for } j = i + 1 \\
&= 0 \qquad\qquad \text{otherwise.}
\end{aligned}\right\} \quad (4.42)$$

Hence, one can express the "evolution equation for $P_{ij}(t)$" during the transient stage of the strain field $\epsilon \in \mathcal{E} \subset X$ (ie. if the experimental observations are the induced strains), as follows:

$$\frac{dP_{ij}^{\epsilon}(t)}{dt} = -bP_{ij}^{\epsilon}(t) + \frac{a}{t}P_{ij}^{\epsilon}(t) + bP_{i,j-1}^{\epsilon}(t) - \frac{a}{t}P_{i,j-1}^{\epsilon}(t), \tag{4.43}$$

$$\text{for } t_1 \leq t \leq t_2$$

The solution of this Kolmogorov differential equation is given by:

$$\left. \begin{aligned} P_{ij}^{\epsilon}(t) &= \frac{[b(t-t_1) - a\ln(t/t_1)]^{(j-i)}}{(j-i)!}(t/t_1)^a e^{-b(t-t_1)}; \quad j \geq i \\ &= 0, \qquad\qquad\qquad\qquad\qquad\qquad\qquad\qquad \text{for } j < i \end{aligned} \right\} \tag{4.44}$$

showing again that $P_{ij}^{\epsilon}(t)$ gives a proper measure for all i,j and is a contraction, i.e. $\| P_{ij}^{\epsilon}(t) \| \leq 1$.

(ii) The inelastic behaviour of multi-component solids:

To illustrate the "limiting process" above, the stochastic analysis is now applied to a Multi-component or MC-system (high-temperature resistant composite) shown previously in Fig. (7b) of Chapter 2. This material is essentially a binary structure consisting of a dispersion of Tungsten particles (2-4 μm) in a Cobalt matrix and belongs to the class of high-temperature composites like BN, SiN_4, $SiC(\alpha,\beta)$ etc. Samples of these structures were tested in the Micromechanics Lab. up to 2000° C. Micrographs ($\times 1500$) by the use of transmission and scanning-electron microscopy were produced for the onset and successive stages during this transiency. The results of some of the tests are given in Fig. 29.

The transient stage referred to as stage II in Fig. (28a) is shown here for a $WC\text{-}Co$ structure in Fig. (29). This material has been tested (curve (a)) similar to those illustrated by Fig. (28) at a temperature of 1250° C under a compression load of 7.2 ton/in^2. The required strain measurements were obtained at time intervals of 5–10 min. The experimental data compared with the theoretical results obtained from an evaluation of the general Poisson equation (4.31) and its solution given in (4.32) are shown in Fig. 30. It may be of interest to note that the stochastic analysis of the transition from the elastic to the inelastic response of a "multi-component or MC-system" can also be carried out by the use of "partially observed Markov jump processes" that have been discussed earlier in Chapt. 3 and in ref. [282]. Alternatively the transiency has been studied more recently in terms of "finite point" processes, as will be shown in a subsequent section.

The stochastic analysis of the second type of transiency or the "final stage" of the material response (Fig. 28a) requires however a more general formulation since it is associated with the "break down" of the microstructure and its ultimate failure. In the conventional representation of this final stage of

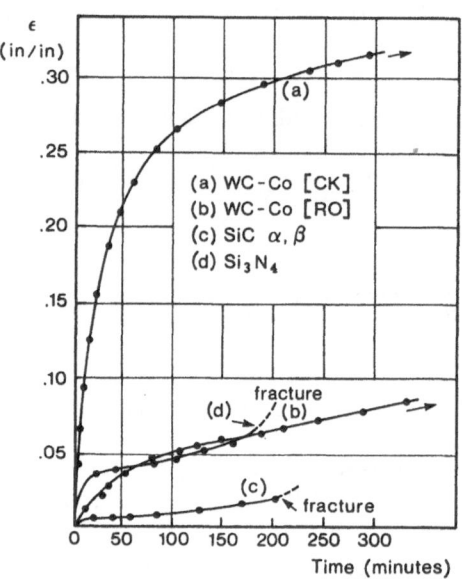

Fig. 29. Strain-time diagrams of MC-systems.

the response, the concepts of "fracture mechanics" are usually employed. This rapidly expanding area of research in solid mechanics is solely concerned with the "crack initiation" and propagation in discrete solids and has only been recently approached on the basis of stochastic mechanics (see refs. [23], [201]). In accordance with the concepts of stochastic mechanics, it may be conjectured that the description of this final transiency should be based on a more general or secondary stochastic process in which second or higher order transition probabilities would be involved. Such a process is then a function of the stochastic deformation process $\mathbf{u}(t)$ or $\phi\{\mathbf{u}(t)\}$. Whilst in general such a process may not satisfy the Markov property or the Kolmogorov relation, there are nevertheless certain conditions that make it possible to satisfy the latter. Since the random deformation process has been defined by $\mathbf{u}_t(\mathbf{X}) \in [\mathcal{U}_\infty, \mathcal{F}_\infty^u, \mathcal{P}^u]$ in which $\mathcal{U}_\infty$ is the ∞-fold product of $\mathcal{U}$ and $\mathcal{F}_\infty$ is the ∞-fold product of $\mathcal{F}$, the "final stage process $\phi\{\mathbf{u}(t)\}$" will be such that $\mathbf{u}(t) \in \mathcal{U}_\infty \backslash \mathcal{U}_s$. Hence, it will be valid only in the subspace $\mathcal{U}_\infty \backslash \mathcal{U}_s$. If $\mathcal{U}_s$ is the steady-state deformation space or the space of $\mathbf{u}(t)$ prior to the final stage, then $\mathcal{U}_\infty \backslash \mathcal{U}_s$ is associated with $\phi\{\mathbf{u}(t)\}$ or the breakdown of the microstructure. Hence it may be called the "fracture-space $\mathcal{U}_f \subset X$" of the discrete solid. It is evident, that an in depth study of this final stage of deformation would be extensive and lies beyond the scope of the

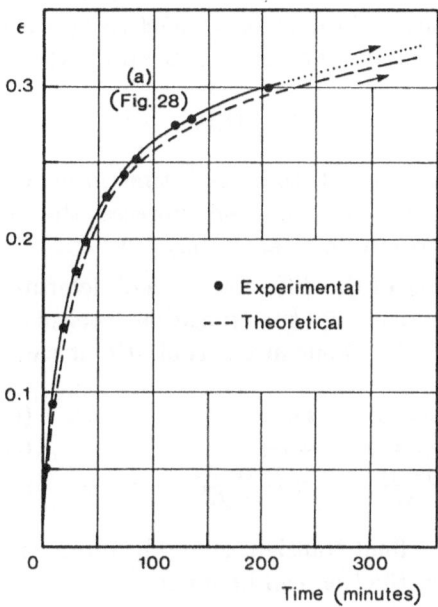

Fig. 30. Comparison of experimental observations with theoretical results (Eqn. 4.32)

present analysis. It may be stated however, that this deformation stage can also be visualized as a "growth process" with regard to the developing cracks and their propagation through time within the microstructure, the representation of which may be carried out by means of an "exponential Markov chain" analogous to the treatment of growth processes given in refs. [269].

The significance of the transition intensities in the representation of the various stages of the general deformation process has become apparent. It has been shown with regard to the transient material behaviour that in a general stochastic deformation process two types of transient stages may be distinguished. The first type shows the tendency of $q_{ij}(t) \rightarrow q_{ij}|_s$ as the variable time $t \rightarrow t_s$, where the latter is the instant of time when the steady-state deformations are first observed. In accordance with the inequality stated in (4.21) and on the assumption that the transition probability $P_{ij}(t)$ is t-continuous on $X(\mathcal{U})$ for all $(t, s) \in \mathbb{R}^+$ a limit will be reached such that:

$$\lim_{t \rightarrow s}[\underline{P}(t) - \underline{P}(s)]\mathcal{P}^u(0) \rightarrow 0 \text{ for all } s > 0. \tag{4.45}$$

The above characteristic will prevail during the steady-state deformation stage until a time instant $t < t_f$ is reached. The time t_f corresponds to the beginning of the second transiency, ie. the onset of the microstructural breakdown. It is of interest to note that the stable stage preceded by the first transient can also be represented analytically in form of a Cauchy sequence $\{P(t_n), n = 1, 2 \ldots\}$ in $L(0, 1)$ for which the following convergence property must be required:

$$|P_{ij}(t_{n+1}) - P_{ij}(t_n)| < \epsilon, \text{ for n} > N(\epsilon) \tag{4.46}$$

in which $0 < \epsilon < 1$ corresponds to a small time interval. By comparison with eqn. (4.21), it is seen that this condition expresses the same requirement for a stable deformation to occur, ie. that $0 < q_{ij} \leq 1$.

The distinct features of the different types of deformation can be described by the two basic quantities, ie. the transition intensity matrix $Q(t)$ and the transition probability $P(t)$. Thus in terms of $Q(t)$ it can be stated that:

$$Q(t) \rightarrow \left\{ \begin{array}{ll} q_{ii} = 0, \ q_{ij} = 0, j = i + 1 & \text{(elastic)} \\ q_{ii} = -\lambda, q_{ij} = \lambda, \ j = i + 1 & \text{(steady-state)} \\ q_{ii} = \frac{-\lambda t_2(t-t_1)}{t(t_2-t_1)}, \ q_{ij} = \frac{\lambda t_2(t-t_1)}{t(t_2-t_1)}, \ j = i+1 & \text{(transient)} \end{array} \right\} \tag{4.47}$$

by the exclusion of the final transient stage. In terms of the transition probabilities the following distinction can be made:

$$\left. \begin{array}{l} P_{ij}(t) = \delta_{ij} \ , \ \ P(t)P(s) = P(t+s) \ ; \ \text{(elastic)} \\ P_{ij}(t) = f(t) \ , \ \ P(t)P(s) = P(t+s) \ ; \ \text{(steady-state)} \\ P_{ij}(t) = g(t) \ , \ \ P(t)P(s) \neq P(t+s) \ ; \ \text{(transient)} \end{array} \right\} \tag{4.48}$$

in which the form of the functions $f(t)$ and $g(t)$ are given by equations (4.37) and (4.44), respectively.

From the above formulated general stochastic deformation theory of discrete solids and the considerations given to the microstructural stability the following conclusions may be drawn:

(i) A random deformation process of discrete solids in a probability space $[\mathcal{U}, \mathcal{F}^u, \mathcal{P}^u]$ is a sequence of the random variables $\{u(X, t)\}$ or $\{u_t\}$ each member of this sequence being generated from the previous one by a random endomorphism.

(ii) The general stochastic deformation process can be subdivided into two "stable stages" that correspond to the purely elastic and steady-state response of the solid. The intermediate and final stage in the deformation process are "transient", whereby the former tends to approach the conditions for stability of the microstructure and the latter is rather unstable leading to the ultimate failure of the material.

(iii) The most important characteristic in the representation of the process is the transition probability and the associated transition intensity. The transition probabilities may be regarded as the kernels of evolution operator on the deformation space as a subspace of X.

(iv) For stability to exist the transition probabilities must satisfy the semi-group property. This property is not satisfied for the two transient stages mentioned above.

The division of "stability zones" in the response of a discrete solid finds an immediate application in the analysis of crack propagation in metals.

(iv) General remarks on material operators (constitutive maps):

The general stochastic deformation theory of discrete solids developed above, were concerned with considerations of the change of the probability measure in the deformation space $\mathcal{U}$ or the strain space $\mathcal{E}$ of X, respectively. Such considerations also apply in the stress space $\sum \subset X$, provided that the random process in that space is assumed to be Markovian. Thus, for instance, for two Borel sets E_r, E_{r+1} in $\mathcal{F}^\sigma$ (σ-algebra of the stress space) corresponding to the time instants, t_r, t_{r+1} on the real line $\mathbb{R}^+$, one can express the probability distribution or measure on $\sum$ as follows:

$$\mathcal{P}^\sigma\{E_{r+1}, t_{r+1}\} = \mathcal{P}^\sigma\{E_{r+1}, t_{r+1}|E_r, t_r\}\mathcal{P}^\sigma\{E_r, t_r\}. \tag{4.49}$$

Assuming that the stochastic process is homogeneous, one obtains analogously to eqn. (4.14):

$$\mathcal{P}^\sigma\{t\} = \mathcal{P}\{\sigma(t)\} = \underline{P}^\sigma(t)\mathcal{P}\{\sigma(0)\} = \underline{P}^\sigma(t)\mathcal{P}^\sigma\{0\} \tag{4.50}$$

where the measure $\mathcal{P}^\sigma(t)$ can be identified with the transition probability of the Markov process in the stress-space resulting in the matrix Kolmogorov differential equation in the form of:

$$\frac{d}{dt}\underline{P}^\sigma(t) = \underline{Q}^\sigma \underline{P}^\sigma(t); \ \underline{P}^\sigma(0) = \underline{I}. \tag{4.51}$$

Evidently in order to establish "constitutive maps" or the relations between the deformation and stress-space a connection between the equations (4.14) and (4.50) is needed. This connection is supplied by the notion of a "material functional" or operator as defined in Chapt. 2. Hence in accordance with the fundamental concepts of stochastic mechanics, the material operators generally replace the conventional relations known as "constitutive laws" in continuum mechanics. These operators contain those stochastic variables or functions of the variables, which are representative of a specific discrete medium, and can be constructed from the knowledge of their distributions due to experimental measurements. In this manner a material operator aims at a realistic representation of the material characteristics of a given discrete medium.

From a mathematical point of view a "micro operator" is an operator that refers to a structural element α ; $(\alpha = 1 \ldots N)$ of the discrete solid and is a mapping between the stress and deformation space (see definitions in section (2.3(iii))) such that:

$$^\alpha m\colon \Sigma \to \mathcal{U} \;;\; ^\alpha m^{-1}\colon \mathcal{U} \to \Sigma \;;\; \sigma \in \Sigma \,,\; \mathbf{u} \in \mathcal{U},$$

where in general $^\alpha m$ may be non-linear. More specifically, considering the micro stress $^\alpha \sigma \in \Sigma$ and the micro deformation $^\alpha \mathbf{u} \in \mathcal{U}$ for a domain $\mathcal{D} \subset \Sigma$ and a range $\mathcal{R} \subset \mathcal{U}$ the following mapping can be given:

$$^\alpha \mathbf{u}(t) = {}^\alpha m[^\alpha \sigma(t)] \;;\; (\alpha = 1 \ldots N). \tag{4.52}$$

In order to obtain the necessary macroscopic relations from the above microscopic ones, particularly by observing the introduced intermediate length scale (meso-domain) the definitions of the corresponding operators, ie. micro and macro-operators must be observed. Thus following the Def. 3 of sect. 2.3(iii) Chapt. 2, the meso operator $^M m(t)$ for instance is determined by the average of the micro-operators or:

$$^M m(t) = E\{^\alpha m(t)\} = < {}^\alpha m(t) > = \sum_1^N {}^\alpha m(t) p\{^\alpha m(t)\} \tag{4.53}$$

where $p\{^\alpha m(t)\}$ is the probability density of $^\alpha m(t)$. It has been shown in earlier work [22], [23] that $^\alpha m(t)$ consists of two linear transform operators, ie.:

$$^\alpha m(t) = m\{^\alpha A(t),\, ^{\alpha\beta} B(t)\} \tag{4.54}$$

in which the transformation $^\alpha A\colon {}^\alpha \Sigma \to {}^\alpha U$; $^{\alpha\beta} B\colon {}^{\alpha\beta} \Sigma \to {}^{\alpha\beta} U$ refer to the mapping of the microstress $^\alpha \sigma$ within each microelement α and the generalized stress $^{\alpha\beta} \sigma$ at the boundary of an element to the associated deformation subspaces $^\alpha U$ and $^{\alpha\beta} U$, respectively. However in this representation the problem of "invertibility" arises as mentioned before in Chapt. 2. Thus, if the operator $^\alpha m$ is "strictly monotone" for some dense subset of the stress and deformation space designated here by Σ_d and $\mathcal{U}_d$, respectively, one has:

$$
\begin{aligned}
&^\alpha \mathbf{u} = {}^\alpha m\{^\alpha \sigma(t)\}, \text{ or by the inverse} &&\text{(a)} \\
&^\alpha \sigma(t) = {}^\alpha m^{-1}\{^\alpha \mathbf{u}(t)\}, &&\text{(b)}
\end{aligned}
\left.\vphantom{\begin{aligned}&\\&\end{aligned}}\right\} \tag{4.55}
$$

where $^\alpha \sigma \in \Sigma_d$ and $^\alpha \mathbf{u} \in \mathcal{U}_d$. It can be shown, that a non-degenerate linear form $< \cdot, \cdot >$ between the elements of Σ_d and $\mathcal{U}_d$ with respect to the operator $^\alpha m$ can always be established, if $< [m^k(\sigma) - m(^\ell \sigma)], [^k \sigma - {}^\ell \sigma] >$ is greater than zero and where $^k \sigma, {}^\ell \sigma$ are two specific values of the microstress $^\alpha \sigma \in \Sigma_d$. Hence this operator is "strictly monotone" and invertible (Tonti [266], Dolezal [142]). For the application of the general stochastic deformation theory however a weaker condition is sufficient, ie. that of strong monotonicity, but where the invertibility still holds. By considering the form (4.54) of the micro-operator, its distribution over a meso-domain can also be written as:

$$\mathcal{P}\{{}^{\alpha}m(t)\} = f[\mathcal{P}\{{}^{\alpha}A(t), {}^{\alpha\beta}B(t)\}] \tag{4.56}$$

where explicit forms of the function $f[.,.]$ will be discussed subsequently by considering the elastic response and the mechanical relaxation behaviour of polycrystalline solids. It may be seen from the form in (4.55) that the second relation is an inverse mapping from the deformation space to the stress-space, e.g. that the probability measure on the stress-space is linked to that on the deformation space by the material operator at the appropriate scale, or more precisely by its measure. In particular, considering the inverse operators and the micro deformations as statistically independent random variables generally the following relation between the distributions will exist:

$$\mathcal{P}\{\boldsymbol{\sigma}(t)\} = \mathcal{P}\{m^{-1}(t)\}\mathcal{P}\{\mathbf{u}(t)\} \tag{4.57}$$

where for simplification of the notation the superscripts have been omitted with the understanding that relation (4.57) holds for the three measuring scales of the stochastic mechanics theory and is compatible with the given definitions. The concepts of statistical independence and that of the "product measure" (eqn. 4.57) together with the assumption of the random processes in the deformation and stress-space to be Markovian leads to differential equations representing the "evolution of the response" of discrete solids. It involves however the distribution of the inverse operator or $\mathcal{P}\{m^{-1}(t)\}$. The solution of such differential equations may not always be possible in a closed form, but may give an insight into the probability distribution of the microscopic or local operator ${}^{\alpha}m(t)$ at any time during the deformation process. The notion of this operator permits further the inclusion of interaction effects between structural element which otherwise cannot be achieved. It is to be noted that the outlined procedure may not hold generally, if, for instance, the probabilistic and topological structure of the stress-space is unknown. In that case on the basis of relation (4.57) and a linearity assumption for the inverse operator, the construction of a "non-degenerate bilinear form" with respect to the inverse operator is still possible, which then includes a proper topological structure of the stress-space $\sum$ (see for instance Moreau [135]). In continuum mechanics the relevant field quantities associated with the deformational behaviour of "homogeneous" or idealized solids are specified in terms of "point functions". The resulting mathematical relations are then considered as "field equations". Such equations together with the "constitutive laws" and the stress equations of motion are unified in the description of the motion of a "material point" of the continuum. Thus the outstanding characteristic of continuum models is the consideration of point functions. This notion is however of no significance in stochastic mechanics, since the latter regards the field variables as random variables and their measures, which are "set functions". Furthermore, due to the basic concept that the general deformation process is representable in form of a stochastic process of a special type (Markov, Poisson, etc.) the corresponding formulations is mainly concerned with the "evolution of the probability distributions" or measures and the resulting differential equations can be regarded as "governing equations" of the discrete medium. Such governing equations consist rather of a set of differential equations involving the probability measures of the relevant stochastic variables. It can be readily seen with reference to equations (4.14) and (4.50) that the following relation can be obtained:

$$\underline{P}^{\sigma}(t)\mathcal{P}\{\boldsymbol{\sigma}(0)\} = \mathcal{P}\{m^{-1}(t)\}\underline{P}^{u}(t)\mathcal{P}^{u}\{\mathbf{u}(0)\} \tag{4.58}$$

involving the distribution of the material operator $m^{-1}(t)$. In order that equation (4.57) be satisfied at all time instants, one obtains for $t = 0$, the following initial relation:

$$\mathcal{P}\{\boldsymbol{\sigma}(0)\} = \mathcal{P}\{m^{-1}(0)\}\mathcal{P}\{\mathbf{u}(0)\} \tag{4.59}$$

which together with (4.58) yields:

$$\underline{P}^{\sigma}(t)\mathcal{P}\{m^{-1}(0)\}\mathcal{P}\{\mathbf{u}(0)\} = \mathcal{P}\{m^{-1}(t)\}\underline{P}^{u}(t)\mathcal{P}\{\mathbf{u}(0)\} \tag{4.60}$$

Thus, if the initial distribution $\mathcal{P}\{\mathbf{u}(0)\}$ is specified by means of some experimental observations and has a non-zero value, it may be concluded that:

$$\underline{P}^{\sigma}(t)\mathcal{P}\{m^{-1}(0)\} = \mathcal{P}\{m^{-1}(t)\}\underline{P}^{u}(t) \tag{4.61}$$

which relates the probability distribution of the operator $m^{-1}(t)$ at any time t to that at $t = 0$, in terms of the transition probabilities $\underline{P}^{\sigma}(t)$ and $\underline{P}^{u}(t)$. In the terminology of the stochastic mechanics theory the above relation (4.61) can be regarded as one form of the governing equations in terms of the probabilities $\underline{P}^{u}$ and $\underline{P}^{\sigma}$. To illustrate this remark consider the "steady-state" deformation process and the one-step transition probability $\underline{P}^{u}(t)$ given earlier (eqn. 4.40). If, in this case the time instant t_2 is considered as the "initial" one, ie. if $t_2 = 0$ and $\underline{P}(t_2)$ is equal to $\underline{I}$, one has:

$$\mathcal{P}\{m^{-1}(t)\} = e^{\lambda t}\mathcal{P}\{m^{-1}(0)\}. \tag{4.62}$$

In the more general case, e.g. when the two transition probabilities are arbitrary functions of time one can differentiate (4.61) to give:

$$\mathcal{P}\{m^{-1}(0)\}\frac{d\underline{P}^{\sigma}(t)}{dt} = \frac{d\mathcal{P}\{m^{-1}(t)\}}{dt}\underline{P}^{u}(t) + \mathcal{P}\{m^{-1}(t)\}\frac{d\underline{P}^{u}(t)}{dt}. \tag{4.63}$$

This relation is quite general, since it includes all the possible stages of the deformational behaviour of a discrete solid with the exclusion of the final stage, ie. the breakdown of its microstructure leading to fracture. By considering elastic deformations only, the above relation reduces to:

$$\frac{dP\{m^{-1}(t)\}}{dt} = 0 \tag{4.64}$$

since both transition probabilities $\underline{P}^u(t)$ and $\underline{P}^\sigma(t)$ can be regarded as constant due to the fact that during an elastic deformation, the material characteristics are fairly constant. Similarly it can be shown that in the steady-state deformations, relation (4.63) reduces to:

$$\frac{dP\{m^{-1}(t)\}}{dt} - \lambda P\{m^{-1}(t)\} = 0 \tag{4.65}$$

which has the solution given in eqn. (4.62). Thus relation (4.63) can be looked upon as the key governing equation of a discrete solid. If the transition probabilities $\underline{P}^u(t), \underline{P}^\sigma(t)$ are given in terms of the earlier Kolmogorov relations one can also write for (4.63) the following expression:

$$P\{m^{-1}(0)\}\underline{Q}^\sigma(t)\underline{P}^\sigma(t) = \frac{dP\{m^{-1}(t)\}}{dt}\underline{P}^u(t) + P\{m^{-1}(t)\}\underline{Q}^u(t)\underline{P}^u(t) \tag{4.66}$$

Hence it may be concluded that, if the probabilistic structure of the deformation and stress-space are well defined, then by an appropriate choice of the stochastic process representing the deformation process for a given discrete solid as a whole or part of it, the evolution of the probability measures on these spaces can be formulated. Since both these spaces are related to each other by the constitutive mapping or the material operators that in themselves contain stochastic parameters, the evolution of the measure on $\mathcal{U}$ and Σ is coupled to that of the operator. This view is expressed above in the key governing equations for a discrete solid (eqn. 4.61). In certain problems, it may become necessary to include "body forces" in the deformation analysis. In this case, one can advantageously use the representation of the deformational behaviour as discussed here and that part of the material operator, which permits the inclusion of body forces on the same scale as the effects caused by a bonding potential.

Since the mapping between the deformation and stress-space is intimately related to a given "boundary value" problem, the significance of the micro, meso and macro-material operators should be emphasized further. In order to show the application of these operators in the classical forms of the boundary value problems, it may be recalled from the theory of elasticity that:

(i) the Dirichlet problem states:

given a linear operator m on the deformation space $\mathcal{U}$ (Hilbert space), a specified function (body force) $\mathbf{f} \in F$ and a prescribed deformation on the boundary $\mathbf{g}$, then the problem consists of finding the deformation $\mathbf{u} \in \mathcal{U}$ such that:

$$^{\mathcal{M}}m\mathbf{u} = \mathbf{f} \text{ in the macrodomain “}\mathcal{M}\text{” of the solid} \qquad (4.67)$$

$$\text{and } \mathbf{u} = \mathbf{g} \text{ on the boundary “}\partial\mathcal{M}\text{”} \qquad (4.68)$$

(ii) the Neumann problem states:

given a linear operator $^{\mathcal{M}}m$ on $\mathcal{U}$ (Hilbert-space), a specified function (body force) $\mathbf{f} \in F$ and a prescribed normal derivative $\mathbf{n} \in \partial\mathcal{M}$, the problem consists of finding the deformation $\mathbf{u} \in \mathcal{U}_m$ such that:

$$^{\mathcal{M}}m\mathbf{u} = \mathbf{f} \text{ in “}\mathcal{M}\text{” (macrodomain)} \qquad (4.69)$$

$$\text{and } \frac{\partial\mathbf{u}}{\partial\mathbf{n}} = \mathbf{s} \text{ in “}\partial\mathcal{M}\text{” (boundary of } \mathcal{M}) \qquad (4.70)$$

Evidently the operator $^{\mathcal{M}}m$ in both these problems in a deterministic formulation has its counterpart in the stochastic theory.

To obtain the probabilistic forms one may consider for simplicity of the analysis the discrete medium as a "continuous random medium" on the assumption that:

(i) the discrete structure can be replaced by a "continuous random structure".

(ii) the physical domain of an ensemble of microelements in the actual medium to be an equivalent domain in the random continuum provided the statistical properties are preserved.

(iii) the relevant field quantities can be taken as in the classical case to be attached to the "points x_i of the random continuum", $x_i \in X$.

As a consequence by the use of the "mass points" $\mathbf{x}_i \in X$, the stress-equations of motion will be given by:

(elasto-static case): $\sigma_{ij,j} + \rho(\mathbf{x})f_i = 0 \qquad (4.71)$

(elasto-dynamic case): $\sigma_{ij,i} + \rho(\mathbf{x})f_i = \dfrac{\partial^2 u_i}{\partial t^2} \qquad (4.72)$

where $\mathbf{x}_i \in X$, f_i are the body force vectors, σ_{ij} the Cauchy-stress tensor and the mass density $\rho(\mathbf{x})$ is a random variable. These forms are also used in the discretization of the random continuous medium, when $x_i \in \mathbb{R}^d$ (lattice construct). In this context, a similar form has been considered by Stein, Zhang and König [270] in the micro mechanical modelling and simulation of the elastic-inelastic behaviour of structured solids. In their work an "overlay" model has been considered in which each material point $\mathbf{x}$ has been assumed to be composed of a dense spectrum of "subelements" indexed by a scalar variable ($\xi = \{0, 1\}$). In this model the stresses and strains are then defined separately for the material points $\mathbf{x}$ as well as for the subelements and where the macro stress field satisfies eqn (4.71).

Using the notion of the material operator $\mathcal{M}m$ in the phenomenological sense one has:

$$\mathcal{M}m(\mathbf{x}) : \Sigma \to \mathcal{U}_m : \mathcal{M}m(\mathbf{x}) \in \mathcal{L}(\Sigma, \mathcal{U}) \tag{4.73}$$

so that

$$\mathcal{M}m(\mathbf{x})\sigma(\mathbf{x}) = \mathbf{u}(\mathbf{x}) \; ; \; \mathbf{x} \in \mathcal{M} \subset X. \tag{4.74}$$

Thus, if $\mathcal{M}m(\mathbf{x})$ denotes a bounded linear operator for the macroscopic purely elastic response of the random structure and where its inverse exists, one has:

$$\mathcal{M}m^{-1}(\mathbf{x}) = E(\mathbf{x}) \cdot \mathrm{grad} = E_{ijk\ell}\frac{\partial}{\partial x_i} \tag{4.75}$$

in which $E_{ijk\ell}$ is the standard modulus of elasticity of the solid. By inserting the above relation (4.75) into (4.73) one obtains the operational form of the deformation equations of motion as:

$$\mathcal{M}m(\mathbf{x}) + \rho(\mathbf{x})\mathbf{f} = 0 \tag{4.76}$$

where

$$\mathcal{M}m(\mathbf{x}) = \mathrm{div}(\underline{E} \cdot \mathrm{grad}) = \frac{\partial}{\partial x_i}(E_{ijk\ell}\frac{\partial}{\partial x_i}). \tag{4.77}$$

However, to formulate the problem (i) and (ii) stated earlier, it is necessary to introduce the following function spaces:

(i) $\mathcal{U}_m = \{\mathbf{u} \in \mathcal{U}; \mathcal{M}m(\mathbf{x})\mathbf{u} \in F\}$; $F \subset X$

(ii) $\partial\mathcal{U}$ = space of boundary values of $\mathbf{u}$ at the boundary $\partial\mathcal{M}$ of the macro domain $\mathcal{M}$.

(iii) ∂G = space of boundary values for $\partial u/\partial n$ at $\partial\mathcal{M}$, $\mathbf{n}$ = outer normal to the boundary surface.

(iv) ∂G^* = the dual of space ∂G.

Hence the stochastic forms of problems (i) and (ii) can be stated as follows:

(i) the Dirichlet problem for $\mathcal{M}m(\mathbf{x})$:

given $\mathbf{f} \in F \subset X$ and $\mathbf{g} \in \partial\mathcal{U}$, find $\mathbf{u} \in \mathcal{U}_m$ such that:

$$\mathcal{M}m(\mathbf{x})\mathbf{u} = \mathbf{f} \; ; \; \mathbf{u}|_{\partial\mathcal{M}} = \mathbf{g} \tag{4.78}$$

(ii) the Neumann problem for $\mathcal{M}m(\mathbf{x})$:

given $\mathbf{f} \in F \subset X$ and $\mathbf{s} \in \partial G^*$, find $\mathbf{u} \in \mathcal{U}_m$ such that:

$$^{\mathcal{M}}m(\mathbf{x})\mathbf{u} = \mathbf{f} \; ; \; \frac{\partial \mathbf{u}}{\partial \mathbf{n}}|_{\partial\mathcal{M}} = \mathbf{s}. \tag{4.79}$$

Thus relations (4.78, 4.79) represent the stochastic forms of the classical ones, in the theory of elasticity. However, if the discrete solid exhibits a complete random structure, a generalization of the random operators and corresponding relations may be required. In this case an approach by the use of "random operators" on Banach spaces may be necessary (see also Bharucha-Reid [129, 271], Kannan [272], Adomian [273]).

4.2 The Response of Polycrystalline Solids:

A large and important class of discrete media are the polycrystalline solids shown earlier in Fig. 17 (undeformed configuration). This figure also indicates the arrangement of the individual crystals or grains in the kinematic subspace $K \subset X$. The corresponding deformation kinematics were considered in earlier work [22, 23]. Polycrystalline solids consist of grains the lattices of which are often imperfect due to defects such as dislocations, the presence of vacancies, impurity atoms that frequently penetrate the lattice structure, etc. The mechanical response of a discrete solid is in general different from, although depending upon, the response of its individual crystals. Hence, it is essential to correlate the response of the aggregate with the response of the individual elements of the structure. An approach of this kind leads then to a modification of the conventional "phenomenological laws" and to a possible elimination of undesirable modes of deformations through the design of new materials. It is evident however, that without some simplifying assumptions these modes cannot be identified due to the randomness of the structure of a real polycrystalline solid. So far as the purely elastic response of a discrete solid is concerned, two effects are briefly discussed below, namely that of the occuring dislocations within the lattice of each element and the influence of existent grain boundaries of a given polycrystalline solid.

(i) The elastic response including interactions:

To illustrate the application of the stochastic mechanics theory discussed so far, it may be convenient to consider first the dislocation effects on the crystals themselves and then to analyse the grain boundary influences on the overall response. The analytical model of the former has been given in ref. [175] under rather restrictive assumptions, that:

(i) The crystal matrix whether free from or containing imperfections responds to an applied stress according to the conventional constitutive equations of a "linear isotropic elastic" material. These relations, as is well-known, contain two independent material characteristics, ie. the shear modulus "G" and the Poisson ratio "ν".

(ii) the density of line dislocations $^{\alpha}\rho_d$ (unit length per unit volume) in the αth crystal or micro element is inversely proportional to its "average radius $^{\alpha}\overline{r}$". This density is taken as:

$$^{\alpha}\rho_d \equiv \rho_d : \rho_d = \frac{6}{|\mathbf{b}|^{\alpha}\overline{r}}$$

in which $\mathbf{b}$ is the Bürger vector (see also [274]).

(iii) the effect of fluctuations in the line dislocation density on the mechanical response of the solid is neglected.

(iv) a ratio "κ" of the line dislocations within a crystal is used. A typical value of this quantity for a representative crystal of average radius $^{\alpha}\overline{r} = 10^5|\mathbf{b}|$ is taken to be 10^{-4}.

(v) Out of the large number of possible slip systems within the αth crystal there is only one slip direction, ie.:

$$^{\alpha}O_{\beta j} \; ; \; \beta = 2, 3$$

for the motion of mobile dislocation, where $^{\alpha}O_{\beta j}$ represents the direction cosines between the internal frame $^{\alpha}Y_{\beta}$ and the external frame j. Thus $^{\alpha}Y_2$ is the direction of the bow-out of the inactivated Frank-Read source [274], while $^{\alpha}Y_3$ designates the normal to the "slip-plane" (see Fig. 31).

(vi) The direction cosines $\underline{O}$ have a random distribution over the crystals within a meso domain M.

(vii) dislocations only interact with an externally applied stress field and not among themselves.

(viii) and finally the overall deformation of a micro element is considered completely reversible.

The above assumptions indicate that this model used for establishing the linear response characteristics of a polycrystalline solid visualizes a large number of closed-packed crystals of equal volumes ($^{\alpha}v \; ; \; \alpha = 1 \ldots N$). Each micro element or crystal in turn contains the same density "ρ_d" of anchored semi-mobile dislocation segments (see Fig. 31 and refs. [175, 264]), which can bow-out in one preferred slip direction on one slip-plane, both of which vary from crystal to crystal in a random fashion. Contributions due to the interaction between dislocation segments are neglected.

Within the framework of these assumptions, one can formulate an "internal micro stress $^{\alpha}\sigma^{(i)}$" as the components of the overall micro stress $^{\alpha}\sigma \in \Sigma \subset X$ and an induced internal micro strain "$^{\alpha}\epsilon^{(i)}$" corresponding to other components

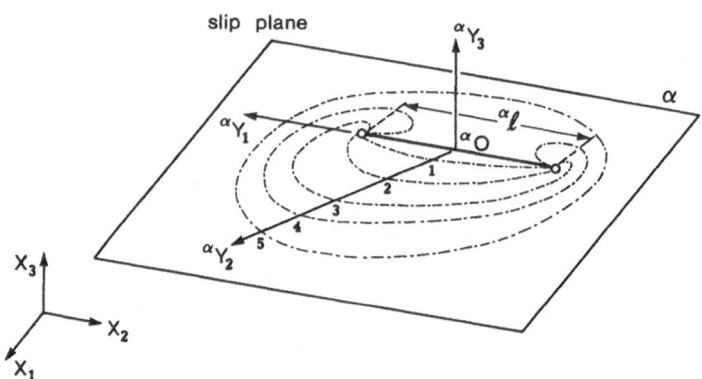

Fig. 31. Dislocation effect in a polycrystalline solid.

of the state-vector ${}^\alpha \mathbf{z}_i$ of the micro element. By using the primitive form of the elastic stress-strain relations, ie.:

$$ {}^\alpha \sigma = {}^\alpha A {}^\alpha \epsilon \tag{4.80} $$

in which the operator ${}^\alpha A$ is identified with the elastic modulus ${}^\alpha E$ of the micro element α. Hence the effect of dislocations on the basis of the above model modifies this relation so that:

$$ \left. \begin{array}{l} {}^\alpha \sigma = {}^\alpha A_D {}^\alpha \epsilon \\ \text{and} \\ {}^\alpha A_D : {}^\alpha E + \eta {}^\alpha \Gamma, \end{array} \right\} \tag{4.81} $$

where the quantity "$\eta {}^\alpha \Gamma$" accounts for the presence of dislocations. It is well-known from continuum mechanics that ${}^\alpha E$ (in the usual tensor notation), is given by:

$$ {}^\alpha E_{ijk\ell} = \frac{2G\nu}{1 - 2\nu} \delta_{ij} \delta_{k\ell} + 2G \delta_{ik} \delta_{j\ell} \tag{4.82} $$

where G is the shear modulus of the pure crystalline solid, "ν" its Poisson ratio and δ the Kronecker delta. Thus the operator ${}^\alpha A_D$ can be written as:

$$ {}^\alpha A_{ijk\ell} = \frac{2G\nu}{1 - 2\nu} \delta_{ij} \delta_{k\ell} + 2G_{ik} \delta_{j\ell} + \eta {}^\alpha \Gamma_{ijk\ell}. \tag{4.83} $$

where the last term on the right-hand side is given by:

$$\eta = \frac{2}{3\pi} G(1-\nu)^{\alpha}\ell(\kappa\rho_d) \qquad \text{(a)}$$
$$^{\alpha}\Gamma_{ijk\ell} = {}^{\alpha}O_{3i}{}^{\alpha}O_{2j}{}^{\alpha}O_{3k}{}^{\alpha}O_{2\ell} \qquad \text{(b)}$$

$$\left.\right\} \text{(4.84)}$$

describing the influence on the deformation by the dislocations anchored by the "characteristic lenght $^{\alpha}\ell$" between pinning points of an edge dislocation on a slip plane the normal of which has the direction cosines $^{\alpha}O_{3j}$ and where "$\kappa\rho_d$" is the mobile dislocation density. The coordinate transformation $^{\alpha}\Gamma$ suitably combines the transformation vector $^{\alpha}O_{3i}$ and $^{\alpha}O_{2j}$ between the local and external reference frames. The quantity η stands for the work done by the force:

$$|^{\alpha}\mathbf{f}| = Gb^{\alpha}\epsilon_{32} \qquad (4.85)$$

which acts on a dislocation segment of length $^{\alpha}\ell$ to displace it for the distance $^{\alpha}\overline{Y}_2$, multiplied by the number "κ" of mobile dislocation segments in the αth crystal (see also ref. [175]). Of greater interest however is the extension of this approximation to a larger material domain. Thus the total energy ^{M}U stored in a meso-domain M due to an applied load is equal to the sum of the strain energy in the elastic matrix and the increase in the self-energy of the anchored dislocation segments in each crystal summed over "$1\dots N$" micro elements so that:

$$^{M}U = \frac{1}{2}\sum_{\alpha=1_{\alpha}}^{N}\int_{v} {}^{\alpha}A_D \cdot {}^{\alpha}\epsilon^{\alpha}\epsilon d^{\alpha}v \qquad (4.86)$$

where $^{\alpha}A_D$ is given by (4.81), and $^{\alpha}v$ is the volume of a single crystal. Evidently by differentiating the integral in (4.85) with respect to $(^{\alpha}\epsilon^{\alpha}v)$ gives the microstress $^{\alpha}\sigma$ in the αth crystal in accordance with (4.81), ie.:

$$^{\alpha}\sigma = {}^{\alpha}A_D{}^{\alpha}\epsilon : [E + \eta^{\alpha}\Gamma]^{\alpha}\epsilon \qquad (4.87)$$

expressing the "local" constitutive relation with the inclusion of the dislocation effect. Since in the stochastic mechanics theory the microstrains $^{\alpha}\epsilon$ are random variables they also depend in this case on the "orientation $\underline{O}$" of the single crystal. For a more detailed discussion reference is made to [23, 175, 275].

It has been mentioned previously, that for the inclusion of "grain boundary" effects in the analysis of the deformational behaviour of polycrystalline solids a probabilistic "surface molecular coincidence lattice" model can be used. The schematics of a typical "undeformed coincidence or unit cell" in the grain surface has been shown in Fig. 18(a, b) (Chapt. 2) and will be further discussed here. In general, the surface between any two crystals α, β will take up such a position that the crystals are "idealized" to form two interpenetrating mathematical translation lattices disregarding the space filling atoms. On the assumption that lattice 1 is fixed and lattice 2 is changing (Fig. 18b), i.e. undergoing a translation and rotation, the latter will be translated in such a way that one point coincides with a point of lattice 1. This point is called the "lattice coincidence site" (see Bollman [177]). Due to the periodicity of the two lattices, a finite number of such points are brought about. The number of points form another lattice termed "coincidence lattice". A measure of the coincidence lattice is given by the surface of its "unit cells" which in turn depend on the lattice parameter "a" of the given crystal and the relative orientation. A somewhat different view has been taken by Goux et al. [178], who considers the two crystals to be separated by an "amorphous" layer of a certain thickness and who uses a minimum energy principle in its representation. However the "interfacial energy" only partially characterizes the grain boundary, since other phenomena must also be considered. It is apparent, that the model used in the stochastic mechanics theory is a combination of Bollman's coincidence lattice site and Goux's amorphous layer model. Since the present model considers the orientations $^{\alpha}\underline{Q}, {}^{\beta}\underline{Q}$ as random parameters, the coincidences areas designated by $^{\alpha}\Delta^{q}S = {}^{\beta}\Delta^{q}S$ (see also Fig. 18(b)) where $q = 1 \dots p$, and p =number of coincidence cells in the surface $^{\alpha}S, {}^{\beta}S$ respectively, will be random functions of the combination of the orientations $^{\alpha}\underline{Q}, {}^{\beta}\underline{Q}$. Their expected values and correlations can be obtained from the distributions by means of Bollman's and Goux's method. Thus by following the deformation kinematics discussed earlier, a significant correlation is the "distance" between the C.M. of two adjacent crystals, ie. $\nu = {}^{\beta}\mathbf{X} - {}^{\alpha}\mathbf{X}$, where the latter are the position vectors with respect to an external reference frame. One obtains therefore the mean value and correlation function for the orientation for instance, as:

$$
\begin{aligned}
E\{^{\alpha}\underline{Q}\} &=< {}^{\alpha}\underline{Q} >_{N} ; \ (\alpha = 1 \dots N) &\text{(a)} \\
R_0\{\nu\} &=< {}^{\alpha}\underline{Q}(^{\alpha}\mathbf{X})^{\alpha}\underline{Q}(^{\alpha}\mathbf{X} + \nu) >_{N} &\text{(b)}
\end{aligned}
\left.\right\} \quad (4.88)
$$

and for the "grain size" measured in terms of the elementary volume $^{\alpha}v$, as:

$$
\begin{aligned}
E\{v\} &=< {}^{\alpha}v >_{N} ; \ (\alpha = 1 \dots N) &\text{(a)} \\
R_v(v) &=< {}^{\alpha}v(^{\alpha}\mathbf{X})^{\alpha}v(^{\alpha}\mathbf{X} + \nu) >_{N} &\text{(b)}
\end{aligned}
\left.\right\} \quad (4.89)
$$

Since "R" in the above relations is considered as a non-isotropic, statistically homogeneous correlation function of the relevant quantity, one can establish in this approximation, the number "p" of coincidence cells and the corresponding coincidence cell areas $^{\alpha}\Delta^{q}S$; $(q = 1 \dots p)$ for the polycrystalline solid.

In general, the "binding energy" existing in the interface between crystals α, β can be formulated as shown earlier in form of a "potential" by means of the distance vector $\mathbf{d}$, i.e. $\phi\{|^{\alpha\beta}\mathbf{d}|\}$. This potential may be of the central force type, parabolic, etc. Similar to the internal energy U of a crystal, the specific form of ϕ is assumed to be known and will have a minimum value when $|^{\alpha\beta}\mathbf{d}|$ is the equilibrium distance. Since the coincidence cell areas $^{\alpha}\Delta^{q}S = {}^{\beta}\Delta^{q}S, q = 1 \dots p$,

(p being the number of coincidence cells), are dictated by the specific structure of the crystals, they will be random functions and without loss of generality "ϕ" can be specified as a random function of the orientations $^\alpha Q, {}^\beta Q$ and the elemental volume $^\alpha v$. Hence in this case ϕ may also be looked upon as a potential function of the coincidence cell areas rather than a pair-potential in the strict sense. In any event the choice of a suitable potential is not essential since the actual type of an "interatomic potential" from a micromechanics point of view has little effect on the structural behaviour of the crystalline solid. Thus one may choose for instance a "Morse function" type potential (see also [22], [23]) of the form:

$$\left.\begin{aligned}
&\phi\{|^{\alpha\beta}\mathbf{d}|\} = \phi_0\{1 - \exp[-b|^{\alpha\beta}\mathbf{d}|]^2\} \\
&\text{where} \\
&{}^{\alpha\beta}\mathbf{d} = |^{\alpha\beta}\mathbf{d}|\mathbf{e} = < |^{\alpha\beta}\mathbf{d}| >_N \mathbf{e},
\end{aligned}\right\} \quad (4.90)$$

$\mathbf{e}$ being the unit vector in the direction of $^{\alpha\beta}\mathbf{d}$, ϕ_0 is the resultant equilibrium potential at the coincidence cell and "b" a material constant both being determined from spectroscopic data. Since ϕ takes into account the size of the coincidence cell "$\Delta^q S$", which in turn, depends on the relative orientations of the crystals and the lattice vector $\mathbf{a}$, the function $\phi\{\mathbf{a}, {}^\beta Q - {}^\alpha Q, \Delta^q S\}$ becomes a dependent stochastic function. Assuming that it has a normal distribution, it will be determined by its expected value and the correlation function, ie.:

$$E\{\phi\} = < \phi >_N; \ (\alpha = 1 \dots N); \ \text{and} \ R_\phi(|\nu|), \ (\text{eqn. } 4.88(b)),$$

where $|\nu|$ is the previously defined correlation distance. Following the representation of interatomic forces given by Yvon [117] among others, the force acting between the grain boundary "coincidence sites" can be formulated from the spatial derivative of this potential in general, as follows:

$$^{\alpha\beta}\hat{\mathbf{f}} = \frac{d\phi\{|^{\alpha\beta}\mathbf{d}|\}}{d|^{\alpha\beta}\mathbf{d}|}\mathbf{e}. \quad (4.91)$$

The discrete force $^{\alpha\beta}\hat{\mathbf{f}}$ acts on each coincidence cell contained in the interface between the αth and its contiguous β^{th} crystal. For the overall interaction effect a summation of these discrete forces is required. This can be carried out by adopting the notion of a "generalized surface force" based on the concept of generalized functions (see for instance Gelfand and Vilenkin [55], and [23]). Hence the generalized surface force can be formally written as:

$$^{\alpha\beta}\tau(^{\alpha\beta}\mathbf{d}) = \delta[^{\alpha\beta}\mathbf{d} - {}^{\alpha\beta}\hat{\mathbf{d}}]^{\alpha\beta}\hat{\mathbf{f}} \quad (4.92)$$

where the "$\hat{\ }$" sign again indicates the discreteness of the vector $^{\alpha\beta}\mathbf{d}$ and δ is the 3-dimensional Dirac-Delta function.

In order to relate this "generalized surface force" to the micro stress interior to each crystal, one can for simplicity assume the latter to be an elastic continuum so that the balance of forces is expressed by:

$$\int_{\Delta^q S} \delta[^{\alpha\beta}\mathbf{d} - {}^{\alpha\beta}\hat{\mathbf{d}}]^{\alpha\beta}\hat{\mathbf{f}}dS = \int_{\Delta^q S} {}^{\alpha\beta}\tau dS = \int_{\alpha\Delta^q S} {}^{\alpha}\boldsymbol{\sigma}^{(i)} \cdot {}^{\alpha}\mathbf{n}dS = \int_{\beta\Delta^q S} {}^{\beta}\boldsymbol{\sigma}^{(i)} \cdot {}^{\beta}\mathbf{n}dS \quad (4.93)$$

in which the interior micro stress for simplicity is regarded as a Cauchy stress and where it is assumed that the outward directed normal vectors of each coincidence cell, $^{\alpha}\mathbf{n} = -^{\beta}\mathbf{n}$ (see also Fig. 18) throughout the deformation.

The "internal energy" of individual crystals with the inclusion of dislocation effects and its summation over a meso-domain of the polycrystalline solid has been given by (4.85). Similarly one can formulate the contribution to the overall energy due to the presence of grain boundaries by considering first the contribution per coincidence cell, ie.:

$$^q U(\phi) = {}^{\alpha\beta}\hat{\mathbf{f}} \cdot {}^{\alpha\beta}\hat{\mathbf{d}}\Delta^q S. \quad (4.94)$$

Summing over all coincidence cells gives then the effect of grain boundaries on the energy for a meso-domain as follows:

$$^M U = \frac{1}{2}\sum_{\alpha=1}^{N}\sum_{q=1}^{p} {}^{\alpha\beta}\mathbf{f} \cdot {}^{\alpha\beta}\hat{\mathbf{d}}\Delta^q S = \frac{1}{2}\sum_{\alpha=1}^{N}\int_{\alpha S} {}^{\alpha\beta}\boldsymbol{\tau} \cdot {}^{\alpha\beta}\hat{\mathbf{d}} \cdot d^{\alpha}S \quad (4.95)$$

in which $q = 1\ldots p$ is the total number of coincidence cells surrounding an individual crystal (α) and $^{\alpha}S$ its surface. It is of interest to note, that the introduction of a "pair-potential" acting on the coincidence sites of a cell, results in a similar form to (4.95). In this case the internal energy pertaining to a meso-domain can be expressed by:

$$^M U = \sum_{\alpha=1}^{N} < \sigma, \epsilon >_{\alpha} + C\sum_{\alpha=1}^{N}\sum_{\beta=1}^{N} \phi(^{\alpha}\mathbf{x}, {}^{\beta}\mathbf{x}) \quad (4.96)$$

in which $^{\alpha}\mathbf{x}, {}^{\beta}\mathbf{x}$ denote here the position vectors to the coincidence sites of the coincidence cells in the lattices 1,2, respectively, as indicated in Fig. 18 and where the bilinear form in the first term of (4.96) can be brought into a quadratic form of the internal strains by the use of the "micro operator $^{\alpha}m$". The response of a polycrystalline solid in the case of an elastic, isothermal deformation can be represented as follows. For any prescribed loading in the elastic range the "macroscopic" strain energy stored in the medium with a specific geometry can be stated by:

$$^M U = \int_{\partial S} \mathbf{T} \cdot {}^{M}\mathbf{u}d\partial S , \quad (4.97)$$

in which the boundary conditions on ∂S or $\partial \mathcal{M}$ (boundary of the macroscopic domain) have been previously discussed. In order to delineate the contribution of dislocations and grain boundaries by equating the external work exerted on a material specimen to that induced in the micro elements of the structure, it should be recognized that for each meso-domain M:

$$\int_{MS} {}^{M}\mathbf{T}^{M}\mathbf{u}\, d^{M}S = \frac{1}{2}\Big\{\sum_{\alpha=1}^{N} {}^{\alpha}A_{D}{}^{\alpha}\epsilon^{\alpha}\epsilon d^{\alpha}v + \sum_{\alpha=1}^{N}\int_{\alpha S} {}^{\alpha\beta}\boldsymbol{\tau}\cdot{}^{\alpha\beta}\mathbf{d}\,d^{\alpha}S\Big\} \qquad (4.98)$$

so that the "macroscopic energy" can be equated to:

$$\int_{\partial S} \mathbf{T}\cdot{}^{\mathcal{M}}\mathbf{u}\, d(\partial S) = \sum_{p=1}^{P}\int_{MS} {}^{M}\mathbf{T}\cdot{}^{M}\mathbf{u}\, d^{M}S$$

$$\text{for } p = 1\dots P \text{ mesodomains.} \qquad (4.99)$$

In terms of a characteristic energy functional or the material operator "M" for a specific meso-domain, one also has:

$$^{P}M = E\{{}^{\alpha}m\} = \sum_{1}^{N} {}^{\alpha}m\Delta\mathcal{P}\{{}^{\alpha}m\} \; ; \; \alpha = 1\dots N \; ,$$

where the micro-operator ${}^{\alpha}m$ and its distribution over the domain has been discussed earlier. Since the constitutive map can also be stated by the use of the "macroscopic operator $\mathcal{M}$" one has:

$$\mathcal{M}: {}^{M}\Sigma \to {}^{M}\mathcal{U}; \; {}^{M}\Sigma, {}^{M}\mathcal{U} \subset X; \; {}^{M}\Sigma = \cup_{p=1}^{P}{}^{M}\Sigma_{p}; \; {}^{M}\mathcal{U} = \cup_{p=1}^{P}{}^{M}\mathcal{U}_{p} \quad (4.100)$$

which is obtained as an element of the space of continuous bounded linear "$\mathcal{L}$", i.e.:

$$\mathcal{M} = \mathcal{L}[{}^{M}\Sigma, {}^{M}\mathcal{U}], \; {}^{M}\Sigma, {}^{M}\mathcal{U} \subset X; \; \mathcal{L} = \cup_{1}^{P}P_{M}. \qquad (4.101)$$

Evidently, for a numerical evaluation of the elastic response of a polycrystalline solid, it becomes necessary to determine the material operators at all three scales, i.e., the micro, meso and macro operator. By using an "idealized random structure", a numerical calculation of these operators can be carried out (see also ref. [22]). For experimental purposes and in order to validate the theory, a random structure was chosen which consisted of pure aluminium "mono crystals" that were embedded in an epoxy-resin matrix. The application of a new test procedure involving the combination of "Holographic Interferometry" and "X-ray Diffraction, Back reflection (Laue)" technique (see also [277, 278]), allowed the establishment of the distributions of two translations and one rotation of the mono crystals. Since the external load to a thin flat specimen has been applied parallel to the $(x_1 x_2)$-plane of the embedded crystals with the dimensions $2 \times 2 \times 2\ mm^3$, the latter were pre-oriented so that the $(0,0,1)$ crystallographic axes always coincided with the x_3-direction. Due to this experimental configuration and some required computational simplifications only the class of polycrystalline solids such as copper and aluminium were investigated (see also [278, 279] for the numerical evaluation and the determination of operators and microstress distributions). Although the test procedure has been applied to these two types of polycrystalline solids, it can be readily used for the experimental determination of microstress distributions and or deformations of other binary structures (metal-polymer composites) or multi-component systems.

(ii) The inelastic behaviour of multi-component systems:

A general stochastic deformation theory and the deformational stability of discrete solids were discussed in the foregoing sections. To illustrate this theory particularly with regard to the significant "transient stage" of the deformation, a multi-component or MC-system has been chosen (Fig. 2 of Chapt. 2), the transient behaviour of which has been formulated before in terms of a "limiting Markov" process. It is the main purpose of this section to present an alternative analysis based on the general theory of "point processes". Recalling Def. 2 of section 2.1 (Chapt. 2), it is more convenient in the case of MC-systems to employ a "micro domain Γ" rather than a micro element of the structure as a primitive element in the analysis. As a consequence the r-dimensional state-vector is defined with respect to the micro domain Γ as follows:

$$
{}^{\Gamma}\mathbf{z} = \begin{pmatrix} \boldsymbol{\sigma}(t) \\ \mathbf{u}(t) \\ \vdots \\ \boldsymbol{\epsilon}(t) \end{pmatrix} = [\boldsymbol{\sigma}(t), \mathbf{u}(t), \boldsymbol{\epsilon}(t) \ldots]^T \in Z \equiv X \tag{4.102}
$$

and the set $\{{}^{\Gamma}\mathbf{z}_i\}$; $(\Gamma = 1 \ldots N, i = 1 \ldots r)$ corresponds to a meso domain of the material. The components of ${}^{\Gamma}\mathbf{z}$ may be scalar or higher-dimensional functions. For convenience and simplification of the stochastic analysis, one may use one or more components of ${}^{\Gamma}\mathbf{z}$. In this case attention must be given to the corresponding "subspaces of $Z \equiv X$" and the "event structure" is to be identified accordingly together with an appropriate probability measure. For the MC-system under consideration, the "overall strain" and/or "strain-rate" vectors are the only "observables" (under isothermal conditions) and hence will be considered subsequently together with their respective subspaces, i.e. $\mathcal{E}^E, \mathcal{E}^{In} \subset Z \equiv X$ (see also the state-space representation of Fig. 12, Chapt. 2).

The state of a microdomain Γ containing the constituents of the MC-system is then characterized by $\mathbf{z} < {}^{\Gamma}\mathbf{z} < \mathbf{z} + \Delta\mathbf{z}$, where ${}^{\Gamma}\mathbf{z}$ is a specific value for the microdomain Γ and $\Delta\mathbf{z}$ the range of accuracy of experimental observations.

Thus the event sets E become subsets of X or by considering them as "open spheres":

$$E_n = \{\mathbf{z}_n < {}^\Gamma \mathbf{z}_n \leq \mathbf{z}_n + \Delta \mathbf{z}_n\} \; ; \; \cup_n E_n = X, \; E_n \cap E_k = \emptyset, \; n \neq k.$$

If X is locally compact, then the subsets are also compact and bounded under closure. Since X is a measurable space, the subspaces $\mathcal{E}^E, \mathcal{E}^{In} \subset \mathcal{E}$ will also be measurable and hence one can introduce the probability measure $\mathcal{P}^\epsilon$ on $\mathcal{E}$ with the properties:

$$\left.\begin{array}{l}
\text{(i) } 0 \leq \mathcal{P}^\epsilon\{E_n\} \leq 1 \text{ for } \forall E_n \in \mathcal{F}^\epsilon, n = 1, 2 \ldots \\[2mm]
\text{(ii) } \mathcal{P}^\epsilon\{E_1 \cup E_2\} = \mathcal{P}^\epsilon\{E_1\} + \mathcal{P}^\epsilon\{E_2\} - \mathcal{P}^\epsilon\{E_1 \cap E_2\}, \\[1mm]
\quad \text{if } E_1, E_2 \in \mathcal{F}^\epsilon \text{ and } E_1 \cap E_2 \neq \emptyset. \\[2mm]
\text{(iii) } \mathcal{P}^\epsilon\{\cup_n E_n\} = \displaystyle\sum_n \mathcal{P}^\epsilon\{E_n\}, \text{ if } E_n \cap E_m = \emptyset, n \neq m, \\[3mm]
\qquad\qquad\qquad\qquad\qquad\qquad\qquad n, m = 1, 2 \ldots \\[2mm]
\text{(iv) } \mathcal{P}^\epsilon\{\mathcal{E}\} = 1.
\end{array}\right\} \quad (4.103)$$

The measure $\mathcal{P}^\epsilon$ is also the distribution function of the micro strains in the domain Γ and the corresponding probability space is represented by the triplet $[\mathcal{E}, \mathcal{F}^\epsilon, \mathcal{P}^\epsilon]$. If this space is identified with the space of all $\mathcal{P}^\epsilon$-regular measurable functions $\epsilon \in \mathcal{E}$ and correspondingly with $\dot{\epsilon} \in \mathcal{E}$ (see also [45], [280]), it leads by recognizing the duality with the stress or stress-rate space, respectively, as a subspace of X to the establishment of constitutional relations in operational form ([23], [281] and [282]).

For the alternative description of the "transient behaviour" of a MC-system (Fig. 32a,b) as mentioned earlier by means of point processes, it is to be recognized that generally point processes are concerned with random variables representing the number of "entities" involved in a process. In the present case these entities refer to the configurations of the constituents of the MC-system in the configuration space $\mathcal{C} \subset X$. They are distributed corresponding to a finite number of states of the system characterized by the "state vector $\mathbf{z} \in Z \equiv X$" or its components. Thus the collection of random points $\{x_1, x_2 \ldots x_n\} \in \mathcal{C}$ can be referred to as a "point process φ" on $\mathbb{R}^d \subseteq E(\mathbb{R}^n) \equiv X$. Mathematically φ can also be regarded as a random variable taking values in a measurable space $[Z, \mathcal{F}^z]$. For the simplification of the analysis and because of the possible observations of strains, one may choose the subspace $\mathcal{E}$ of $Z \equiv X$ and hence $[\mathcal{E}, \mathcal{F}^\epsilon]$ as the measure space. The strain-space $\mathcal{E}$ (Fig. 12) is then the family of all subsets φ_ϵ of $\mathbb{R}^d$ that will have to satisfy the following conditions:

(i) the subset φ_ϵ is "locally finite", i.e. each subset of $\mathbb{R}^d$ is bounded and contains only a finite number of the $\{x_n\}$ so that:

$$\varphi = \{x_1, x_2 \ldots x_n\} \in \mathcal{C} \Rightarrow \{\epsilon_1, \epsilon_2 \ldots \epsilon_n\} = \varphi_\epsilon \in \mathbb{R}^d \qquad (4.104)$$

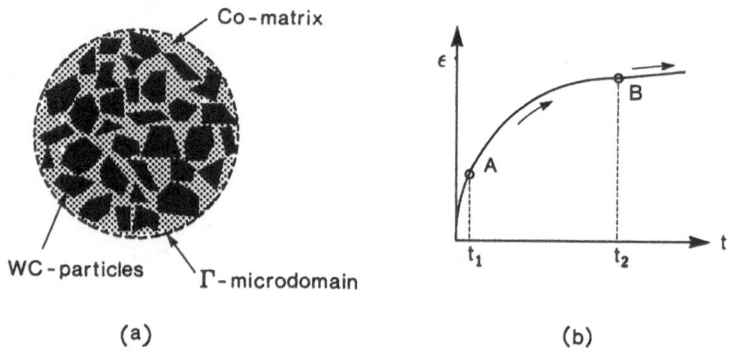

Fig. 32. Transient stage of a MC-system.

(ii) the set φ is simple, i.e. $x_i \neq x_j$, if $i \neq j$.

(iii) satisfies the nearest neighbour conditions: $x_1 \in \varphi_1$, $x_2 \in \varphi$, iff

$$\| x_1 - x_2 \| = 1.$$

For convenience one can designate the point process by $\varphi(x) = \{x_n\} \in X$ or equivalently by $\varphi_\epsilon = \{\epsilon\} \in \mathcal{E}$, where conditions (i), (ii) above are satisfied, or where each φ_ϵ in $\mathcal{E} \subset Z \equiv X$ is a closed set of $\mathbb{R}^d$ and where (iii) refers to a constraint of φ_ϵ due to the interactions between the constituents to be discussed below.

It is to be noted that there are two ways of considering a point process φ, i.e. either as a "random sequence $\varphi_\epsilon(\epsilon) = \{\epsilon_1, \epsilon_2 \dots \epsilon_n\}$" or as a "random measure" for each Borel set E (event). Thus the probability measure $\varphi(E) \equiv \mathcal{P}^\epsilon$ is the number of points of $\varphi_\epsilon(\epsilon)$ that fall in the set E. Since the random pattern is locally finite, all points fall into a given material domain of the MC-system, the set function $\varphi(E) \equiv \mathcal{P}^\epsilon$ or measure is also locally finite and a σ-additive Borel measure. Thus, if E is a Borel set, $E \cap \varphi(x)$ or $\varphi_\epsilon(\epsilon)$ is the "random set" of points of $\varphi(x)$ that also belongs to the set E.

The basis for the description of a finite point process can be expressed by the following conditions:

(i) the random points are located in $X = \mathbb{R}^d$, which is assumed to be a completely separable metric space.

(ii) the distribution $P_n = \{p_n\}, (n = 0, 1, 2 \dots)$ determines the total number of points in the set $\{x_n\}$ and correspondingly in the set $\varphi(\epsilon) = \{\epsilon_n\}$.

(iii) for each integer $n \geq 1$ the probability distribution P_n is given in the Borel sets of the "product space" $X^n = X \times \ldots \times X$ and determines the joint distributions of the points of $\varphi(x)$ or correspondingly of $\varphi(\epsilon)$, where the total number of points is "n".

Implicit in the above statements is the assumption that either of the set $\{x_n\}$ or $\{\epsilon_n\}$ is an "unordered" set of points (see also refs. [241] and [282]). The distribution P_n should be symmetric however, in order to consider the point process in terms of a theory of unordered sets or in other words P_n should be given equal weight to all $n!$ permutations of the locations $\{x_1, x_2 \ldots x_n\}$. Hence writing the distribution in a symmetrized form, viz.:

$$P_n^s(E_1 \times \ldots E_n) = \frac{1}{n!} \sum P_n(E_{i_1} \times \ldots \times E_{i_n}) \qquad (4.105)$$

where the sum is taken over all $n!$ permutations $(i_1 \ldots i_n)$ of the integers $(1 \ldots, n)$ and $1/n!$ is the normalizing factor.

Alternatively, for the symmetrization of P_n one can use the measure discussed earlier as the "Janossy measure" (see also Srinivasan [283]) in such a manner that:

$$J_n(E_1 \times \ldots \times E_n) = p_n n! P_n^s(E_1 \times \ldots \times E_n) \qquad (4.106)$$

Thus, if $X = \mathbb{R}^d$, the density $j_n(x_1 \ldots x_n)$ of J_n with respect to the Lebesgue measure on $X^{(n)}$ with $x_i \neq x_j$ for $i \neq j$, is the probability that there are n-points in $\varphi(x)$ or $\varphi(\epsilon)$ respectively, in each of the n distinct regions $x_i, x_i + dx_i$ or that: $j_n(x_1 \ldots x_n)dx_1 \ldots dx_n$. The normalization condition, i.e. $\sum p_n = 1$ in terms of the Janossy measure becomes therefore:

$$\sum_{n=0}^{\infty} \frac{1}{n!} J_n(X^{(n)}) = 1 \qquad (4.107)$$

with $J_0(X^{(0)}) = p_0$ and for $n \geq 1$:

$$J_n(X^{(n)}) = p_n n! \qquad (4.108)$$

Indicating that by the use of any family of symmetric measures $J_n(\cdot)$ satisfying (4.108), one can determine a probability distribution $\{p_n\}$ or P_n and a set of symmetric probability measures $\{P_n^s\}$. To establish a "global probability measure $\mathcal{P}^\epsilon$" on the Borel sets E of the countable union of X, i.e.:

$$X^{\cup} = X^0 \cup X^{(1)} \cup X^{(2)} \cup \ldots \qquad (4.109)$$

in which $X^{(0)}$ is an isolated point, one then has the "canonical probability space $[X^{\cup}, \mathcal{P}]$ of the point process (see also Moyal [284]). Hence for the given measure $\mathcal{P}^\epsilon$, $1/n! J_n$ is a "restriction of $\mathcal{P}^\epsilon$" to the components of the product space $X^{(n)}$.

For a more detailed analysis of the transient behaviour of the MC-system, it is necessary to employ a special class of point processes, i.e. the "Gibbs point processes" in a bounded region of the material specimen. Such regions are the microdomains Γ (Fig. 32(a)). It is further important to recognize the well established equivalency of Markov random fields with Gibbs random fields (see also Dobrushin [285], Spitzer [286] and Preston [287, 288]) in order to represent the possible interactions between the components of the MC-system. Since point processes are unordered sets of points, where $\varphi(x)$ or $\varphi(\epsilon)$ represent closed sets of $\mathbb{R}^d$, one can define a probability measure in such a manner that to each "configuration (state)" $x_i \in \{x_n\} \in \Gamma \subset C$ or X an energy is assigned in Γ, as follows:

$$U^\Gamma = \sum_{i=1}^{n} u_i(x_i) + C \sum_{i,j} \phi(x_i, x_j) \tag{4.110}$$

in which the first term refers to the potential energy of the system in Γ due to the external force field and the second term represents a "pair-potential" or interaction potential (see also ref. [23]). The parameter C in (4.110) is a constant or a material property that reflects whether the interaction potential is attractive, i.e. $C > 0$ or repulsive for $C < 0$. Hence for a given configuration of the components of the MC-system in the Γ domain one can express the total energy by:

$$U^\Gamma(x_1 \ldots x_n) = \sum_{i=1}^{n} u_i^{ext}(x_i) + C \sum_{i=1}^{n-1} \sum_{j=i+1}^{n} \phi(x_i, x_j) \tag{4.111}$$

where it is assumed that the nearest neighbour condition in (iii) holds for $x_i \neq x_j$. By the well-known principle of statistical mechanics, the probability distribution of a particular configuration of the domain in the state of equilibrium can be stated as:

$$P^\Gamma(x_n) = G^{-1} \exp^{-\frac{1}{\kappa\theta} \cdot U^\Gamma\{x_n\}} \tag{4.112}$$

where this probability measure is called a "Gibbs measure" (see also [287]), κ is a universal constant, θ the temperature and G a normalizing constant referred to as the "partition function". The latter is given by:

$$G = \sum_{x_i} e^{-\frac{1}{\kappa\cdot\theta} \cdot u(x_i)}. \tag{4.113}$$

Alternatively by using the Janossy measure introduced in (4.106) or its density j_n, one can also write that:

$$j_n(x_1 \ldots x_n) = G \exp\{-\theta U^\Gamma(x_1 \ldots x_n)\}. \tag{4.114}$$

Considering now a set of microdomains for the MC-system over the entire "macrodomain" of the material, where $(\Gamma_1 \ldots \Gamma_n)$ represents a finite partition of the probability space X, one can define the probability of finding exactly n_i points in $\Gamma_i, (i = 1, \ldots k)$ by means of the Janossy measure $(n_1 + \ldots + n_k = n)$ in the following manner:

$$P_k\{\Gamma_1 \ldots \Gamma_k; n_1 \ldots n_k\} = (n_1! \ldots n_k!) J_n(\Gamma_1^{(n_1)} \times \ldots \times \Gamma_k^{(n_k)})$$

$$= p_n \begin{pmatrix} n \\ n - n_k \end{pmatrix} P_n^s(\Gamma_1^{n_1} \times \ldots \times \Gamma_k^{(n_k)}) \tag{4.115}$$

in which the multinomial coefficient indicates the grouping of the n points that correspond to the locations of the components of the MC-system when n_i lies in $\Gamma_i; (i = 1 \ldots k)$. The sets Γ_i are disjoint and have $\cup_{i=1}^{k} \Gamma_i = X$. The joint distributions of the random variable $N(\Gamma_k)$, i.e. the total number of points in $\{\Gamma_k\}$ can be readily established. For the requirements of the finiteness of the distributions and their existence reference is made to Daley and Vere-Jones [241]. One can also define a "counting measure" from the fact that each realization or outcome of the point process is representable in terms of the state vector $\mathbf{z} \in Z = X$ for some $n \geq 0$ such that:

$$N(\Gamma) = \#\{i: z_i \in \Gamma \subset Z\} \tag{4.116}$$

where z_i are the components of $\mathbf{z} \in Z$. It is to be noted however, that in the present case only one "measurable component" is available, i.e. the strain field $\epsilon \in \mathcal{E} \subset Z$. Since $\mathbf{z}$ induces a mapping from $X^{(n)}$ into the space of all counting measures in X, it is measurable and holds for every $n \geq 1$ so that the process for the MC-system as a whole is a "finite point process" with the global measure $\mathcal{P}^\epsilon$ with the restriction given by eqn. (4.108). It is significant that for the consideration of the integral geometrical aspects of the transient behaviour of the MC-system as well as for the "simulation of the process", it becomes necessary to discriminate between the individual states of the microdomain Γ. Thus the local energies $u_i \in U^\Gamma$ (first term of eqn. 4.110) correspond to the positions or occupations of the components of the MC-system in the set Γ, i.e. $\varphi(x_n)$ or equivalently φ_ϵ are closed random sets in $\mathbb{R}^d$.

Since in the short elastic or reversible stages before the transiency takes place, the stress-space Σ is dual to the strain-space $\mathcal{E}$, i.e. $(\sigma \in \Sigma, \epsilon \in \mathcal{E})$ both being subspaces of Z or X, one can replace u_i by the inner produce $< \sigma, \epsilon >$ so that eqn. (4.110) also reads:

$$U^\Gamma = \sum_{i=1}^{n} < \sigma, \epsilon > + C \sum_{i=1}^{n-1} \sum_{j=1}^{n} \phi(x_i, x_j) \tag{4.117}$$

in which $< ., . >$ represents again the local potential energy at equilibrium due to the external force field. Evidently by neglecting interaction effects this relation reduces to the first term only. It is apparent that for the whole point process in Γ, one has to consider the elastic as well as the "inelastic" contributions of the components of the system or:

$$\psi = \varphi_\epsilon \circ \psi_{\bar{\epsilon}} \subset \mathbb{R}^d \tag{4.118}$$

where the process ψ by neglecting to a first approximation the interaction effects, will represent the composition of the elastic and inelastic contributions and where $\psi_{\bar{\epsilon}} = \{\bar{\epsilon}_1, \bar{\epsilon}_2 \ldots\} \in \mathbb{R}^d$ in general are "time-dependent" strains. For a better understanding of this process one can use "conditional probabilities" in the sense of Preston [287], that are referred to as "local specifications", i.e.:

$$\pi^\Gamma_{BC\Gamma}(\psi_{\bar{\epsilon}}|\varphi_\epsilon) = P\{\psi \cap B \in \Gamma \subset \mathbb{R}^d | \psi \cap B^c \in \Gamma = \varphi_\epsilon\} \tag{4.119}$$

in which B is any set in Γ and B^c its complement. The above equation indicates that up to a certain strain level corresponding to $\varphi_\epsilon = \{\epsilon_1, \epsilon_2 \ldots \epsilon_n\} \in \mathbb{R}^d$, (Gibbsian equilibrium states) and $\psi_{\bar{\epsilon}} \in \mathbb{R}^d$ associated with the inelastic response, the quantity π^Γ or conditional probability of having $\psi_{\bar{\epsilon}}$ additional to the points $\varphi_\epsilon \in \mathbb{R}^d$ can be referred to as a "local specification". This probability can also be written in exponential form in accordance with the Gibbs measure defined in equations (4.112) and (4.113) so that:

$$\pi^\Gamma_{BC\Gamma}(\cdot,\cdot) = \int 1_\Gamma[((\psi_{\bar{\epsilon}} \cap B) \cap B) \cap \varphi_\epsilon] \exp[-\overline{U}(\psi_{\bar{\epsilon}} \cap B, \varphi_\epsilon]W \, d\psi_{\bar{\epsilon}}/G(B, \varphi_\epsilon) \tag{4.120}$$

where 1_Γ is the indicator function for any subsets of the point process ψ in Γ, and $\overline{U}$ the energy increase due to the addition of the "inelastic" set of points $\psi_{\bar{\epsilon}}$ in Γ such that: $\overline{U}: \Gamma \times \Gamma \to \mathbb{R}^1$, where in the sense of Stoyan [236], Kendall [240], it is assumed that the local energies have the additivity property, i.e.:

$$\overline{U}(\psi_{\bar{\epsilon}}, \varphi_\epsilon) = \overline{U}(\bar{\epsilon}_1, \varphi_\epsilon) + \overline{U}(\bar{\epsilon}_2, \varphi_\epsilon \cup \{\bar{\epsilon}_1\}) + \ldots$$

$$\ldots + \overline{U}(\bar{\epsilon}_n, \varphi \cup \{\bar{\epsilon}_1, \ldots \bar{\epsilon}_{n-1}\}) \tag{4.121}$$

It is significant to interpret these energy levels in terms of a so-called "excursion set" (Adler [289]), which can be written in the present case as follows:

$$A_{\overline{U}} = \{\pi^\Gamma(\psi, \Gamma)\} = \left\{ \begin{array}{l} \pi^\Gamma(\psi, \Gamma) \colon U^\Gamma + \overline{U}^\Gamma \geq U^\Gamma_{t=t_1}, \\ t \in (t_1, t_2) \in T \end{array} \right\} \tag{4.122}$$

indicating that the point process ψ made up of the elastic and inelastic energy contributions of the components of the MC-system can be described as shown above. It is of interest to note that the approach by means of point processes and a counting measure (eqn. 4.116) can also be carried out by the use of "martingale dynamics" that equally permits the characterization of the transient response of the MC-system.

(iii) Dynamics of structured solids:

(a) Stochastic Models of wave propagation:

Earlier work by Kampé de Feriét [290], Keller [291], Frisch [292] and others dealt with the wave mechanics of random continuous media. Their work was concerned with the formulation of the "wave propagation" in random media and the solution of the corresponding random differential and or integral equations. The aspect of discreteness of the microstructure other than that of a perfect lattice did not enter into the analysis. Before discussing the stochastic formulation of the wave phenomena in structured solids (see [23], [293]), it is convenient to introduce apart from the employed spatial scales (i.e., postulate P.1 of the theory) also two "time-scales" in the analysis. Hence, generally one has a "macro-time" for the passage of a wave through the entire material body "$\mathcal{B}$" and a "micro-time" with reference to the passage of the wave through an "individual element α" of the microstructure. In the present case in order to simplify the analysis, the microstructure is considered to be that of a "cubic type" polycrystalline solid, whereby each element or grain α has an elastic modulus $^{\alpha}E$ and a mass density $^{\alpha}\rho$. The probability distributions of these parameters are assumed to be homogeneous in space. The intrinsic field quantity is chosen to be the wave velocity vector $^{\alpha}\mathbf{v} \in \mathcal{V} \subset X$, which is identified with the deformation rate $^{\alpha}\dot{\mathbf{u}}$ of an element. Since from the onset the interfaces between contiguous elements are admitted, although they may be small, one needs two additional field parameters defined as follows:

$$\left. \begin{array}{l} \text{(i) a "reflection coefficient": } C_r \triangleq \dfrac{|^{\alpha}\mathbf{v}_r|}{|^{\alpha}\mathbf{v}_i|} \\[3mm] \text{(ii) a "transmission coefficient": } C_{tr} \triangleq \dfrac{^{(\alpha+1)}|\mathbf{v}_{tr}|}{^{\alpha}|\mathbf{v}_i|} \end{array} \right\} \quad (4.123)$$

in which $^{\alpha}\mathbf{v}_i$, $^{\alpha}\mathbf{v}_r$ and $^{(\alpha+1)}\mathbf{v}_{tr}$ are the incident, reflected and transmitted wave velocity vectors, respectively. Another significant parameter is the "passage time" or the time required of the wave front to pass through an individual element α of the structure. It is defined by:

$$^{\alpha}\tau \triangleq {}^{\alpha}d/^{\alpha}c \qquad (4.124)$$

where $^{\alpha}d$ is the grain size in the direction of the wave propagation and $^{\alpha}c$ the wave propagation velocity through α, if for simplicity the element is regarded as a continuum, i.e., without any defects. The above definitions make it possible to account for two important effects arising in the wave propagation through structured solids, i.e.:

(i) the "transmitted" wave velocity vector $^{(\alpha+1)}\mathbf{v}_r$ is different
from the "incident" wave velocity vector $^{\alpha}\mathbf{v}_i$,
due to the existence of the $(\alpha, \alpha + 1)$ grain boundary,

(ii) the wave propagation velocity $^{\alpha}c$ is a random quantity
for every micro element α.

$$(4.125)$$

It is seen that $^{\alpha}\tau$ defined by (4.124) introduces a time-scale due to the admittance of the existing microstructure that depends on the size of the element itself. Since the wave length of a specific pulse given to the semi-infinite bar $\mathcal{B}$ may be smaller, equal or larger than the grain size $^{\alpha}d$, this specifies in a sense the wave propagation in the discrete solid more accurately than the otherwise used "correlation length" of disturbances due to local fluctuations of the wave propagation usually stipulated by a continuum approach.

It should be noted, that the distributions of C_{tr} and $^{\alpha}\tau$ can be obtained from corresponding experimental investigations concerning single or bi-crystals, respectively. In absence of such data these distributions can be found by the application of a simulation technique such as the Monte-Carlo method to be discussed subsequently.

Thus, if an arbitrary pulse is given to the front of the semi-infinite bar $\mathcal{B}$, (Fig. (33a)), the wave propagation in the space domain will occur within a certain "dispersion cone", the projection of which is indicated in Fig. (33b). One can distinguish three possible cases for the modelling of the wave propagation. Thus an one-dimensional model, which is considered as an ensemble of k non-intersecting sequences S_k; $(k = 1 \ldots K)$ (Fig. 33a) consisting of "cubic crystals" $(\alpha = 1 \ldots N)$. A 2D-model, which is regarded as an ensemble of k sequences S_k that interact in the $X_1 X_2$-plane (plane-wave propagation) and a 3D-model where the ensemble of k sequences interact in all three directions of the $\mathbb{R}^3$-space. In general a pulse given to the body $\mathcal{B}$ will induce a "longitudinal" as well as a "transverse" wave, each of which will propagate with its own velocity, i.e., c_L and c_T, respectively. In the following stochastic analysis which is valid for either of this type of waves, it will be assumed for simplification of the analysis that the "longitudinal" wave formation is the predominant one, and that the wave velocity vector field is Markovian represented by $\mathcal{V}(\mathbb{R}^1 \times \mathcal{J}) \subset X$. This necessitates however to consider the introduced time scale (belonging to an index set $\mathcal{J} = \mathcal{J}_1 \oplus \mathcal{J}_2$):

(i) if "$\bar{t}$" designates the macroscopic or "average time" required for the passage of a longitudinal wave along an "average path", ie. $\bar{t} \in \mathcal{J}_1$, then an "internal time t" will be necessary for the "sample path" of such a wave where $t \in \mathcal{J}_2$.

(ii) due to the random velocity $^{\alpha}c_L$, the longitudinal wave will arrive anywhere in a time-space graph (projection of dispersion-cone, Fig. 33(b)) with an indeterminate time difference with respect to the macro-disturbance (average path).

It follows that according to (i) a particular random variable can be identified namely the "dispersion time" of the longitudinal wave defined by the non-negative quantity:

$$\tau_L \overset{\Delta}{=} t - \bar{t} \qquad (4.126)$$

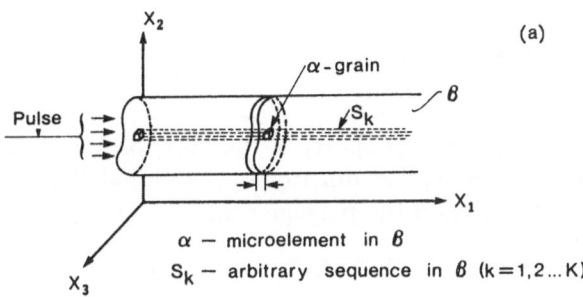

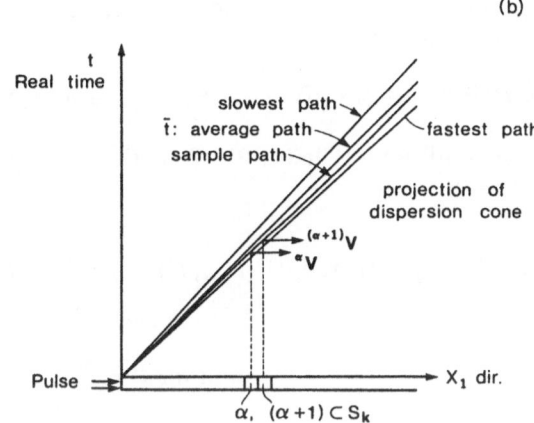

Fig. 33. Model for the wave propagation in a polycrystalline solid with cubic structure.

Although τ_L defined by (4.126), is a random variable intrinsic in the velocity process $\mathbf{v}(\bar{t}, t)$, it is itself a random process parametrized by the average time $\bar{t}$, i.e., $\tau_L = \tau_L(\bar{t})$. Hence one can express the probability of the process $\mathbf{v}(\bar{t}, t)$ as follows:

$$P\{\mathbf{v}(\bar{t}, t)\} = P\{\mathbf{v}(t)\} P\{\tau_L(\bar{t})\} \qquad (4.127)$$

in which both the effects (i) and (ii) of (4.125) have been separated. Evidently the right-hand side of the above relation represents a modulation of the wave velocity vector, and accounts for the dispersion of the wave front in the time-space graph. Moreover (4.127) indicates that the velocity vector $\mathbf{v} \in \mathcal{V}$ evolves independently, when parametrized by the internal and average times. These evolutions being independent of their respective past histories will now be recognized as Markov processes.

Thus considering first the L-wave evolution in terms of the internal time t, one has:

$$\mathcal{P}\{\mathbf{v}(t) \in E\} = \int_{\mathcal{V}} \mathcal{P}\{\mathbf{v}(t) \in E | \mathbf{v}(t_0) = x\}\mathcal{P}(dx) \; ; \; t_0 \leq t \qquad (4.128)$$

which expresses the Markov property of the "$\mathbf{v}(t)$ process". Writing the conditional probability as an explicit function of the probability distributions of the transmission coefficient and the passage time, one obtains:

$$P\{\mathbf{v}(t_0 + {}^{\alpha}\tau) \in E | \mathbf{v}(t_0) = x\} = \mathcal{P}\{C_{tr}: \mathbf{v}(t_0 + {}^{\alpha}\tau) = C_{tr} \cdot \mathbf{v}(t_0) \in E\}. \quad (4.129)$$

Recognizing this Markov process to be a temporally homogeneous one and since $\mathcal{V}$ is a subspace of X and is separable, one can postulate the existence of a transition function in the form of:

$$P({}^{\alpha}\tau, x, E) \overset{\Delta}{=} \mathcal{P}\{\mathbf{v}({}^{\alpha}\tau) \in E | \mathbf{v}(0) = x\}; \; {}^{\alpha}\tau \in \mathcal{J}_2 = [0, \infty) \qquad (4.130)$$

satisfying the usual conditions (Chapt. 2). Thus the corresponding Chapman-Kolmogorov equation becomes:

$$\left. \begin{array}{c} P(t + t', x, E) = \displaystyle\int_{\mathcal{V}} P(t, x, dy)P(t', y, E) \; ; \; t = \sum_{\alpha=1}^{n} {}^{\alpha}\tau \\[4mm] \text{and for } t' = \sum_{\alpha=1}^{n'} {}^{\alpha}\tau \geq 0. \end{array} \right\} \quad (4.131)$$

If $C(\mathcal{V})$ designates a Banach space of all bounded continuous functions $f(x)$ on the velocity space $\mathcal{V}$ endowed with the norm $\| f \| = \sup_{x \in \mathcal{V}} |f(x)|$, one can define an operator T_2 on $\mathcal{V}$, by:

$$T_2(t)[f(x)] \overset{\Delta}{=} \int_{\mathcal{V}} f(y)P(t, x, dy) \; ; \; f \in C(\mathcal{V}) \qquad (4.132)$$

where $\{T_2(t), t \geq 0\}$ is a contraction semi-group of operators on $(\mathcal{V})$ (see also Chapt. 2). Returning to eqn. (4.127) and considering the second term on the right-hand side that involves the L-wave propagation, the corresponding stochastic process: $\mathbf{v}(\bar{t}, t) = \mathbf{v}(\bar{t})$ for a fixed time t, it will be characterized by the following probability function:

$$P\{\mathbf{v}(\bar{t},t)\} = \text{const} \cdot P\{\tau_L(\bar{t})\}. \tag{4.133}$$

As a consequence the evolution of this process will be represented by a linear BD-Markov process at regular intervals in terms of the average passage time $< {}^\alpha\tau >$. This process can therefore be expressed in terms of a one-step transition function, by:

$$P(\bar{t}_0, z, \bar{t}, D_L) \overset{\Delta}{=} P\{\tau_L(\bar{t}) \in D_L | \tau_L(\bar{t}_0) = z\} \; ; \; \bar{t} \in \mathcal{J}_1, \; D_L \in \mathcal{D}_L. \tag{4.134}$$

which is also a function of the distribution of the passage time, i.e., $P({}^\alpha\tau)$ so that:

$$P\{\tau_L(\bar{t}) \in D_L | \tau_L(\bar{t}_0) = z\} = P\{{}^\alpha\tau: {}^\alpha\tau - < {}^\alpha\tau >= \tau_L(\bar{t}) - \tau_L(\bar{t}_0)\} \tag{4.135}$$

where $\bar{t} = \bar{t}_0 + < {}^\alpha\tau >$. The random variable τ_L is an outcome in the "sample space $\mathcal{D}_L$" of the dispersion times (see also Fig. 33b) and D_L a Borel set of $\mathcal{D}_L$ belonging to the σ-field $\mathcal{F}_L$. Assuming that this process is a temporally homogeneous Markov process one can write the following transition function in a reduced form, i.e.:

$$P(\bar{t}, z, D_L) \equiv P(0, z, \bar{t}, D_L) \tag{4.136}$$

satisfying the conditions of a Markov process. The corresponding Chapman-Kolmogorov equation is then:

$$P(\bar{t} + \bar{t}', z, D_L) = \int_{\mathcal{D}_L} P(\bar{t}, z, dw) P(\bar{t}', w, D_L), \quad \bar{t}, \bar{t}' \geq 0. \tag{4.137}$$

Similarly as before, one can introduce a transition operator parametrized by the average time $\bar{t}$, which is given by:

$$T_1(\bar{t})[f(z)] \overset{\Delta}{=} \int_{\mathcal{D}_L} f(w) P(\bar{t}, z, dw) \quad ; \; f \in C(\mathcal{D}_L) \tag{4.138}$$

and where $\{T_1(\bar{t}), \bar{t} \geq 0\}$ is again a contraction semi-group of the operators T_1 on $C(\mathcal{D}_L)$.

For the description of the L-wave propagation in a discrete 1-D solid for example, it becomes necessary to combine the above results. Thus by following the work of Hille and Phillips [38] concerning "multi-parameter semi-groups", one can state the following (see also ref. [294]):

"if $\{T_1(\bar{t}), \bar{t} \in \mathcal{J}_1\}$ and $\{T_2(t), t \in \mathcal{J}_2\}$ are one-parameter semi-groups of contraction operators on $C(\mathcal{D}_L)$, then there exists a two-parameter semi-group $\{T(\bar{t}, t), (\bar{t}_1, t) \in \mathcal{J}_1 \oplus \mathcal{J}_2\}$" such that:

(i) $T(\bar{t}, t)$ is the direct product of two operators T_1 and T_2 or:

$$T(\bar{t},t) = T_1(\bar{t})T_2(t),$$

(ii) $\lim\limits_{(\bar{t},t)\to 0} T(\bar{t},t) = I,$

(iii) $T(\bar{t},t)$ is a contraction: $\| T(\bar{t},t) \| \leq 1.$

(b) Application of the Monte-Carlo simulation:

It is seen from the above brief discussion that the evolution of the longitudinal wave in an 1D-model can be formulated by the use of an abstract dynamical system represented by the quadruplet $[\mathcal{V} \times \mathcal{D}_L, \mathcal{F}^{\mathsf{v}} \times \mathcal{F}_L, \mathcal{P}^{\mathsf{v}} \times \mathcal{P}^L, T(\bar{t},t)]$. In particular the semi group property of the operator $T(\bar{t},t)$ indicates that both the Chapman-Kolmogorov relations, i.e., (4.131) and (4.137) can be jointly used to describe the evolution of the velocity vector field $\mathbf{v}(\bar{t},t)$. So far as the problem of random plane waves in a three-dimensional structured solid is concerned, the same simplification can be made as before, i.e., by considering only a "cubic structure" with random physical properties. Due to these properties of the elements of the structure, the plane wave will evolve in form of a random plane-wave. Although the interactions between contiguous elements occur in all three directions, the dominant interaction will be that in the direction of the wave-propagation. Again by assuming that the macroscopic body $\mathcal{B}$ is made up of a number of sequences each of which contains a denumerable number of micro elements, the wave-propagation will be predominantly a longitudinal wave with slight disturbances due to the passage through the micro elements. It is further assumed that the coupling between the sequences is weak with respect to the wave propagation. Hence similarly to the one-dimensional case, two effects will be significant here, i.e.:

(i) the transmitted wave velocity vector $^{(\alpha+1)}\mathbf{v}_{tr}$ is different from the incident wave velocity vector $^{\alpha}\mathbf{v}_i$ due to the grain boundary denoted by $\partial^{\alpha}\mathcal{D}$ ($^{\alpha}\mathcal{D}$ is the domain of a single grain α).

(ii) the wave propagation velocity $^{\alpha}c$ is random for every element α.

Since the wave velocity vector $^{\alpha}\mathbf{v}$ goes into $^{(\alpha+1)}\mathbf{v}$ after a random passage time, which has to account for the transmission process associated with the interaction with four neighbouring grains (for a cubic structure), the velocity vector $\mathbf{v}$ becomes a function of the three Euclidean coordinates. Evidently the transmission coefficient will be thus a transmission operator, so that:

$$C(\omega): {}^{\alpha}\mathbf{v}(t) \to {}^{(\alpha+1)}\mathbf{v}(t + {}^{\alpha}\tau) ; \quad {}^{\alpha}\mathbf{v}, {}^{(\alpha+1)}\mathbf{v} \in \mathcal{V}. \tag{4.139}$$

where $C(\omega)$ also depends on the surface interactions at the boundary $\partial^{\alpha}\mathcal{D}$ and the velocities in the four neighbouring grains which can be designated by $\gamma_i, (i = 1\ldots 4)$. Hence C can be expressed by:

$$C = C(C_{tr} \text{ on } \partial^{\alpha}\mathcal{D}, {}^{\gamma_i}\mathbf{v} ; i = 1\ldots 4). \tag{4.140}$$

Since the $^{\gamma_i}\mathbf{v}$'s are time dependent random processes, the random operator $C(\omega)$ will also be time-dependent. It has been shown in [23] that $C(\omega)$ is a "random endomorphism" on $\mathcal{V}$ so that for the elastic wave propagation in the $\mathcal{L}(\mathcal{V})$ space (Banach), it will be a Banach space valued random function. On the basis of the above theoretical model, a simulation of the wave propagation can be obtained by the application of the Monte Carlo method.

The simulation can be performed for any number of sequences S_k (Fig. 33a) concurrently, either without considerations of interaction effects or with the inclusion of interactions between individual sequences. It is to be noted, however, that in the 1-D model interactions are accounted for only within an arbitrary sequence. In the 2-D and 3-D models, the interactions also occur between neighbouring sequences in both the longitudinal and transverse directions. Hence to deal with the simulation a slight change in notation is required. Considering the 1D-model first then the transmission operator reduces to a scalar transmission coefficient considered before, ie.:

$$^{(\alpha+1)}\mathbf{v}(t + {}^\alpha\tau) = {}^{(\alpha+1)}C_{tr}{}^\alpha\mathbf{v}(t). \tag{4.141}$$

On the assumption that the grains are perfectly elastic and have a given density, one can derive a so-called "impedance ratio χ" by considering the grain boundary between conjugate crystals to be rigid [295], [296], such that:

$$^{\alpha,(\alpha+1)}C_{tr} = \frac{2}{1 + {}^{\alpha,(\alpha+1)}\chi} \;;\; {}^{\alpha,(\alpha+1)}\chi = \frac{{}^{(\alpha+1)}\rho^{(\alpha+1)}c}{{}^\alpha\rho^\alpha c} \tag{4.142}$$

For the description of the interaction effects in the 2D and the 3D case, a model may be used, consisting of K-sequences of elements arranged parallel to each other in either the (X_1, X_2)-plane or the (X_1, X_2, X_3)-space, respectively. Thus a wave propagating through the system $\mathcal{B}$ in the X_1-direction for instance (2D-model), will pass through the elements for any given sequence in a different manner. In this case every grain will interact with one or two neighbours within $\mathcal{B}$ or the semi-infinite bar, but not on its lateral surfaces. For simplification of the analysis, it may be assumed that these interactions occur independently at the boundaries denoted in a shorter form by $(\alpha\beta)$ and $(\alpha\gamma)$, respectively. However an additional assumption is necessary, namely that the interactions occur in accordance with the conservation of energy principle with respect to the half-spaces of the neighbouring crystals as for instance the upper half-space of "α_2" and the lower half-space of the "β_1" grain, respectively (see also [23], [295]). Moreover, it is to be recognized that the initial condition for entering any two grains in two neighbouring sequences S_k is given by:

$$\Delta\bar{v}_i = {}^{\alpha_2}\bar{v} - {}^{\beta_1}\bar{v} \tag{4.143}$$

where the subscript i designates the initial state and the bar sign refers to the average over the cross-section of an element. Evidently, the assumed boundaries between sequences, which are specified by the given grain size $^\alpha d$ itself, require the continuity of velocities so that:

$$^{\alpha_2}\overline{v}_{tr}(t_0+ <\,^\alpha\tau >) = {}^{\beta_1}\overline{v}_{tr}(t_0+ <\,^\alpha\tau >) \qquad (4.144)$$

where "tr" refers to the final state pertaining to a micro element. This condition together with the energy conservation in terms of a "power flux" [23] in the neighbouring halves can be expressed by the probabilities:

$$^{\alpha_2}\mathcal{P}(t_0) + {}^{\alpha_1}\mathcal{P}(t) = {}^{\alpha_2}\mathcal{P}(t_0+ <\,^\alpha\tau >) + {}^{\beta_1}\mathcal{P}(t_0+ <\,^\alpha\tau >) \qquad (4.145)$$

leading to the transmitted velocities in the following form:

$$^{\alpha_2}\overline{v}_{tr} = {}^{\beta_1}\overline{v}_{tr} = \left\{ \frac{(^{\alpha_2}\overline{v})^2 + {}^{\alpha\beta}\chi(^{\beta_1}\overline{v})^2}{{}^{\alpha\beta}\chi + 1} \right\}^{1/2}. \qquad (4.146)$$

Hence considering the element α, a similar result is obtained for the transmitted velocity $^{\alpha_1}\overline{v}_{tr}$ in the "α_1 part" of α only, and so on. The final velocity $^\alpha\overline{v}_i$ incident upon the $(\alpha, \alpha + 1)$ interface is obtained by the use of the energy conservation as follows:

$$^\alpha\overline{v}_i = \left[\frac{1}{2}\{(^{\alpha_1}\overline{v}_{tr})^2 + (^{\alpha_2}\overline{v}_{tr})^2\}\right]^{1/2}. \qquad (4.147)$$

The transmission at the $\alpha, (\alpha + 1)$ interface can be calculated in the same manner as in the 1D-model by using relation (4.141).

Evidently, the above scheme can be extended to the 3D-model by allowing four half-spaces for the chosen element α in the sequence S_k. In order to obtain numerical results for the simulation the following computer scheme has been adopted:

(i) postulate a uniform initial velocity distribution in the first layer, i.e., $\overline{v}^{(0)} = 1\ m/sec$. The first layer of B is designated by 0 and the subsequent ones by $1, 2 \ldots$,

(ii) calculate the passage time: $\tau^{(0)} = {}^\alpha d/c_L^{(0)}$,

(iii) evaluate the arrival time at the $(0,1)$ interface: $t = \tau^{(0)}$,

(iv) generate two random variables: $\rho^{(1)}, E^{(1)}$ for every micro-element in this layer,

(v) evaluate the propagation velocity: $c_L^{(1)} = \left(\frac{E^{(1)}}{\rho^{(1)}}\right)^{1/2}$,

(vi) evaluate the passage time: $\tau^{(1)} = {}^\alpha d/c_L^{(1)}$,

(vii) evaluate the arrival time at the $(1,2)$ interface: $t = {}^\alpha d/c_L^{(1)}$,

(viii) calculate the transmitted wave velocity: $\overline{v}^{(1)} = C_0\overline{v}$,

(ix) evaluate the velocities due to the interactions at all (α, β) boundaries in the transverse direction according to (4.145) and (4.146),

(x) generate two random variables $\rho^{(2)}, E^{(2)}$ and repeat the procedure starting from (v).

For the numerical evaluation of the above models an f.c.c. metal (copper) was chosen amongst various other polycrystalline solids (see also [296]). The following mean values of the Gaussian distribution of the physical properties were used:

$$\left.\begin{array}{l} < {}^\alpha E >= 14.459 \cdot 10 \; N/m^2 \; ; \; \sigma_E =< {}^\alpha E > /1000 \\ < {}^\alpha G >= 5.46 \cdot 10 \; N/m^2 \; ; \; \sigma_G =< {}^\alpha G > /1000 \\ < {}^\alpha\rho >= 89 \cdot 10^3 /kg/m^3 \; ; \; \sigma_\rho =< {}^\alpha\rho > /200 \end{array}\right\} \quad (4.148)$$

in which ${}^\alpha E$ is the elastic modulus, ${}^\alpha G$ the shear modulus and ${}^\alpha\rho$ the density of a single crystal. The application of the Monte-Carlo method to the semi-infinite bar $\mathcal{B}$ of this material resulted in the velocity and dispersion time distributions as discussed above. A global comparison of the three models is indicated in Fig. 34, which represents the evolution of the given pulse in terms of the Gaussian fit of the probability density $p(v)$ taken at the locations 0.1 m, 0.5 m and 1.0 m from the front end of the bar $\mathcal{B}$ in Fig. (33b).

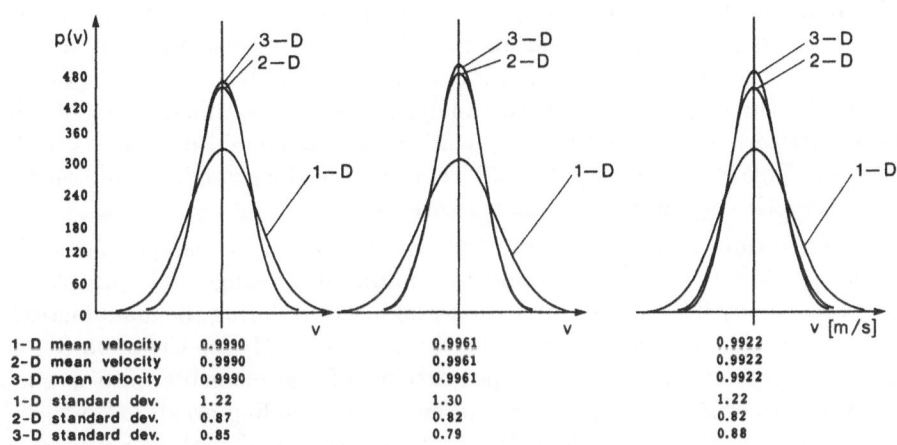

1-D mean velocity	0.9990	0.9961	0.9922
2-D mean velocity	0.9990	0.9961	0.9922
3-D mean velocity	0.9990	0.9961	0.9922
1-D standard dev.	1.22	1.30	1.22
2-D standard dev.	0.87	0.82	0.82
3-D standard dev.	0.85	0.79	0.88

Fig. 34. Probability densities of the velocity for the 1D, 2D and 3D-models of the wave propagation in $\mathcal{B}$ (Monte-Carlo simulation).

It may be seen from the Monte-Carlo simulation that the stochastic analysis considers the dispersion effects due to the discreteness of the microstructure. Although only elementary forms of interactions have been used to simplify the analysis more rigorous forms can be inserted for a more detailed simulation.

4.3 The Stochastic Analysis of Fibrous and Polymeric Networks.

In this section first the response behaviour of "fibrous structures" to a given load will be investigated by the use of stochastic mechanics. For the illustration of the theoretical analysis a cellulosic network with specified properties has been chosen (see also micrograph in Fig. 4 and the "idealized" structure of Fig. 9a). The significant relations linking the microscopic variables to the intermediate ones (mesoscopic) and thus to the macroscopic quantities are again the characteristic material functionals or operators. An operator for the given cellulose structure will be derived in an explicit form so that it becomes possible to numerically evaluate the induced "internal stresses" in the network.

The second application of the stochastic mechanics theory in this section concerns another class of discrete media, namely the behaviour of polymer melts, when subjected to external influences. This class of materials is of considerable interest in chemical engineering practice.

(i) Stochastic mechanics of fibrous structures:

It is apparent that for the description of the response of the network, the behaviour of single fibres and the effect of bonding between them at a "fibre cross-over" within the network is fundamental. It has been shown to be convenient for the simplification of the analysis to define a "micro element" of a fibrous network as a "free fiber segment" that is anchored between two fiber crossings. This is indicated in Fig. 35, representing the geometry and deformation kinematics of a fiber α in the network.

By definition, a meso domain of the network consists of an ensemble of $\alpha = 1 \ldots N$ fibres, and a corresponding number of crossings and "pores". The latter are usually filled by a binding material inserted during the manufacturing of the network, but are neglected in the present analysis. Customarily, the fiber crossings are taken as the "projection" of the elements onto a plane. However, an actual crossing occurs via an actual bonding area "^{b}a", which in general may be time-dependent due to a possible bond motion between the crossing fibres. This will be discussed subsequently after the introduction of geometrical probability concepts. It is readily seen that for the purely elastic response of the fiber induced by an external load, a "micro deformation $^{\alpha}u$" and a corresponding micro stress $^{\alpha}\sigma$, will be given by:

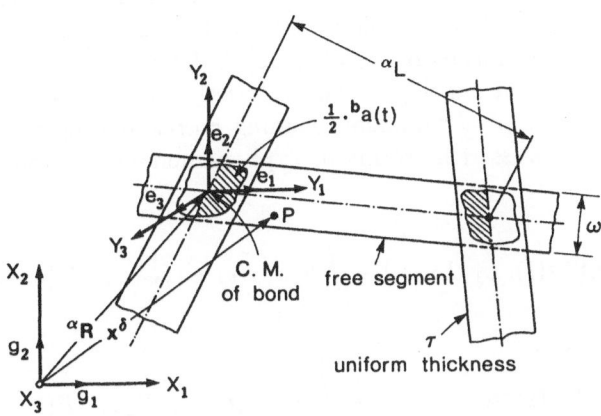

Fig. 35. "Idealized micro-element" of fibrous structures (cellulose network).

$$^{\alpha}u_i = {}^{\alpha}\Lambda_{Ii}{}^{\alpha}u_I \; ; \; {}^{\alpha}\Lambda_{Ii} = \cos(e_I, g_i),$$
$$^{\alpha}\sigma = {}^{\alpha}E\nabla^{\alpha}\mathbf{u} \text{ or } {}^{\alpha}\sigma = {}^{\alpha}E \; {}^{\alpha}\epsilon$$

$$\left.\begin{array}{r}\\ \\ \end{array}\right\} (4.149)$$

where the quantity $^{\alpha}\Lambda_{Ii}$ denotes the transformation matrix from the internal to the external reference frame $(X_1, X_2, X_2)(Fig.35)$, $^{\alpha}E$ the elastic modulus of the fiber and $^{\alpha}\epsilon$ the microstrain.

Using the base vectors of the two frames gives an approximate strain measure for the elastic deformation, if the element is assumed for simplicity to be homogeneous, where ϵ:

$$\epsilon: \epsilon_{IJ} = \frac{1}{2}(\delta_{IK}\nabla_J + \delta_{JK}\nabla_I)u_K. \tag{4.150}$$

However, most fibrous structures exhibit a rheological behaviour and the duration of a purely elastic response is rather short. Thus in order to characterized this behaviour of a single fiber, one can use a phenomenological form in terms of hereditary integrals [297] and the usual tensor notation so that:

$$^{\alpha}\underset{\approx}{\epsilon}(t) = {}^{\alpha}\underset{\approx}{E}^{-1}\{{}^{\alpha}\underset{\approx}{\sigma}(t) + \int\limits_{-\infty}^{t} {}^{\alpha}\underset{\approx}{H}(t-\tau)^{\alpha}\underset{\approx}{\sigma}(\tau)d\tau\}$$

$$^{\alpha}\underset{\approx}{\sigma}(t) = {}^{\alpha}\underset{\approx}{E}\{{}^{\alpha}\underset{\approx}{\epsilon}(t) - \int\limits_{\infty}^{t} {}^{\alpha}\underset{\approx}{R}(t-\tau)^{\alpha}\underset{\approx}{\epsilon}(\tau)d\tau\}$$

$$\left.\begin{array}{r}\\ \\ \\ \\ \\ \\ \end{array}\right\} (4.151)$$

in which the signs "$\approx$" and "$\underline{\underline{}}$" designate second and fourth order tensors, respectively, $^{\alpha}\underline{H}, ^{\alpha}\underline{R}$ the tensorial creep and relaxation functions of the time-dependent micro-strains and micro-stresses, (see also [297], [181]). Experimental observations of fibrous materials [299], indicate that almost all such structures have non-linear rheological characteristics. Hence the above response relations in the case of a "uniaxial" loading corresponding to an available test data, can be expressed in terms of a phenomenological model [297], [22] as follows:

$$^{\alpha}\epsilon(t) = g_1[^{\alpha}\sigma(t), L_1, L_2 \ldots] + \int_{0+}^{t} h_1(^{\alpha}\sigma(t), C_1, C_2 \ldots)H(t-\tau)d\tau, \qquad (4.152)$$

$$^{\alpha}\sigma(t) = g_2[^{\alpha}\epsilon(t), K_1, K_2 \ldots] + \int_{0+}^{t} h_2(^{\alpha}\epsilon(t), B_1, B_2 \ldots)R(t-\tau)d\tau, \qquad (4.153)$$

where the functions $g_1(\cdot)$ and $g_2(\cdot)$ correspond to the non-linear elastic response, while $h_1(\cdot)$ and $h_2(\cdot)$ represent the non-linear hereditary effects. Since the arguments of the functions $g(\cdot)$ and $h(\cdot)$ contain unknown constants, a differential approximation method should be used for their determination (see for instance ref. [300]). Hence assuming that in a first approximation $g_1(\cdot)$ and $g_2(\cdot)$ in the above relations are linear and that in the case of "stress relaxation", $(^{\alpha}\epsilon(t) \equiv ^{\alpha}\epsilon(0^+) = \epsilon_i)$, the ϵ_i's are time independent, one obtains the following approximation:

$$^{\alpha}\sigma_i(t) = ^{\alpha}E\epsilon_i + h_2(\epsilon_i, B_1, B_2 \ldots)\int_{0+}^{t} R(\tau)d\tau \qquad (4.154)$$

To find an expression for the relaxation kernel $R(\tau)$, it may be assumed that the latter will satisfy an nth order differential equation, in the form of:

$$a_0R + a_1R^{(1)} + a_2R^{(2)} + \ldots + a_{n-1}R^{(n-1)} + a_nR^{(n)} = 0 \qquad (4.155)$$

where

$$R^{(i)} = \frac{d^{(i)}R}{dt^{(i)}} \; ; \; (i = 1, 2 \ldots n) \qquad (4.156)$$

with the general solution:

$$R(t) = \sum_{i=1}^{n} D_n \exp(R_i t) \qquad (4.157)$$

and where $R_i; (i = 1, 2 \ldots n)$ are the roots of eqn. (4.154) and the D_n's are constants. Thus (4.154) may be expressed by:

$$^{\alpha}\sigma_i(t) = \,^{\alpha}E\epsilon_i + h_2(\epsilon_i, B_1, B_2 \ldots) \sum_{i=1}^{n} \frac{D_i}{R_i}[\exp(R_i t) - 1] \qquad (4.158)$$

Defining the second term of the above expression by the operator $R^*(\epsilon_i, t)$ or:

$$R^*(\epsilon_i, t) = -h_2(\epsilon_i, B_1, B_2 \ldots) \sum_{i=1}^{n} \frac{D_i}{R_i}[\exp(R_i t) - 1]$$

leads to the constitutive relation in its local form as:

$$^{\alpha}\sigma_i(t) = \,^{\alpha}\mathcal{A}(t)\,^{\alpha}\epsilon_i \text{ with } \,^{\alpha}\mathcal{A}(t) = \,^{\alpha}E - R^*(\epsilon_i, t) . \qquad (4.159)$$

Generalizing this relation to hold for any experiments and all components of the stress and strain tensors, respectively, gives:

$$^{\alpha}\boldsymbol{\sigma}(t) = \,^{\alpha}\mathcal{A}(t)\,^{\alpha}\boldsymbol{\epsilon}(t) \text{ or } \,^{\alpha}\sigma_{ij}(t) = \,^{\alpha}\mathcal{A}_{ijk\ell}\,^{\alpha}\epsilon_{k\ell}(t) . \qquad (4.160)$$

The constitutive relation (4.159) can of course be expressed as shown before in terms of micro stresses and micro deformations so that:

$$^{\alpha}\boldsymbol{\sigma}(t) = \,^{\alpha}A(t) \cdot \,^{\alpha}\mathbf{u}(t) , \text{ where } \,^{\alpha}A(t) = \,^{\alpha}\mathcal{A}(t) \cdot \boldsymbol{\nabla} \qquad (4.161)$$

which renders the operator $^{\alpha}A(t) \equiv \,^{\alpha}m(t)$, (micro-operator) a third order tensor.

In order to formulate the response behaviour of a fibrous system, it is important to include the bond interaction considered earlier in Chapt. 3. In particular these considerations were focused on the molecular structure of cellulose (Fig. 9a) and the fibre-fibre interface model pertaining to hydrogen bonding (Fig. 21). It has been shown that the most significant parameter in a bond motion is the "relative distance vector $^{\alpha\beta}\mathbf{d}$" at the crossing of the fibres α, β. To clarify this parameter further, the model used now assumes "perfect bonding" between the α, β fibres. This however can only occur for a relatively short duration under the influence of an external load. It is more likely that under a sustained loading some bond dissociation as well as formation of new bonds will occur, which can be represented by a stochastic bond deformation process. The latter expresses this behaviour in general by the evolution of the distribution of the number of bonds within a bonding area $^b a(t)$ or by the earlier discussed "BD-process" having a non-linear stochastic intensity. Studies of the molecular structure of cellulose networks have revealed that apart from an "intermolecular" bonding there occurs also an "intramolecular" bonding in order to account for the spatial configuration of the bond (see Fig. 20). X-ray data on cellulosic systems show that cellulose in natural fibres crystallizes partially or completely within the fibres in such a manner that "unit cells" are formed (Fig. 21). This "idealized model" has already been discussed with respect to an effective bonding area between the surface layers of fibres α, β. In order to extract a "force transmission" between these fibres at their crossing one can employ the Bollmann's "coincidence cell" model given previously with regard to the grain boundary configurations in polycrystalline solids. In the present case the 3-dimensional configuration of a "unit cell" within the interface between crossing fibres α, β (Fig. 21a) will be adopted. It is considered that the lattice formed by the cellulose unit cells is such that the length of the "repeating unit" is in the direction of the fibre-axis. The "coincidence sites" in this case are the OH-groups (2–2′), (Fig. 21(b)), that initially are at a distance $^b\Delta$ apart corresponding to the interatomic distance of the bond potential. A mechanical force exerted on the bond in the direction of the fibre axis, i.e. by excluding for simplicity a possible action due to bending or torsion, will result in a relative displacement denoted by $^b\mathbf{d}$ and the OH-groups will move to the configuration 2″ in Fig. 21(b). Whilst this model characterizes, in the simplest way the relative motion for the case of perfect bonding, the actual bond motion will be far more complex. It occurs in a random manner, although the significant parameter $^b\mathbf{d}$ of the bond potential could still be retained (see also [22]). It has also been argued in that reference that the bond potential can be reasonably represented by a "modified Morse potential" function of the following form:

$$\phi(r) \simeq \phi_0\{\exp[-2\nu|r - r_0|] - 2\exp[-\nu|\nu - \nu_0|]\} \tag{4.162}$$

in which ϕ_0 represents the equilibrium potential of the $\phi(r)$-curve at a distance r_0 from the origin and where higher order terms in $|r - r_0|$ have been neglected. Generally for a cellulosic structure $r_0 = 1.7\text{Å}$, $\phi_0 = 4.5$ KCal/mole and the spectroscopic constant ν is of the order 2Å^{-1}. It is assumed in this simplified model that a translation of "matching points" in the hydroxyl groups of the bond will occur that can be accounted for by the relative deformation vector $^b\mathbf{d}$. The latter is associated with a specific configuration during a deformation within the fibre crossing so that:

$$^b\mathbf{d}(t) = {}^b\boldsymbol{\delta}(t) - {}^b\boldsymbol{\Delta} \tag{4.163}$$

which by using the transformation matrix $^\alpha\Lambda_{Ii}$ from before, gives $^b\mathbf{d}(t)$ as:

$$^bd_i(t) = {}^\alpha\Lambda_{Ii}{}^bd_I(t) \tag{4.164}$$

Using this time-dependent vector yields the potential acting at an individual bond as follows:

$$\phi = \phi_0\{\exp[-2\nu|^b\mathbf{d}(t)|] - 2\exp[-\nu|^b\mathbf{d}(t)|]\} \tag{4.165}$$

where it is assumed that the number of bonds existing within the specific area ^ba of the interface (α, β) remains unchanged during the application of a load. Evidently, this simple model excludes the possibility of a partial or complete bond failure. In order to establish an interaction force in the bond from the above potential, the following fundamental relation [22], [23] can be used, viz.:

$$ {}^b\hat{\mathbf{f}} = -\frac{d\phi}{d|{}^b\mathbf{d}|}\mathbf{e} = \mathbf{e}\phi_0\{.,.\} \tag{4.166}$$

in which the " ^ " symbol indicates the "discreteness" of this force and the bracket corresponds to that of eqn. (4.165). To combine this discrete force with the fibre stress, it becomes necessary to generalize the latter so that:

$$ {}^b\tau =< \delta({}^b\mathbf{d} - {}^b\hat{\mathbf{d}}), {}^b\mathbf{f} > \tag{4.167}$$

where δ is the 3-dimensional Dirac-delta function. If ${}^b\mathbf{n}$ denotes the outward normal to the bonding area ${}^b a$, the bond stress ${}^b\boldsymbol{\sigma}$ can then be defined by:

$$ {}^b\boldsymbol{\sigma} = {}^b\mathbf{n}\frac{{}^b\mathbf{f}(t)}{{}^b a}. \tag{4.168}$$

Using the form of the interaction potential given in (4.165) and (4.168) yields thus the time-dependent bond stress:

$$ {}^b\boldsymbol{\sigma}(t) = \frac{2\nu}{{}^b a}\phi_0\{\exp[-2\nu|{}^b\mathbf{d}(t)|] - \exp[-\nu|{}^b\mathbf{d}(t)|]\}\mathbf{g} \cdot {}^b\mathbf{n} \tag{4.169}$$

where the base vector $\mathbf{g}$ is identical to ${}^b\mathbf{d}(t)/|{}^b\mathbf{d}(t)|$. Hence the bond stress in an operational form will be expressed by:

$$ {}^b\boldsymbol{\sigma}(t) = B \cdot {}^b\mathbf{d}(t) \tag{4.170}$$

in which B is the corresponding operator. It is seen that for the characterization of the behaviour of a single fiber two operators A, B are required. An explicit form of these operators will be given subsequently, after dealing with the application of geometric probabilities, that enter the analysis through an evaluation of the test data. These probabilities also arise from the use of an "appropriate mass density" in a 3-dimensional fibrous network. In this context, it becomes necessary to discuss briefly the "micro scale" introduced in the stochastic mechanics theory in conjunction with the possible experimental observations. It has been shown in previous work that in a sample of a fibrous network, surface deformations can be assessed to a high degree of accuracy by the application of "stress-holographic interferometry" [301, 302]. The employed procedure permitted the measurements for a network sample to be carried out at 240 points over the entire surface of the sample. Experimentally such a point is called "observation point (OP)". It has, itself, a dimension of 500 μm in diameter and contains, therefore, a certain number of fibres (on the average 200 fibres for the cellulosic network considered here). It is obvious that the suggested "microscale" pertaining to a single fibre or micro element is for experimental reasons too small and must be adjusted to the area and the volume corresponding to an "observation point (OP)". The situation arising from the experimental observation points is depicted in Fig. 36.

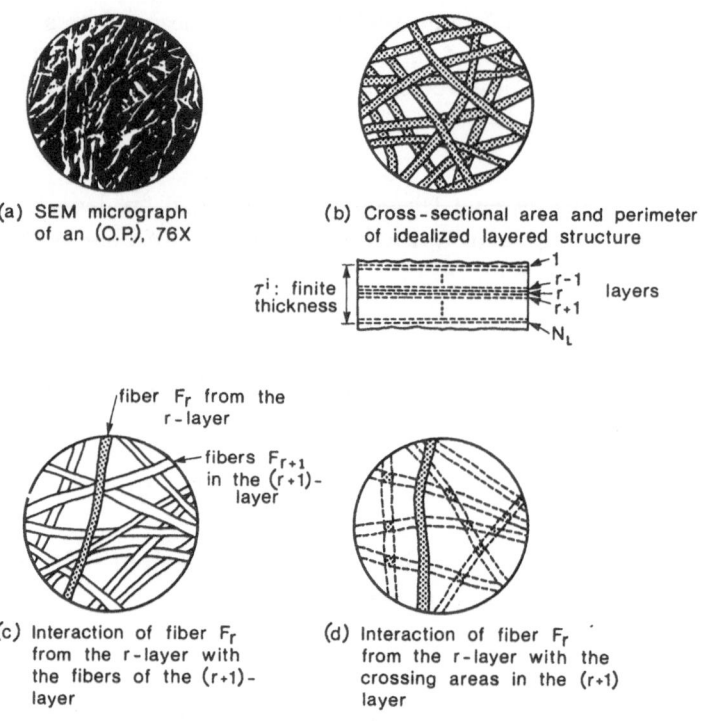

(a) SEM micrograph
of an (O.P.), 76X

(b) Cross-sectional area and perimeter
of idealized layered structure

(c) Interaction of fiber F_r
from the r-layer with
the fibers of the (r+1)-
layer

(d) Interaction of fiber F_r
from the r-layer with the
crossing areas in the (r+1)
layer

Fig. 36. Micro-structure at the observation points (OP).

Thus Fig. 36(a) shows a SEM micrograph of a cellulosic network, clearly indicating the planar structure of the system. It may be observed that in general,
several crossings of one fibre with other fibres may occur and thus it becomes
necessary to introduce geometrical probabilities of fiber crossings, the average
number of fibers, etc. pertaining to an observation point. Moreover, if the conventional definition of "mass density" is to be used, the corresponding mass
and volume of the (OP) has to be established. Thus, the mass density will be
given by $^i\rho = {^iw}/{^iv}$, $i \in Z^+$ for the "idealized plug" as shown in Fig. 36(b).
The volume ^iv of each (OP) is considered to be formed by a number of layers
N_L. Hence $^i\rho$ becomes a random variable. It is assumed, however, in the idealization of the structure, that individual layers are made up of fibers remaining
in the "plane of the layers" so that ^iv contains only a certain number of layers
N_L. A definition for the average number of fibers contained in ^iv can be given
by:

$$< {}^iN_f > = \frac{{}^iw}{\lambda \cdot \rho_f} \tag{4.171}$$

in which $\lambda = \frac{\pi R}{2}$ is an "average fiber length" (Fig. 36(b)), ρ_f the mass density of a single fiber, ${}^iw$ the total mass of fibers in ${}^iv, i \in Z^+$. This average number is experimentally accessible for most fibrous structures and for the given number of layers N_L in the volume ${}^iv$ of an (OP). The average number of fibers per layer can be stated as:

$$< {}^in_f > = \frac{< {}^iN_f >}{N_L} . \tag{4.172}$$

With these basic definitions of the geometrical parameters of the structure and following the discussion of section 3.4 of Chapt. 3 (see also Santaló [225]), one obtains the geometrical probabilities of fiber crossings within the cross-sectional area as follows:

$$P\{F_r \cap F_s \neq \emptyset; r \neq s\} = \frac{2\pi A + \pi L(\mathrm{w}_r + \mathrm{w}_s) + \pi^2 \mathrm{w}_r \mathrm{w}_s}{(L + \pi \mathrm{w}_r)(L + \pi \mathrm{w}_s)} . \tag{4.173}$$

By considering for simplicity that $\mathrm{w}_r = \mathrm{w}_s = \mathrm{w}$ (see Fig. 35) and the probability that only two fibers are crossing, the average number of crossings will be given by:

$$< {}^in_c > = P\{.,.\} \frac{1}{2} {}^in_f({}^in_f - 1)$$

and hence by using (4.172) and (4.173) one has:

$$< {}^in_c > = \frac{{}^in_f}{2}({}^in_f - 1)(1 - \frac{2R^2}{(2R + \mathrm{w})^2}). \tag{4.174}$$

In the above statement, no interaction effects between layers in the volume ${}^iv$ are considered. However, if ${}^iv$ is made up of several layers as indicated, the fibers F_r in the layer r and F_{r+1} in the layer $(r + 1)$ will interact. In this case the number of crossings on the average will be of the following form:

$$< {}^iN_c > = N_L < {}^in_c > + (N_L - 1) < n_c^{r,r+1} > \tag{4.175}$$

in which $n_c^{r,r+1}$ gives the number of "new crossings", whilst the other quantities retain the meaning given previously. To illustrate the possible interaction effects between layers and the fibers contained in them, consider Fig. 36(c) which indicates the situation of a fiber F_r interacting with the fibers in layer $(r + 1)$. It is shown that F_r crosses only two fibers in the layer $(r + 1)$, thus producing one new crossing only. In general, in order to establish an average value of $n_c^{r,r+1}$, one may visualize that the latter is formed by:

$$< n_c^{r,r+1} > = < n_{c_1} > - < n_{c_2} > \tag{4.176}$$

where $< n_{c_1} >$ is the number of crossings of all F_r's intersecting all F_{r+1}'s while $< Nn_{c_2} >$ represents the number of crossings between the F_r's and the crossing areas K_{r+1} in the layer $(r + 1)$ as shown in Fig. 36(d). The difference between $< n_{c_1} >$ and $< n_{c_2} >$ in (4.176) thus excludes the counting of more than one crossing of a fiber with fibers in adjacent layers. Considering the number of fibers to form a convex set, the geometrical probability for $< n_{c_1} >$ becomes:

$$< n_{c_1} >= P\{F_r \cap F_{r+1} \neq \emptyset\} <\,^i n_f >= \left(1 - \frac{2R^2}{(2R + w)^2}\right)(^i n_f)^2 \qquad (4.177)$$

and similarly for $< n_{c_2} >$:

$$< n_{c_2} >= P\{F_r \cap K_{r+1} \neq \emptyset\} <\,^i n_f >$$

so that:

$$< n_{c_2} >= \left[\frac{\Sigma \ell_{(r+1)} + \pi <\,^i n_c > w}{L + \pi w}\right] <\,^i n_f > \qquad (4.178)$$

where the $\ell_{(r+1)}$'s correspond to the circumference of the crossing areas $K_{(r+1)}$ in the layer $(r+1)$. The quantity $\ell_{(r+1)}$ can be determined for a fibrous network by the procedure due to Mack [304], giving:

$$< \ell_{(r+1)} >= \frac{4\pi \lambda^2 \cdot w}{2\pi \lambda w + \lambda + w}. \qquad (4.179)$$

Thus in terms of the geometric probabilities, one obtains by using relations (4.176–4.179), the average number of crossings in the elemental volume $^i v$ of an (OP) as follows:

$$< \,^i N_c >= N_L <\,^i n_c > +(N_L - 1)\Big[\left(1 - \frac{2R^2}{(2R + w)^2}\right)(<\,^i n_f >)^2$$

$$-\left(\frac{4\lambda^2 w + 2\pi \lambda w + \lambda + w}{(2R + w)(2\pi \lambda w + \lambda + w)}\right) <\,^i n_c ><\,^i n_f >\Big]. \qquad (4.180)$$

It is to be noted that all parameters in the above expression are experimentally accessible and have also been discussed in references [22, 303].

So far as the "local" operators pertaining to fibers, their bonding and respective forms are concerned, the following remarks will serve to clarify the micro material operator $^\alpha m(t)$ denoted here by $^i m(t)$ for the (OP) scale. Since the material functional in general is essentially a function of time due to the change with time of an "actual bonding area $^b a(t)$" under the applied stress, one can write at the "micro element" scale that:

$$^\alpha \kappa(t) = \frac{^b a(t)}{\lambda \cdot w}, \qquad (4.181)$$

in which $^{\alpha}\kappa(t)$ is valid for specified fibers of width w, λ its average length and $^{b}a(t)$ an actual time-dependent bonding area between crossing fibers. Hence the "overall micro deformation" of a fibrous element α can be expressed to a first approximation by the form:

$$^{\alpha}\mathbf{u}(t) = \{(1 - {}^{\alpha}\kappa(t)){}^{\alpha}\Lambda^{f}\mathbf{u}(t) + {}^{\alpha}\kappa(t){}^{b}\mathbf{d}(t)\} \qquad (4.182)$$

in which $^{\alpha}\Lambda$ is the orientation matrix that transforms the local components of $"^{f}\mathbf{u}(t)"$ and $^{b}\mathbf{d}(t)"$ to the external frame X_i. It has been shown in [22], that the stress in the fiber segment $^{f}\sigma(t)$ and that in the actual bonding area $^{b}\sigma(t)$ can be combined so that the force transmission in the bond, i.e. from the fiber stress to bond stress can be expressed simply by the ratio in terms of a parameter $^{\alpha}\mu(t)$ such that:

$$^{b}\sigma(t) = \frac{fa}{{}^{b}a(t)}{}^{f}\sigma(t) = {}^{\alpha}\mu(t){}^{f}\sigma(t). \qquad (4.183)$$

in which $^{\alpha}\mu(t)$ can be established by experimental observations. By using the earlier notation and on the assumption that $^{b}\sigma(t)$ is proportional to the stress in the unsupported fibre segment, one can express to a first approximation the overall micro-operator of a fibrous element as follows:

$$^{\alpha}m(t) = \{[1 - {}^{\alpha}\kappa(t)]^{\alpha}\Lambda A^{-1}(t) + {}^{\alpha}\kappa(t){}^{\alpha}\Lambda B^{-1}(t) \cdot {}^{\alpha}\mu(t)\}. \qquad (4.184)$$

In view of the above geometrical probabilities of the fiber arrangement within the finite volume of the experimental points (OP), indexed by $i = 1 \dots N, i \in Z^+$, the corresponding material operator $^{i}m(t)$ has been obtained as follows:

$$\left.\begin{aligned}
{}^{b}a(t) &= \frac{\pi\lambda^2 \cdot \mathrm{w}^2}{2\pi\lambda\mathrm{w} + \lambda + \mathrm{w}} \\
{}^{i}\kappa(t) &= \frac{2\pi\lambda \cdot \mathrm{w} \cdot {}^{i}N_c}{(2\pi\lambda\mathrm{w} + \lambda + \mathrm{w})^{i}N_f} \\
{}^{i}\mu(t) &= \frac{(2\pi\lambda \cdot \mathrm{w} + \lambda + \mathrm{w}) \cdot \tau}{\pi\mathrm{w}\lambda^2}
\end{aligned}\right\} \qquad (4.185)$$

and hence:

$$\left.\begin{aligned}
{}^{i}m(t) = \{&[1 - \frac{2\pi\lambda \cdot \mathrm{w} \cdot {}^{i}N_c}{(2\pi\lambda\mathrm{w} + \lambda\mathrm{w})^{i}N_f}]^{i}\Lambda A^{-1}(t) \\
&+ [\frac{\tau \cdot {}^{i}N_c}{\lambda \cdot {}^{i}N_f}]^{i}\Lambda B^{-1}(t) \cdot {}^{i}\mu(t)\},
\end{aligned}\right\} \qquad (4.186)$$

where τ refers to the thickness of an individual fiber and the transformation matrix $^{i}\Lambda$ refers to the centre of mass of the experimental point (OP) with respect to the external frame.

For the numerical evaluation of the "stress" distribution in the 3-dimensional fibrous network, which is of the main interest here, a cellulosic system has been used with the following material properties:

Young's modulus of elasticity (macroscopic):

$E_{e\ell} = 1.26 \times 10^{13} \ dynes/cm^2$. (measured in the direction of the applied load)

Viscoelastic modulus (macroscopic):

$E_{vis} = 0.9 \times 10^{13} \ dynes/cm^2$.

Average mass density:

$^i\rho = 731 \ kg/cm^3$ (according to measurements in [301, 303].)

Overall sample size: 88 mm (length) $\times 55$ mm (width) $\times 0.085$ mm (thick).

Environmental conditions: $23.5°$ C temperature and 50% relative humidity.

Applied Load (uni-axial): 82 N.

The quantities related to the observation points (OP), $i = 1\ldots 240$ for each surface of a sample (see Fig. 37) in the "Double-sided or DSHI-holographic interferometry method ([301], [303]) were as follows:

Average mass of fibers:	$^iw = 1.96 \times 10^{-7} \ gr$
Average thickness:	$^i\tau = 85.6 \times 10^{-4} \ cm$
Average number of fibers:	$^iN_f = 200$
Average number of fiber crossings:	675
$< {}^\alpha\kappa(t) >= 0.0676$,	$^iu(t) = 0.01535$
Average value of operator:	$A = 165.6 \times 10^{-15} \ dynes/cm^3$
Average value of operator:	$B = 285.3 \times 10^{-22} \ dynes/cm^3$
Transformation matrix Λ:	$\begin{bmatrix} 0.9936 & -0.1132 \\ 0.1132 & 0.9936 \end{bmatrix}$.

With the above basic properties of the material samples and the use of the DSHI-method [302], the micro deformations $^i\mathbf{u}$ (not tabulated here) have been established for the (OP) points of the grid indicated in Fig. 37(a).

The macroscopic deformations of the test samples were obtained by means of "Photonic-Sensor" measurements as described in ref. [301], [302] and resulted in the phenomenological curve given in Fig. 37(b). The use of the material data given above in a FORTRAN computer program led then to the numerical values of the local operator $^im(t)$ for the observation points (OP), $(i = 1\ldots 240)$ on each sample surface and the corresponding relation (4.186). A division of the grid on the material surface into nine meso-domains (Fig. 37(a)) brought about the numerical values of the meso operator as outlined earlier and hence the "internal stress distribution" as indicated in Fig. 38(a,b) below.

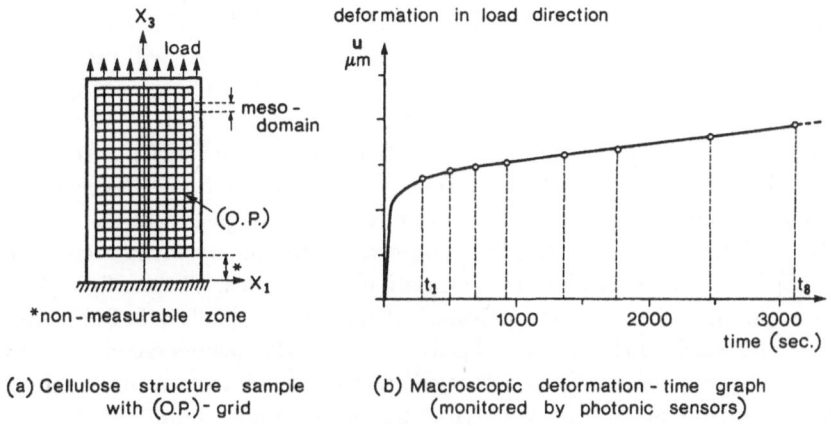

(a) Cellulose structure sample
with (O.P.)- grid

(b) Macroscopic deformation - time graph
(monitored by photonic sensors)

Fig. 37. Uni-axial loading and macroscopic deformation of a cellulose structure.

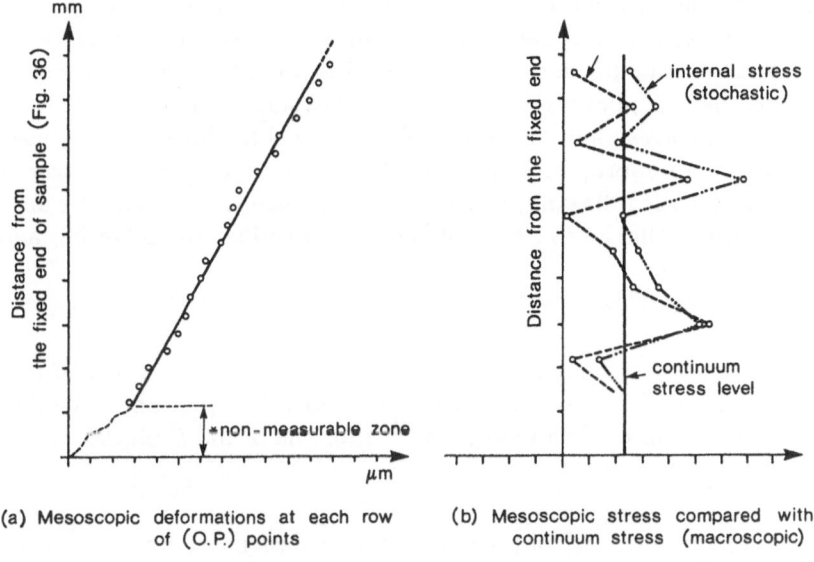

(a) Mesoscopic deformations at each row
of (O.P.) points

(b) Mesoscopic stress compared with
continuum stress (macroscopic)

Fig. 38. Internal stress distribution of a cellulose network sample.

(ii) Stochastic analysis of polymer melts.

This section is concerned with the stochastic dynamics of two immisicible polymer melts that has been considered earlier in refs. [305, 306]. Since the amorphous state of the binary microstructure of the melt consists of chain-molecules of the two polymers, which in general assume the shape of "random coils", the most significant aspect of their behaviour due to an applied external field, is their interaction at a large number of "contact points". Since a specific material domain contains a finite number of chains, the latter will cross each other without interpenetration causing an "entanglement". These entanglements are regarded in the present analysis as "time-dependent or temporary junctions (T.J.)" at which the monomeric or "structural units (S.U.)" are bonded together by a weak interaction of the van der Waals type. Several analytical models for the dynamics of polymers have been suggested in the past (see for instance [307–315]). In particular temporary polymer networks were considered in [318] from the point of view of statistical mechanics. The configurational and chain-dynamics were treated amongst others in refs. [313, 315, 317] and a topological approach to the motion of polymer networks has been considered in [316] and subsequent papers. It is obvious that from a microstructural point of view the description of the overall dynamics of the melt should be based on the motion of the individual macromolecules of the polymers with the inclusion of their interactions. The assumptions made in ref. [305, 306] that the melt-microstructure (amorphous) is of a random nature will be maintained here. The two phases of the polymer melt consist of an α-polymer that is dispersed in a β-polymer (matrix) resulting in a highly non-coherent interface between them. It is also assumed, in accordance with the often occurring case in practice, that the α, β phases have chain molecules such that their average length is $^\beta\ell \gg {}^\alpha\ell$ and corresponds to their molecular weights. Since the stochastic analysis requires an appropriate choice of an "element" of the microstructure it is convenient for the following molecular considerations as well as the description of the "local" behaviour of the polymers to introduce the following additional definitions:

Def. 1:

A "structural or monomeric unit (S.U.)" is defined by a pair of monomers with the bond vector $\mathbf{b}_0$. It has the length $\ell = |\mathbf{b}_0|$ and a conformation angle θ. It will be denoted by $^\alpha m, {}^\beta m$ for the α and β phases, respectively.

Def. 2:

A "chain segment s_k" of the α, β polymers designated by $^\alpha s_k, {}^\beta s_k$ is defined by its length $^\alpha\ell, {}^\beta\ell$, respectively times a set of rotation angles, i.e.:

$$^\alpha s_k = |^\alpha\mathbf{b}_0| \sum_{i=1}^{k-1} \cos {}^\alpha\theta_i \; ; \; {}^\beta s_k = |^\beta\mathbf{b}_0| \sum_{i=1}^{k-1} \cos {}^\beta\theta_i.$$

It is important to note that the maximum length of the chain segments s_k is given by:

$$[s_k]_{max} = \bar{\bar{s}}_k |\mathbf{b}_0| \equiv \text{Dia. of the "unit cell"},$$

where $\bar{\bar{s}}_k$ is the cardinal number of the set $\{s_k\}$ (see also [306]).

Def. 3:

A "unit cell (U.C.)" is defined as the union of a set of α and β segments, i.e.:

$$\{{}^{\alpha}s_k\} \cup \{{}^{\beta}s_k\} \Rightarrow U.C.$$

The topological space $\Gamma \subset \mathbb{R}^d$ which includes the set of "temporary junctions (T.J.)$\equiv \{{}^{\alpha\beta}m\}$" is defined by:

$$(\Gamma \backslash \{{}^{\alpha}s_k\}) \cup (\Gamma \backslash \{{}^{\beta}s_k\}) \cup \{{}^{\alpha}s_k\} \cap \{{}^{\beta}s_k\} \Rightarrow \Gamma \subset \mathbb{R}^d.$$

Since the union of the sets (Fig. 39(a)) are composed of both polymer chains, the "volume fractions" can be introduced, where:

$${}^{\alpha}c + {}^{\beta}c = 1 \; ; \; {}^{\alpha}V = \text{volume of } \{{}^{\alpha}s_k\} \; ; \; {}^{\beta}V = \text{volume of } \{{}^{\beta}s_k\}$$

and hence:

$${}^{\alpha}c = {}^{\alpha}V/V_u \; ; \; {}^{\beta}c = {}^{\beta}V/V_u \; ;$$

in which V_u is the volume of the unit cell (U.C.). Due to the assumption that the average chain length of the two polymers, i.e. ${}^{\beta}\ell \gg {}^{\alpha}\ell$ so that ${}^{\beta}s_k({}^{\beta}\ell) \gg {}^{\alpha}s_k({}^{\alpha}\ell)$, it is convenient to choose the diameter of the unit cell to be the "maximal length" of the chain segments.

It is seen from the above definitions that the "chain structure" of the two polymers of the melt as well as the random coil-segments are regarded as a sequence of "pair-molecules" with their associated "inter-molecular potentials". The individual temporary junctions (T.J.) at which a weak interaction occurs may be regarded in terms of an "intra-molecular" or "pseudo-potential". The latter is usually employed in the analysis of binary compounds (see for instance [170]). A diagrammatic sketch of possible chain configurations in a domain $(\overline{AB})$ is given in Fig. 39.

A sketch of a "unit cell" with two segments of the participating "test chains ${}^{\alpha}c_1, {}^{\beta}c_1$" is shown in Fig. 39(b), whilst Fig. 39(a) indicates the possible entanglements over several segments of the chain ${}^{\alpha}c_k$ with ${}^{\beta}c_1 \in \{{}^{\beta}c_k\}$. Although the intermolecular interaction of a structural unit (S.U.) determines the microstructure of each polymer in the melt, the occurence of macromolecules and the formation or dissociation of the temporary junctions between individual segments ${}^{\alpha}s_k, {}^{\beta}s_k$ add considerably to the complexity of the formulation of the flow dynamics. Hence for the simplification of the analysis the "idealized configurations" indicated in (a), (b) of Fig. 39 are assumed and the notion of "unit cells' will be employed. This excludes however any stereo-chemical arrangements of the participating structural units. Due to the large number of

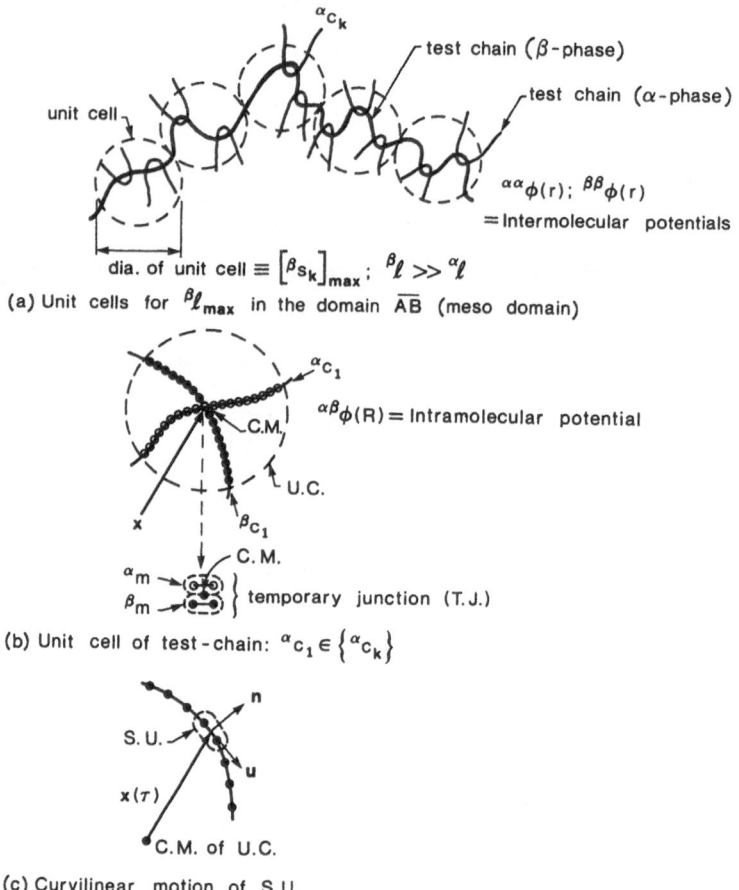

(a) Unit cells for ${}^{\beta}\ell_{max}$ in the domain $\overline{AB}$ (meso domain)

(b) Unit cell of test-chain: ${}^{\alpha}c_1 \in \left\{ {}^{\alpha}c_k \right\}$

(c) Curvilinear motion of S.U.

Fig. 39. Sketch of a possible configuration in the domain (Dia.= $\overline{AB}$, meso-domain) and U.C. in $\overline{AB}$.

possible configurations of both types of chain molecules and their entanglement, the motion including "weak interactions" can be considered as a random phenomenon.

(a) The Poissonian behaviour of entanglements of the polymers:

As shown in ref. [306] the dynamics of the melt can be conveniently studied to a first approximation, on the basis of Markov theory and an abstract dynamical system characterized by the triple $[X, \mathcal{F}, \mathcal{P}]$. The probabilistic function space X can be identified with a state-space Z of the composite material, $\mathcal{F}$ is then the σ-algebra of event sets (Borel) in this space and $\mathcal{P}$ an appropriate probability measure. However, in dealing with the chain behaviour of the polymers by adopting the "monomeric units" as primitive elements, a rigorous analysis would involve quantum chemistry with the corresponding probabilistic forms and an "event structure" that leads together with the Neumann algebra to a Hamiltonian field representation (see Percus [211], Hafner [321] and others). In the present formulation, an event structure is used in accordance with the classical theory of probability which is based on the principles and axiomatic definitions of stochastic mechanics given in Chapt. 2.

By considering the fundamental interactions, i.e., the entanglement between the two polymers as a "random phenomenon", it can be characterized by a family of real random variables or the "number of temporary junctions (T.J.) denoted by "N_t" per unit cell of the melt, and the time parameter $t \in T$ in form of a stochastic process $(N_t, t \in T)$. If the joint distribution functions of the random variables $\{N_{t_1}, N_{t_2} \ldots N_{t_n}\}$ is known for all finite $(n = 1, 2 \ldots)$ and all sets of values $(t_1, t_2 \ldots t_n) \subset T$ and if these distributions are compatible, then there exists a probability field $[X, \mathcal{F}, \mathcal{P}]$. Hence $\{N_t, t \in T\}$ is defined on the probability space X or a subspace of it, (for instance the configuration space $\mathcal{C} \subset X$) for which the measure $\mathcal{P}\{N_{t_1} \leq x_1, \ldots N_{t_n} \leq x_n\}$ is equal to the prescribed distribution for every $(n = 1, 2 \ldots)$ and $(t_1, \ldots t_n) \subset T$. In particular the random process $\{N_t\}$ is Markovian, if:

$$\mathcal{P}\{N_t \leq x | N_{t_1} = y_1, N_{t_2} = y_2 \ldots N_{t_n} = y_n\} = \mathcal{P}\{N_t \leq x | N_{t_n} = y_n\} \quad (4.187)$$

for all $t_1 \leq t_2 \leq \ldots t_n \leq t; (n = 1, 2 \ldots)$ and all possible values of the random variables. This process is uniquely determined by the initial and the conditional distribution functions as follows:

$$\mathcal{P}\{N_0 \leq x\} = P(0, x)$$

$$\mathcal{P}\{N_t \leq x | N_s = y\} = p\{s, y; t, x\} \; ; \; (s \leq t) \quad (4.188)$$

which are the "transition probabilities". Since

$$\mathcal{P}\{N_t \leq x\} = P(t, x) \text{ or } P(t, x) = \int\limits_{-\infty}^{\infty} p(0, y; t, x) d_y P(0, y)$$

it follows by the theorem of total probability that:

$$p\{x, y; t, x\} = \int\limits_{-\infty}^{\infty} p\{u, z; t, x\} d_z p\{s, y; u, z\}; \; (s \leq u \leq t) \subset T \quad (4.189)$$

expressing the Chapman-Kolmogorov relation of Markov theory. The random process of chain entanglements, i.e., the creation and destruction of temporary junctions is in the present case a special Markov process. For the simplification of the analysis, it is assumed that the process $\{N_t\}$ is a "homogeneous Poisson process", or a process with independent increments in which the transition probabilities depend only on $(x - y)$ apart from t and s and that during a short time interval $(0, t) \in T$ only "one event" of "entanglement" occurs. This can be expressed by:

$$P\{N_t \equiv N\} = P_N(t) \text{ for } t \geq 0 \text{ ; } (N_0 = 0) \tag{4.190}$$

or in general for the homogeneous process:

$$P\{N_t - N_s = N\} = P_N(t - s). \tag{4.191}$$

For such a process there exists a constant $\lambda > 0$ such that:

$$P_N(t) = \frac{(\lambda t)^N}{N!} e^{-\lambda t} \text{ ; } (N = 1, 2 \ldots) \tag{4.192}$$

Hence the distribution function $P_N(t)$ will be determined for a given initial distribution by a system of differential equations, viz.:

$$\frac{dP_N(t)}{dt} = -\lambda P_N(t) + \lambda P_{N-1}(t) \text{ ; } (N = 1, 2 \ldots) \tag{4.193}$$

According to (4.192) the transition probability is a step function of x, i.e.:

$$p\{s, y; t, x\} = \sum_{N=0}^{(x-y)} \frac{\lambda^N (t - s)^N}{N!} e^{-\lambda(t-s)}, \tag{4.194}$$

if y is a non-negative integer $y \leq x$ and $s \leq t$. Equivalently, one can consider the stochastic process of entanglement as a finite Markov process in terms of a finite number of "entanglement states" corresponding to the random events of their occurence. These states are part of the more general "microstate", representable by a state-vector (multi-dimensional case), as an element of the state-space $Z \subset X$. Thus, if the number of entanglements at a given instant of time t, $N \equiv \xi$; designates the state "j" of the temporary junction, the transition probability is given by:

$$P\{\xi_t = j | \xi_s = i\} = p_{ij}(s, t) \text{ ; } (s \leq t) \tag{4.195}$$

and the conditional distribution becomes:

$$p\{s, y; t, x\} = \sum_{j \leq x, i \leq y} p_{ij}(s, t) \tag{4.196}$$

so that the Chapman-Kolmogorov relation in matrix form is expressed by:

$$P_{ik}(s, t) = \sum_j P_{ij}(s, u) P_{jk}(u, t) \text{ ; } (s \leq u \leq t). \tag{4.197}$$

As stated earlier, under certain conditions [138], one can obtain the evolution of the transition probabilities from relations (4.197) by the use of Kolmogorov's first and second system of differential equations. However, in formulating the flow dynamics of the polymer melt, the occuring entanglements consist not only of the formation but also of the destruction of the temporary junctions. This random phenomenon is therefore better described by a Birth-Death or (BD) Poisson process. Again, to simplify the analysis and considering the one-dimensional case only, if the number $\xi_k \equiv N$ of temporary junctions per unit cell at time t is associated with the "microstate N" or $\{^{\alpha\beta}m\}$ at this time, then for one more (T.J.) to occur within the time interval $(t, t + \Delta t)$, one has the probability of "formation $\lambda\Delta t + O(\Delta t), \lambda > 0$". Similarly, for the occurence of the destruction of a (T.J.) in this time interval, one has the probability "$\mu\Delta t + O(\Delta t), \mu > 0$", independently of the other existing T.J.'s. This leads to an evolution equation for $P_N(t)$ in the form of:

$$\frac{dP_N(t)}{dt} = -N(\lambda + \mu)P_N(t) + (N-1)\lambda P_{N-1}(t) + (N+1)\mu P_{N+1}(t) \quad (4.198)$$

with the initial conditions:

$$P_N(0) = \begin{cases} 1 & \text{for } N = 1 \\ 0 & \text{for } N \neq 1. \end{cases} \quad (4.199)$$

The solution of the above relation can be obtained by the method of generating functions. Eqn. (4.198) in fact is based on the assumption that there exists a continuous function $c_j(t)$ holding uniformly in t so that the transition of interaction of the α, β polymers within $(t, t + \Delta t)$ occurs with $c_j(t + O(\Delta t))$ and where for simplification of the analysis, it is assumed that:

$$c_j(t) \equiv c_j \; ; \text{ and } c_N \equiv c(^{\alpha\beta}m) = N(\lambda + \mu) \qquad \text{(a)}$$

which in the present case are expressed by:

$$c_N P_{N,N+1} = N\lambda \; ; \; c_N P_{N,N-1} = N\mu \qquad \text{(b)}$$

$$\left. \right\} (4.200)$$

It is apparent that the N_t-process of entanglement between the polymers of the melt will have in the dynamic case variable intensities or $\lambda(t)$ and $\mu(t)$. Hence it signifies the use of a wider class of random processes, ie. the "random point processes". Moreover, the formation and destruction of the T.J.'s will occur at "micro-times" which themselves are random variables. Hence by considering the given complete probability space $[X, \mathcal{F}, \mathcal{P}]$ with a distinguished family $(F = \{F_t\}, \; t \geq 0)$ of right-continuous sub-σ-algebras of $\mathcal{F}$ and letting $T = \{\tau_n\}, \; n \geq 1$ be a sequence of "Markov times" with respect to the system F such that with P.a.s., one has:

(i) $\tau_1 > 0$

(ii) $\tau_n < \tau_{n+1} \; ; (\tau_n < \infty)$

(iii) $\tau_n = \tau_{n+1} \; ; \; (\tau_n = \infty)$

$$\left. \right\} (4.201)$$

The random sequence $T = \{\tau_n\}, n \geq 1$ is fully characterized by a "counting process" or:

$$N_t = \sum_{n \geq 1} I_{\{\tau_n \leq t\}} \; ; \; t \geq 0. \tag{4.202}$$

In this sense, the analysis of the sequence $T = \{\tau_n\}, n \geq 1$ is equivalent to that of the process $N = N_t, t \geq 0$. The sequence of Markov times T satisfying (4.201) is then a "random point process" and N_t defined by (4.202) is also a point process.

It is to be noted that the latter is a special case of a "multivariate point process (M.P.P.)". In relation to the N_t-process, one can define for each $t \geq 0$ a sub-σ-field of $\mathcal{F}$ or $\mathcal{F}_t^N$ by:

$$\mathcal{F}_t^N = \sigma(N_s, s \in [0, t]) \tag{4.203}$$

which is generated by the family of random variables $N_s, s \in [0, t]$, where $[0, t]$ is a closed time interval. The family $\mathcal{F}_t^N$ is often referred to as the "internal history of the N_t-process". A family of sub-σ-fields $\mathcal{F}_t$ of $\mathcal{F}$ is called a history, if it is increasing, i.e., $\mathcal{F}_t \geq \mathcal{F}_s$ whenever $s \leq t$. The above consideration forms a linkage to the "martingale" theoretical analysis [47].

Apart from the representation of the entanglement between the α and β polymers, it is also necessary to consider their interactions as well as the "local balance" relations that arise due to the application of an external field. For this purpose it is convenient to use the "configuration space $\mathcal{C} \subset X$" as a subspace of the stochastic state-space. It may be noticed from Fig. 39(b) that the position vectors pertaining to the structural units of the polymers and that of the T.J.'s with respect to the C.M. of a "unit cell" revert in the following sets on the domain Γ of a U.C. to:

$$\left. \begin{array}{l} \{^{\alpha}x_i\} = [^{\alpha}x_1, {}^{\alpha}x_2 \ldots {}^{\alpha}x_n] \Rightarrow \{^{\alpha}m\} \subset \Gamma \\ \{^{\beta}x_j\} = [^{\beta}x_1, {}^{\beta}x_2 \ldots {}^{\beta}x_m] \Rightarrow \{^{\beta}m\} \subset \Gamma \\ \{^{\alpha\beta}x_\ell\} = [^{\alpha\beta}x_1, {}^{\alpha\beta}x_2 \ldots {}^{\alpha\beta}x_N] \Rightarrow \{^{\alpha\beta}m\} \subset \Gamma. \end{array} \right\} \tag{4.204}$$

It is apparent that the chain segments consisting of many structural units are kept together by bond potentials or intermolecular potentials $^{\alpha}\phi$ and $^{\beta}\phi$, respectively. Analogously the "weak interactions" at the temporary junctions of two chain segments can be visualized as a "pseudo-potential" of the following form:

$$^{\alpha\beta}\phi(^{\alpha}x_\ell, {}^{\beta}x_\ell) = \begin{cases} ^{\alpha\beta}\phi(0, |^{\alpha}x_\ell - {}^{\beta}x_\ell|), & \text{if } |^{\alpha}x_\ell - {}^{\beta}x_\ell| \leq \Delta R, \\ & (\ell = 1 \ldots N) \\ 0 & \text{if } |^{\alpha}x_\ell - {}^{\beta}x_\ell| \geq \Delta R \end{cases} \tag{4.205}$$

where "ΔR" denotes the range of weak interactions between the participating structural units "$^{\alpha}m$" and "$^{\beta}m$", as shown in Fig. 40(a) and Fig. 39(b,c).

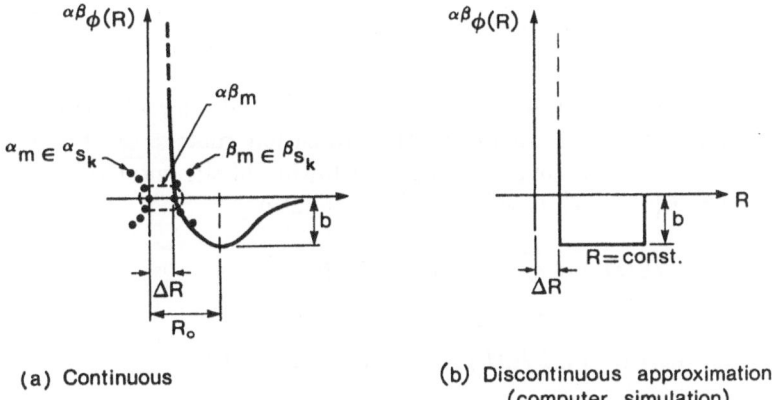

(a) Continuous

(b) Discontinuous approximation
(computer simulation)

Fig. 40. "Pseudo-potential $^{\alpha\beta}\phi(R)$.

It is assumed that $^{\alpha\beta}\phi$ has the usual properties of symmetry, homogeneity and the nearest neighbour characteristic given earlier. For an instantaneous dynamical equilibrium within a unit cell (U.C.), the equilibrium distance denoted by "$\overset{\circ}{R}$" is given by:

$$\overset{\circ}{\mathbf{R}}_\ell = {}^\alpha\overset{\circ}{\mathbf{x}}_\ell - {}^\beta\overset{\circ}{\mathbf{x}}_\ell \; ; \; \ell = 1 \ldots N,$$

and similarly for the intermolecular potentials $^\alpha\phi(r), {}^\beta\phi(r)$:

$$^\alpha\overset{\circ}{\mathbf{r}}_i = {}^\alpha\overset{\circ}{\mathbf{x}}_i - {}^\alpha\overset{\circ}{\mathbf{x}}_{i-1} \; ; \; {}^\beta\overset{\circ}{\mathbf{r}}_j = {}^\beta\overset{\circ}{\mathbf{x}}_j - {}^\beta\overset{\circ}{\mathbf{x}}_{j-1}.$$

In the above form of $^{\alpha\beta}\phi(R)$, (4.205) the representation follows that of weak binding in molecular solids, when the latter is considered per atom as a fraction of the characteristic atomic or molecular energies. If "ΔR" is taken as the range of the strongly repulsive region at short distances, then the longer range corresponds to the range of the familiar dispersion for the van der Waals attraction. Fig. 40(b) indicates an approximation of $^{\alpha\beta}\phi(R)$ as a non-smooth or discontinuous function of R usually employed in computer calculations.

The application of an external flow field (extensional or shear flow are most commonly considered), designated by the velocity $\mathbf{v}(\mathbf{x}) \in \mathcal{V}$, where $\mathcal{V}$ is a subspace of Z, induces a change of the position vectors $^\alpha\mathbf{x}_\ell, {}^\beta\mathbf{x}_\ell$, respectively within the range of interactions "ΔR". This should be recognized in the formulation of the interactions and particularly in the representation of the entanglement process N_t. As previously stated the transition probabilities (eqn. 4.197) on the assumption of a time-homogeneous Markov process satisfy the systems of Kolmogorov equations. The latter can also be written as follows:

$$P'_{ik}(t) = P_{ik}(t)q_k + \sum_{j \neq k} P_{ij}(t)q_k, \text{ (forward)}, \; i,k \geq 0, \qquad \text{(a)}$$

$$P'_{ik}(t) = -q_i P_{ik}(t) + \sum_{j \neq i} q_{ij} P_{jk}(t), \text{ (backward)}, \; i,k \geq 0 \qquad \text{(b)}$$

$$\left. \right\} \quad (4.206)$$

in which the q_{ij}'s are known as the "infinitesimal characteristics" of the N_t-process and are the elements of the transition matrix Q of the Chapman-Kolmogorov relation. Denoting the distribution of the entanglements at time t by: $P_k(t) = P[N_t = k]$, it will satisfy the relation:

$$P'_k(t) = -P_k(t)q_k + \sum_{j \neq k} P_j(t)q_k \; ; \; k \geq 0. \qquad (4.207)$$

The initial distribution: $P_0(k) = P^e_k$ will also satisfy:

$$P^e_N q_k = \sum_{j \neq k} P_j q_{jk} \; ; \; k \geq 0 . \qquad (4.208)$$

Evidently $P_k(t) = P^e_N$ holds for all $t \geq 0$ and all k and hence it is called the "equilibrium distribution" of the N_t-process. Regarding the latter as a BD-Markov process in which the Q-matrix has non-zero elements, one can write that:

$$q_{N,N+1} = \lambda_N \; ; \; q_{N+1,N} = \mu_N \; ; \; N \geq 0 \qquad (4.209)$$

resulting for $\lambda_N \mu_{N+1} > 0$ for all $N > 0$ in an equilibrium distribution given by:

$$P^e_N = K r_N$$

where

$$r_0 = 1 \; ; \; r_N = \frac{\lambda_0 \dots \lambda_N}{\mu_1 \dots \mu_{N+1}} \; ; \; N \geq 1 \; ; \; K = \frac{1}{\sum\limits_{j=0}^{\infty} r_j} < \infty. \qquad (4.210)$$

It is evident that within a unit cell Γ the behaviour of the entanglements $\{^{\alpha\beta}m\}$ changes with time although this set is bounded according to the definition of Γ (Def. 3). Hence a change must occur in such a manner that the "total number" of the participating structural units of both the α, β polymers in Γ remains the same. This is reflected in the earlier assumption of the N_t-process to be homogeneous and stable where all microstates of the T.J.'s, $\ell \in \mathbb{R}^+$ and $q_\ell < \infty$ as well as $q_\ell = \sum\limits_{k \neq \ell} q_{k\ell}$ are conserved. As a consequence one can write for the potential energy per unit cell, that:

$$\Gamma U = \sum_{i=1}^{|\alpha N|} {}^{\alpha}U_i + \sum_{j=1}^{|\beta N|} {}^{\beta}U_j + {}^{\alpha\beta}U_N , \tag{4.211}$$

in which here $|^{\alpha}N|, |^{\beta}N|$ denote the cardinality of the sets $\{^{\alpha}m\}$, $\{^{\beta}m\}$, and $^{\alpha}U_i$, $^{\beta}U_j$ the energies corresponding to the quilibrium values of the intermolecular potentials $^{\alpha}\phi, {}^{\beta}\phi$ respectively. A rigorous calculation of these potentials as well as that of the interaction energy is possible by the methods of quantum chemistry of molecular complexes (see for instance Hobza and Zahradnik [170, 171]). It follows from relation (4.211) that the interaction energy per unit cell Γ is given by:

$$\alpha\beta U_N = {}^{\Gamma}U - \sum_{i=1}^{|\alpha N|} {}^{\alpha}U_i - \sum_{j=1}^{|\beta N|} {}^{\beta}U_i . \tag{4.212}$$

The potential energy of an individual temporary junction at equilibrium in terms of the pseudo-potential $^{\alpha\beta}\phi$ is then:

$$\alpha\beta U_i = r_\ell {}^{\alpha\beta}\phi(0, |^{\alpha}\mathbf{x}_\ell - {}^{\beta}\mathbf{x}_\ell|) \tag{4.213}$$

where the parameter r_ℓ is defined by relation (4.210). This means that at an equilibrium microstate of the junction $^{\alpha\beta}m \in \{^{\alpha\beta}m\}$ this parameter is given by:

$$r_\ell \triangleq \frac{\lambda_\ell}{\mu_\ell} = 1 \; ; \; (\ell = 1\dots N) \tag{4.214}$$

and where $^{\alpha\beta}U_\ell$ is at its minimum, while the entropy s_ℓ is at its maximum for $R_\ell \equiv \overset{\circ}{R}_\ell$.

It is of interest to note that a simplification of this formulation can be achieved by regarding the destruction of the T.J.'s or the "disentanglement" as the "primary process" and the formation of new entanglements as a "secondary process" (see Kröner et al. [318, 322]). In this case the BD-Poisson process of the entanglement becomes a simple D-Poisson process with the single intensity μ. It is possible by the use of the above forms (4.211, 4.213) to formally state the "free energy" per unit cell Γ, i.e.:

$$G_\ell = U_\ell - Ts_\ell$$

and

$$\Gamma G = (|^{\alpha}N| - n_u)^{\alpha}\phi_{min} + (|^{\beta}N| - n_u)^{\beta}\phi_{min} + n_u [{}^{\alpha}\phi_{min} + {}^{\beta}\phi_{min} + r_\ell {}^{\alpha\beta}\phi_{min}]$$
$$-T\{n_u({}^{\alpha}s_\ell + {}^{\beta}s_\ell + \Delta^{\alpha\beta}s_\ell) + {}^{m}s(|^{\alpha}N| + |^{\beta}N| - 2n_u)\} \tag{4.215}$$

in which ^{m}s refers to the entropy of the n_u or non-participating monomeric units in the BD-process. The above form (Gibbs free energy) has been explored rigorously by Müller and Wilmanski [323] in considerations of the transition of crystalline polymers to the amorphous state of the melt.

(b) The local balance relations and flow dynamics:

It has been shown above, that the entanglements between the α, β polymers of the melt can be represented by a Poisson process. For the required flow dynamics of the melt, it is necessary to consider the "local balance" relations first. Most analytical investigations concerned with the dynamics of polymer blends or polymers in solution employ the well-known reptation (or tube) model and frequently the linear "Rouse model". More recently a rigorous analysis using the projection method of statistical mechanics has been given by Hess [324] (see also [325]) in accounting for the longitudinal and lateral motion of a "test chain" in the polymer. Polymer solutions were studied by Öttinger [326] in terms of the Langevin equation approach. The present analysis considers the motion of chain segments belonging to either of the α or β polymers (Fig. 39(c)) in which a "curvilinear path" is formed at any instant of time by the "local tangent vector" to the chain segment. This can be expressed by:

$$\mathbf{u}_k(s_k, t) = \frac{\partial}{\partial s_k} \mathbf{x}_u(s_k, t) \; ; \; |\mathbf{u}| = 1 \qquad (4.216)$$

in which $\mathbf{x}_u$ is the position vector of a structural unit of the chain segment with respect to the C.M. of the unit cell Γ. It is also necessary to consider the "contour length ℓ_k" of the segment s_k at time t, which is defined by:

$$\ell_k \triangleq \int_0^{s_k} ds'_k |\frac{d\mathbf{x}_u(s'_k)}{ds'_k}| \qquad (4.217)$$

where $\ell_k = \bar{s}_k |\mathbf{b}_0|$, (see Def. 2) corresponds to the maximum path length of the segment. These relations (4.216, 217) hold for both the α and β polymer chain segments, respectively. It is to be noted that ℓ_k is also equal to the ratio of the molecular weight M_{s_k} of the segment to that of the monomeric unit M_u, which is often required for numerical calculations. To determine the resistance to the induced motion by an external field, the following forces must be taken into account.

First, there are the forces due to the given intermolecular potentials (eqn. 4.165) that are inherent in either chain of the α, β-polymer. They are defined by their respective bond vectors of the structural units, ie.:

$$\left. \begin{aligned} {}^\alpha\mathbf{f} &= -\frac{\partial^\alpha \phi({}^\alpha\mathbf{r}_i)}{\partial^\alpha r}{}^\alpha\mathbf{u} = -k_1 |{}^\alpha\mathbf{b}_0|{}^\alpha\mathbf{u} \\ {}^\beta\mathbf{f} &= -\frac{\partial^\beta \phi({}^\beta\mathbf{r}_j)}{\partial^\beta r}{}^\beta\mathbf{u} = -k_2 |{}^\beta\mathbf{b}_0|{}^\beta\mathbf{u} \end{aligned} \right\} \qquad (4.218)$$

in which ${}^\alpha\mathbf{u}, {}^\beta\mathbf{u}$ are here the unit vectors tangential to the curvilinear motion and ${}^\alpha\mathbf{r}_i = {}^\alpha\mathbf{x}_i - {}^\alpha\mathbf{x}_{i-1}; {}^\beta\mathbf{r}_j = {}^\beta\mathbf{x}_j - {}^\beta\mathbf{x}_{j-1}$ are the arguments in the molecular potentials $^{\alpha}\phi, ^{\beta}\phi$, respectively. A second contribution to the local forces originates from the T.J.'s due to the weak interaction potential $^{\alpha\beta}\phi(R)$, which also causes a slight lateral motion. It is equivalent to the derivative of the potential in the direction of $^{\alpha\beta}R$ multiplied by the unit tangent vector and an "average friction coefficient ξ_0", valid for a fixed number of "n_u" of a chain segment so that:

$$^{\alpha\beta}\mathbf{f} = -n_u\xi_0\frac{\partial^{\alpha\beta}\phi(R)}{\partial R} \cdot \frac{^{\alpha}\mathbf{u} \times ^{\beta}\mathbf{u}}{|^{\alpha}\mathbf{u} \times ^{\beta}\mathbf{u}|}. \tag{4.219}$$

in which it is assumed that the conformation angles of the structural units of both polymers are approximately the same. The introduction of an external flow field, i.e., $\mathbf{v}(\mathbf{x}) \in \mathcal{V} \subset Z$ induces a change in the position vectors $^{\alpha}\mathbf{x}_i$ and $^{\beta}\mathbf{x}_j$ within the range "ΔR" of the interaction and hence causes a sliding of the T.J.'s. Thus in terms of a relative velocity $\mathbf{w} = ^{\alpha}\dot{\mathbf{x}}_i - ^{\beta}\dot{\mathbf{x}}_j$ and an average viscosity η_0, this force may be expressed as follows:

$$^{v}\mathbf{f} = n_u\eta_0\mathbf{w} \; ; \; \mathbf{w} \in \mathcal{V} \subset Z.$$

The contribution of the interaction force normal to the tangent of the curvilinear motion, i.e., the lateral force component at the T.J. by recognizing from Fig. 39(c) that:

$$\mathbf{n} = \frac{\mathbf{x}^n(\tau)}{|\mathbf{x}^n(\tau)|} \; ,$$

where in general:

$$\mathbf{x}(\tau) = \mathbf{x}^n(\tau) + \mathbf{x}^u(\tau) \; ; \; \mathbf{x}^u(\tau) = (\mathbf{x}(\tau) \cdot \mathbf{u}(\tau))\mathbf{u}(\tau)$$

$$\mathbf{x}^n(\tau) = \mathbf{x}(\tau) - (\mathbf{x}(\tau) \cdot \mathbf{u}(\tau))\mathbf{u}(\tau) \tag{4.220}$$

gives $^{v}\mathbf{f}$ as follows:

$$^{v}\mathbf{f} = -\frac{\partial^{\alpha\beta}\phi(R)}{\partial|\mathbf{x}^n(\tau)|} \cdot \mathbf{n}, \tag{4.221}$$

where in relations (4.220) and 4.221) the time-dependent position vectors to the C.M. of the T.J.'s (Fig. 39(c)) is given for $\tau > 0, \tau \in T \equiv \{\tau_N\}$ in accordance with eqn. (4.201).

It is of interest to note that the polyatomic structure of the α, β polymers which results in the assumed flexible random coil structure also induces some randomness in the relative directions. However, the directional correlation between bonds of the structural units more than one bond length $|\mathbf{b}|$ apart becomes rather negligible in the stochastic mechanics of the melt. In particular in the "melt state" the chain segments interpenetrate or become strongly "entangled", which in the present analysis has been assumed to be only "weak

interactions". Although each chain interacts directly with a large number of
others, the forces transmitted can produce a rather large reversible deformation in each chain. However over short time intervals of the set $\{\tau_N\}$, the melt
responds "elastically" to mechanical distortions. The "relaxation times" for
disentanglement is usually of the order of micro seconds. Depending on the
process of mixing and the choice of chain lengths of the polymer as well as
their conformations, one obtains a different response behaviour for shorter and
longer time scales. The latter are in the context of the entanglement of the
polymers associated with the earlier considered set of Markov times.

By taking into account relations (4.218–4.221), the local forces pertaining
to a segment "s_k" of either polymer can be obtained by neglecting (4.221) for
simplification as follows:

$$\mathbf{f}_{sk} = -{}^\alpha n_u \frac{\partial^\alpha \phi({}^\alpha\mathbf{r})}{\partial^\alpha r}{}^\alpha\mathbf{u} - {}^\beta n_u \frac{\partial^\beta \phi({}^\beta\mathbf{r})}{\partial^\beta r}{}^\beta\mathbf{u} - n_u\xi_0 \frac{\partial^{\alpha\beta}\phi(R)}{\partial R}{}^{\alpha\beta}\mathbf{u} + n_u\eta_0\mathbf{w} \quad (4.222)$$

Hence by introducing the total number "$\mathcal{N}$" of segments contained in a unit
cell results in the following force balance relation:

$$-\frac{\{{}^\alpha\bar{\bar{s}}_k\}}{\mathcal{N}}{}^\alpha n_u \frac{\partial^\alpha \phi({}^\alpha\mathbf{r})}{\partial^\alpha\mathbf{r}}{}^\alpha\mathbf{u} - \frac{\{{}^\beta\bar{\bar{s}}_k\}}{\mathcal{N}}{}^\beta n_u \frac{\partial^\beta \phi({}^\beta\mathbf{r})}{\partial^\beta\mathbf{r}}{}^\beta\mathbf{u}$$

$$+\{{}^{\alpha\beta}\mathbf{f}^f + {}^{\alpha\beta}\mathbf{f}^v\} < N(t) > -\mathbf{f}^{ext} = 0$$

or equivalently in briefer form:

$$- < {}^\alpha\mathbf{f} > - < {}^\beta\mathbf{f} > +\{{}^{\alpha\beta}\mathbf{f}^f + {}^{\alpha\beta}\mathbf{f}^v\} < N(t) >= \mathbf{f}^{ext} \quad (4.223)$$

in which according to the given definition (Def. 3):

$$\frac{\{{}^\alpha\bar{\bar{s}}_k\}}{\mathcal{N}}{}^\alpha n_u = {}^\alpha c \; ; \; \frac{\{{}^\beta\bar{\bar{s}}_k\}}{\mathcal{N}}{}^\beta n_u = {}^\beta c$$

and $\mathbf{f}^{ext}$ the average force acting on the unit cell caused by the external field.
It is also assumed in the above relation that the motion of the α,β-chains
is a collective one, i.e., that only the process of "disentanglement" from the
instantaneous equilibrium is predominant for the deformation process (see also
Kröner [318]). Hence the Poissonian dissociation process applies so that eqn.
(4.198) reduces to:

$$\frac{dP_N(t)}{dt} = -\mu N(t)P_N(t) + \mu[N(t) + 1]P_{N+1}(t) \quad (4.224)$$

where at equilibrium ${}^{\alpha\beta}\phi$ has its minimum ($t = t_0, \lambda = \lambda_0 = 0$) and only
disentanglement of the T.J.'s occur with $N_0 = N(t_0)$. At any other instant of
time, i.e., $t \geq t_0$ the process $N(t) \leq N_0$ with $c_N > 0$ (in eqn. 4.199). Thus the
change with time of the probability of disentanglement has the solution:

$$P_N(t) = \binom{N_0}{N} e^{-\mu N_0 t} [e^{\mu t} - 1]^{N_0 - N} \tag{4.225}$$

so that the average value of the entanglement per unit cell from the equilibrium state onwards and its variance can be given as follows:

$$E\{N(t)\}_{t \geq t_0} = <N(t)> = N_0 e^{-\mu t} , \ t \geq t_0,$$

$$\sigma^2(t) = N_0 e^{-\mu t} [1 - e^{-\mu t}], \tag{4.226}$$

in which the disentanglement process is seen as the primary, while the new formations of the T.J.'s, as a secondary process.

Although, it may be assumed for simplification that the energy density $^\Gamma E$ per unit cell will approximately be the same for all unit cells, i.e., that it is conserved, the application of an external flow field characterized by the velocity $\mathbf{v}(\mathbf{x}) \in \mathcal{V} \subset Z$ induces a change so that:

$$^\Gamma E = U_i + \frac{1}{2}\mathbf{v}^2$$

and hence

$$\frac{D}{dt} \int_\Gamma (U_i + \frac{1}{2}\mathbf{v}^2) dV_u = \int_\Gamma \mathbf{f}_u \cdot \mathbf{v} dV_u \tag{4.227}$$

The operator D/Dt however can be brought under the integral, since the microdomain Γ is fixed, viz:

$$\int_\Gamma [\frac{D}{Dt}(v_i + \frac{1}{2}\mathbf{v}^2) - \mathbf{v}_u \cdot \mathbf{v}] dV_u = 0. \tag{4.228}$$

If the flow is represented by a potential ψ, then it follows from (4.228) that:

$$\frac{dv_i}{dt} = \mathbf{v}_u \ \text{grad} \ \psi - \text{grad} \ \psi \frac{d}{dt} \ \text{grad} \ \psi , \tag{4.229}$$

showing that the evolution of the energy $^\Gamma E$ per unit cell can be related to the microstructure of the polymer melts and the entanglement process. It is apparent, that for numerical calculations, it becomes necessary to establish corresponding "macroscopic relations" characterizing a particular flow field (in most practical applications this field is an extensional or shear flow). The required step by using an intermediate material domain (according to the given definitions of Chapt. 2) can then be taken as follows. Denoting by $\mathbf{X}(t)$ the position vectors of individual unit cells with respect to the C.M. of a meso-domain and correspondingly the configurational distribution function by $P(\mathbf{X}_1, \mathbf{X}_2 \ldots \mathbf{X}_m, t)$ yields an evolution equation for this distribution of the Fokker-Planck type, i.e.:

$$\frac{\partial P(\mathbf{X}, t)}{\partial t} = -\sum_{k=1}^{m} \frac{\partial}{\partial \mathbf{X}_k} [\mathbf{v}(\mathbf{X}, t) + \sum_{k=1}^{m} \mathbf{f}_u(t)] P(\mathbf{X}, t) \qquad (4.230)$$

in which $\mathbf{v}(\mathbf{X}, t)$ is the applied velocity field and $\mathbf{f}_u(t)$ the time-dependent overall force vector by relations (4.223). The solution of the above stochastic differential equation is discussed amongst others in refs. ([129, 131, 153]). The stochastic analysis of polymer melts can be extended to a "state-space" analysis [23, 153], in which the unit cells are regarded as bounded random sets. The evolution of the distribution function can then be formulated in terms of a Gibbs point process in the state-space [287].

4.4 Simple Fluids and the Flow in Fully Saturated Porous Media.

It has been mentioned in section (2.2) of Chapt. 2 that the molecular dynamics and lattice models of discrete fluids are based on the statistical mechanics of particle systems with the main aim to relate the macroscopic observable quantities to the details of the occurring intermolecular effects between the elements of a fluid on the microscopic level. In the analytical development of fluid dynamics, one can distinguish two formalisms for the description of the flow phenomena of discrete fluids. The first formulation known as "molecular dynamics" introduces the classical Newtonian equations of motion of an ensemble of elements (atoms, molecules), which are solved numerically and then integrated to yield the evolution of the configurational and velocity distributions. These calculations position a system of N-elements within a cell of fixed volume and a set of velocities is given to this ensemble selected in such a manner as to make the net momentum equal to zero. Subsequent trajectories of the elements are calculated by step-wise numerical integration of the equations of motion with the inclusion of interaction effects. In contrast to this completely deterministic approach, the second type of formalism introduces stochastic elements in the analysis by using computer simulation techniques. The most frequently used technique is the Monte-Carlo method already discussed earlier. By the use of the latter method a set of molecular configurations is generated by allowing the elements of an ensemble to undergo random displacements. This computational procedure requires, however, a criterion for the acceptability or rejection of a new configuration so that after a certain number of transitions a configuration will occur with a probability proportional to the Boltzmann factor. In this section the class of discrete fluids known as "simple liquids" will be treated from the point of view of stochastic mechanics. Such fluids are characterized by spherically non-saturating interactions and are commonly modelled by a Lennard-Jones potential. More recently it has been shown that non-equilibrium molecular dynamics, which were initially developed to establish the flow properties of simple fluids, can be extended to polymeric liquids [317], and in particular to polymer melts discussed in the foregoing section. Since the molecular dynamics approach and the simulation method are both of a stochastic nature, the dynamics of discrete fluids can be treated on the basis of probabilistic concepts. Thus the fundamental concepts of the stochastic mechanics of fluids are the same as for solids. In the former the field quantities assigned to a "microelement" (atom, molecule) are again regarded as random variables or functions of such variables and the axiomatic definitions (section (2.3) Chapt. 2) remain valid.

(i) The dynamics of discrete fluids.

Here only those quantities pertaining to the molecular dynamics of fluids are briefly re-stated. In general an element "α" is taken as the primitive base and designates a molecule of the discrete fluid. A smaller unit (atom), if considered, is again defined by $^{\alpha}a \overset{\mathrm{df}}{=}$ an element of $\alpha \in \{\mathcal{A}\}$. A "configuration" is the image of α at time t : $\mathbf{r}(\alpha, t) \in \mathbb{R}^3$ or briefly $^{\alpha}\mathbf{r} \in \mathbb{R}^3$ and a "motion" of α is $\alpha \overset{\mathrm{df}}{=} \{^{\alpha}\mathbf{r}; -\infty < t < \infty\}$ in the discrete and time continuous case. These motions are regarded as a stochastic process $\{x(t)\}$ for each α. A meso-domain M is defined as a countable set of molecules, i.e., $M = \{\alpha\}$. The macro-domain of the fluid body $\mathcal{M} = \{M\}$. The volume of $\alpha \in \{\mathcal{A}\}$ at time t is defined by:

$$^{\alpha}v \equiv {}^{\alpha}v(\alpha, t) = \int\limits_{\mathbf{r}(\alpha, t)} d^3\mathbf{r} \; ; \; {}^{\alpha}v_0 \equiv {}^{\alpha}v(\alpha, 0) = \int\limits_{\mathbf{R}(\alpha, 0)} d^3\mathbf{R} \qquad (4.231)$$

$\mathbf{r}, \mathbf{R}$ being the current and initial position vectors of α in the Euclidean frame, respectively and where the notation of Chapt. 2 has been used. An intersecting system of molecules $\alpha, \beta \ldots \in \{\mathcal{A}\}$ has the volume $v(\alpha \cap \beta) = 0$. Evidently, this definition excludes the more complex case of strong interactions such as bonding to occur. By admitting a finite volume of the elements, the mass density is accordingly somewhat different from that given previously for the "particle density" at an arbitrary point $\mathbf{r}$ in the medium (compare with relation in Chapt. 2). Thus the mass density is now defined by: $^{\alpha}\rho = {}^{\alpha}\mu / {}^{\alpha}v$, where $^{\alpha}\mu$ is a scalar element of a set M representing the mass of a molecule $\alpha \in \{\mathcal{A}\}$. For a particular meso domain, one has therefore:

$$^{M}\rho = \int\limits_{R_\rho} {}^{\alpha}\rho d\mathcal{P}(^{\alpha}\rho) \; ; \; \int\limits_{R_\rho} d\mathcal{P}(\rho) = 1 \; , \qquad (4.232)$$

where the Lebesgue integration extends over the subspace $R_\rho \subset X$ on the set M that is embedded in the general probabilistic function space X and $\mathcal{P}(^{\alpha}\rho)$ is the Lebesgue-Stieltjes measure on R_ρ. Moreover, considering the domain M

on R_ρ as an open sphere of radius A or $\{^M\mathbf{r} : |\mathbf{R} - {}^M\mathbf{r}| < A\}$ with respect to the Euclidean frame, where $^M\mathbf{r}$ denotes the position vector to the C.M. of the meso-domain. Thus the volume of such a domain will be:

$$^Mv \sim v(M,t) \triangleq \int\limits_{\mathbf{r}(M,t)} d^3\mathbf{r} \quad \text{and} \quad {}^Mv_0 \equiv v(M,0) \triangleq \int\limits_{\mathbf{R}(M,0)} d^3\mathbf{R}. \qquad (4.233)$$

Analogously to the formulation of solids by recognizing the "events" as the basic elements in the stochastic analysis and that they form a Borel σ-algebra $\mathcal{F}$ on X gives together with the probability measure $\mathcal{P}$ of these events an abstract dynamical system $[X, \mathcal{F}, \mathcal{P}]$. In a given experiment with a finite range $\Delta\mathbf{z}$ of the state-vector $\mathbf{z} \in Z \equiv X$ defines then the events by:

$$E = \{^\alpha\mathbf{z} : \mathbf{z} < {}^\alpha\mathbf{z} < \mathbf{z} + \Delta\mathbf{z}\}$$

yielding an experimental σ-algebra $\mathcal{F}_{exp}$ so that the convergence of results in terms of $\mathcal{P}$ and $\mathcal{P}_{exp}$ has to be achieved. It is assumed that the components of $^\alpha\mathbf{z}$ are real-valued functions of the geometrical and physical properties of the fluid. To represent a "configurational" meso-domain of the fluid body that conceptually corresponds to the notion of a "control volume" in hydrodynamics, one has to use a set of state vectors or $\{^\alpha\mathbf{z} : {}^\alpha\mathbf{z}_i, \alpha = 1, 2 \ldots N, i = 1, 2 \ldots r\}$ and an appropriate function space. Analogously to the treatment of solids, it is also convenient in the considerations of fluids to use subspaces of the state-space X. Thus, if the fluid is modelled by hard spheres and a Lennard-Jones potential, one can use for instance:

$$^\alpha\mathbf{v}(t) \in \mathcal{C}, \; \mathcal{C} \subset X \; ; \; {}^\alpha\mathbf{v}(t) \in \mathcal{V}, \; \mathcal{V} \subset X \qquad (4.234)$$

where $\mathcal{C}$ is the configuration space and $\mathcal{V}$ the velocity space. With the above given event structure, one obtains subsets $E \subset X$ that include the state of an element "α". Similarly in order to analyse the random behaviour of single elements of the fluid or their collective mode of motion an appropriate measure on these sets must be chosen so that $\mathcal{P}$ for the events E is bounded. The notion of a measure is related to the fact that a "dynamical variable" of the flow is a random variable, characterized by its probability distribution function $P\{E\}, 0 \leq P\{E\} \leq 1$ for all $E \in \mathcal{F}$ satisfying the properties of a regular measure. Thus by choosing the velocity space $\mathcal{V} \subset X$ and a corresponding measure $\mathcal{P}^v$ one has $[\mathcal{V}, \mathcal{F}^v, \mathcal{P}^v]$ for the representation of the discrete fluid. The velocity vector $\mathbf{v} \in \mathcal{V}, \mathcal{V} \subset X$ is then defined as a $\mathcal{P}^v$-regular measurable function in $\mathcal{V}$. The expectation or mean value of the velocity vector then becomes:

$$E\{^\alpha\mathbf{v}\} = <{}^\alpha\mathbf{v}> = \int\limits_{\mathcal{V}\subset X} {}^\alpha\mathbf{v} d\mathcal{P}^v \qquad (4.235)$$

where the above integral is understood in the Lebesgue-Stieltjes sense. Similarly the standard deviation is given by:

$$D^{\mathbf{v}} = \left\{ \int ({}^{\alpha}\mathbf{v} - <{}^{\alpha}\mathbf{v}>)^2 d\mathcal{P}^{\mathbf{v}} \right\}^{1/2} \tag{4.236}$$

which makes the velocity space $\mathcal{V}$ a Hilbert space with the inner product $({}^{\alpha}\mathbf{v}, {}^{\beta}\mathbf{v}) = E\{{}^{\alpha}\mathbf{v}, {}^{\beta}\mathbf{v}\}$. If the expected value $E\{\mathbf{v}\} = 0$, then the norm $\| \mathbf{v} \|$ is equal to the standard deviation or:

$$\| \mathbf{v} \| = ({}^{\alpha}\mathbf{v}, {}^{\beta}\mathbf{v})^{1/2} = D^{\mathbf{v}}. \tag{4.237}$$

In general $D^{\mathbf{v}}$ may satisfy the properties of a norm or in certain cases that of a semi-norm only. This would give rise to a specific topology of the velocity space $\mathcal{V}$. On the other hand, one can consider the topology in the configuration space $\mathcal{C} \subset X$. In this case a "distance function": ${}^{\alpha\beta}d = |{}^{\alpha}\mathbf{r} - {}^{\beta}\mathbf{r}|$ between two neighbouring elements (molecules) α, β can be used, which is as a metric on the configuration space, the topology of which is that of a Hilbert space. Hence by considering the velocity space as a subspace of X and a function space $[\mathcal{V}, \mathcal{F}^{\mathbf{v}}, \mathcal{P}^{\mathbf{v}}]$ at any particular time t, one has for the entire duration of the flow of a single element "α" a set of these function spaces or a product space. In particular for $\mathbb{R} = [0, \infty]$ and each time instant $t_r \in \mathbb{R}, r = 1, 2 \ldots n$ there will be a triple $[\mathcal{V}, \mathcal{F}^{\mathbf{v}}, \mathcal{P}^{\mathbf{v}}]$ leading to an n-fold product space in which the velocity $\mathbf{v}({}^{\alpha}\mathbf{r}, t)$ is a measurable function. If for convenience the n-fold product space is extended to infinity, i.e. $[\mathcal{V}_{\infty}, \mathcal{F}_{\infty}, \mathcal{P}]$, the velocity function $\mathbf{v}(\mathbf{r}, t)$ becomes a continuous random function. As shown earlier this can also be formulated by means of an operator T_t:

$$T_t : \mathcal{V} \to \mathcal{V} \text{ for all } t_r \in \mathbb{R}^+ = [0, \infty) \text{ ; } \mathbf{v}_t(\mathbf{r}) \in \mathcal{V} \tag{4.238}$$

so that the velocity field is described in terms of a random function generated by the endomorphism T_t for all $t \in \mathbb{R}^+$.

It is assumed in the kinematics of a simple fluid, that its behaviour is Newtonian and that it consists of a collection of "undistinguishable" particles (molecules $\alpha, \beta \ldots$). These assumptions are frequently made in molecular fluid dynamics. One can regard the motion of a single particle or the collective mode of motion of an ensemble of particles $\{\alpha\} = M$ as a random endomorphism represented by the function $\{\mathbf{r}(\alpha, t)\} = \mathcal{A}_t \subset \mathcal{C}$ giving the position vector to its C.M. at various times.

The collective motion of a set of position vectors at time t can be identified as a one-parameter family of "configurational meso-domains", i.e., $\mathcal{A}_t = \{\mathbf{r}(\alpha, t) \equiv {}^{\alpha}\mathbf{r}(t)\}_{\alpha=1}^N \subset \mathcal{C} \subset \mathbb{R}^3$. In this sense fluid motion (uniform) means that the random endomorphism G_t is such that:

$$\mathcal{A}_t = G_t \mathcal{A}_0, \text{ where } \mathcal{A}_0 \equiv \mathcal{A}_t \text{ at time } t = 0.$$

Hence the sets $\mathcal{A}_0, \mathcal{A}_t$ are bounded point sets in the Euclidean space $\mathbb{R}^3$ and the corresponding domain will be occupied by the "same" fluid elements at time t in $\mathbb{R}^3$. The concept of a Lagrangian description has been used by G.I. Taylor [327] and the notion of a "geometrical transformation "G_t" in mathematical fluid mechanics has been introduced by R.E. Meyer [328]. This transformation corresponds in probability theory to the above mentioned random endomorphism. In the description of the (steady) fluid motion, it is assumed that $\mathbf{r} = \mathbf{r}(\alpha, t)$ is continuous with respect to t on the domain of definition and that an inverse of G_t exists. This notion means implicitly that the bounding surface of the respective fluid domain consists of the same type of elements (molecules) and that they are regarded as structureless entities. Hence in this approximation the effect of intramolecular forces is only to maintain the geometry of the flow. It excludes therefore the case for example of "shear" induced distortions of the distribution functions of molecular fluids. In this context the modelling of interactions has more recently been extended to "non-spherical" and "extended particles", i.e. (macromolecules) as treated in the preceding section.

In conventional hydrodynamics it is not necessary to consider an explicit form of $\mathbf{r}(\alpha, t)$ in the description of the fluid motion. The latter can be characterized by the velocity field. Hence all corresponding random function can be defined for a fluid domain as a family of associated random transformations during a time interval such that:

$$G_t\{{}^\alpha \mathbf{r}(0)\} = \{{}^\alpha \mathbf{r}(t)\},$$

for instance, on the assumption that there exists a class $C^n(\mathcal{C})$ with continuous derivatives of order $\leq n$ with respect to $\mathbf{r}$ and t for a chosen interval: $t_1 < t < t_2$.

It is well known that the Lagrangian represenation in terms of $\mathbf{r}(\alpha, t)$ in the hydrodynamic theory is the solution of the deterministic differential equation of the form:

$$\frac{\partial^\alpha \mathbf{r}(t)}{\partial t} = {}^\alpha \mathbf{v}({}^\alpha \mathbf{r}, t) \tag{4.239}$$

with the initial condition: ${}^\alpha \mathbf{r}(0) = \mathbf{r}_0$ in $\mathcal{A}_0$ and the velocity vector $\mathbf{v} \in C^\infty(G_t \mathcal{A}_0)$. Evidently the connection of this description to the Eulerian one is given by:

$$h({}^\alpha \mathbf{r}, t) \equiv f[{}^\alpha \mathbf{r}(\alpha, t)] \tag{4.240}$$

where

$$Df/Dt \equiv \frac{\partial}{\partial t} h({}^\alpha \mathbf{r}, t) = [\frac{\partial}{\partial t} + \mathbf{v} \cdot grad]f \tag{4.241}$$

or Dt/Dt being the material derivative of f. Since the rate of change with time of characterizing the motion of $\alpha \in M$ passing through the position of α at time t, the derivative of $\mathbf{r}(\alpha, t)$ differs generally from $\partial f/\partial t$ and the material derivative need to be taken. Equation (4.241) in stochastic mechanics is, however, a stochastic differential equation with the same initial conditions and for which a solution exists. The motion of fluid elements will be considered in the next section on the basis of Markov theory.

So far as the interactions between neighbouring elements are concerned, it has been shown that the former have both attractive and repulsive components. Simple fluids are commonly modelled by hard-spheres, square-well and more often by the Lennard-Jones potential $\phi(r)$ designated usually by:

$$\phi(r) = 4\epsilon[(\frac{r_e}{r})^m - (\frac{r_e}{r})^n]$$ (4.242)

or also referred to as the (12,6) potential, since the exponents $m = 12, n = 6$ are commonly adopted and where ϵ is the "well-depth", r the intermolecular distance, respectively. Accordingly, the hard-sphere potential is specified by:

$$\phi(r) = \left\{ \begin{matrix} \infty, & 0 < r < r_e \\ 0, & r > r_e \end{matrix} \right\}$$ (4.243)

in which "r_e" correponds to the intermolecular distance at equilibrium.

As mentioned earlier in Chapt. 2, direct structural information concerning the molecular arrangement in liquids is obtained from X-ray and fast neutron-scattering experiments. Theoretical results of the molecular dynamics analysis are frequently checked by "computer simulations" of which the Monte-Carlo method is the most often used. The interaction potential between molecules α, β or $^{\alpha\beta}\phi$ is a function of the intermolecular distance $^{\alpha\beta}d = |^{\alpha}\mathbf{r} - ^{\beta}\mathbf{r}|$. It is evident, that it will be accurate only for short-range interactions. Hence by assuming a "hard sphere" model for the simple fluid, it can be regarded as an intrinsic or reference potential. In actual fluids, where bond association with other molecules cannot be excluded within a certain neighbourhood of the molecule, one can express an "overall potential" in the form of:

$$^{\alpha\beta}\phi(d) = {}^{\alpha\beta}\phi_0 + {}^{\beta}\phi$$ (4.244)

in which $^{\alpha\beta}\phi_0$ is the intrinsic potential and $^{\beta}\phi$ is the part accounting for bonding with surrounding molecules. This part of $^{\alpha\beta}\phi$ will be non-zero only for a small range of relative portions and orientations of the molecules. Since the reference potential $^{\alpha\beta}\phi_0$ contains a strongly repulsive part that prevents two molecules from coming too close to one another, the distance $|^{\alpha\beta}d|$ corresponds to that of a short-range order or a "minimum distance". In the present analysis, it is seen as a distance function, i.e., $d(\alpha, \beta) \equiv d(^{\alpha}\mathbf{r}, {}^{\beta}\mathbf{r})$ or as a metric on the topological configuration space $\mathcal{C} \subset X$. It should be mentioned here, that one can distinguish various "density regimes" in real fluids. Thus, by using the "number density $n(\mathbf{r}, t)$" introduced in Chapt. 2 and a characteristic length "d_c" for the range of the interaction potential, it may be stated that for the low density regime the condition $n(\mathbf{r}, t)d_c^3 \ll 1$ holds. In the case of dense gases or at liquid densities this condition is however approximately $n(\mathbf{r}, t)d_c^3$ greater than 0.1 (see for instance Batchelor [329]). Since stochastic mechanics employs the notion of meso-domains, which for an idealized discrete fluid is conceptually synonymous to the control volume of hydrodynamics, a characteristic length L of such a fluid domain is important. It must be chosen such that $L \gg \ell_d$ or the diffusion length associated with diffusion of micro element (molecule) out of a particular meso domain [23]. Correspondingly various time scales of the fluid may be considered: a characteristic "macro-time" $\tau_\mu \sim L^2/6K$ (where K is the conventional diffusion constant) and a characteristic "micro-time τ_m" at which the particle (molecule) velocity distribution occurs. Evidently the macro-time scale is much larger than the micro-time scale (compare also with the statements of section (4.2 iii).

(ii) Markov theory in the mechanics of discrete fluids:

Analogous to the mechanics of discrete solids, Markov processes provide the simplest models of the dependent random behaviour and thus can be used in the analysis of the flow of molecular fluids. The evolution of the velocity field $\mathbf{v} = \mathbf{v}(\mathbf{r}, t)$ for instance is then representable by the triple $[\mathcal{V}, \mathcal{F}^v, \mathcal{P}^v]$ with an associated operator $T_t : \mathcal{V} \to \mathcal{V}$ parametrized by $t \in \mathbb{R} = [0, \infty)$ so that $\mathbf{v}_t(\mathbf{r}) \in \mathcal{V}$ is a stochastic process.

It may be assumed that for a uniform (steady) flow condition, a stable generating mechnism of the Markov process will exist, which is thus characterized by its transition function $P(t, x, E)$. The latter is $\mathcal{F}$-measurable and is a function of x for each set or events $E \in \mathcal{F}$ and a probability measure $\mathcal{P}$ on the Borel field $\mathcal{F}$ for each $x \in X$. $P(t, x, E)$ represents the probability that an outcome at time t_{r+1} of the Markov process falls in the set E given that at time t_r the observation x was made. Hence considering the velocity field $\mathbf{v}(\mathbf{r}, t)$ and using conditional probabilities as shown for discrete solids, then to each T_t there corresponds a conditional probability measure $\mathcal{P}^v\{E_{r+1}|E_r\}$ such that whenever

$$T_t E_r = E_{r+1}$$

$$\mathcal{P}^v\{E_{r+1}\} = \mathcal{P}^v\{E_{r+1}|E_r\}\mathcal{P}^v\{E_r\}$$

which similarly as before generalizes to a set of conditional probability measures, i.e.:

$$\mathcal{P}^v\{E_n\} = \mathcal{P}^v\{E_1\} \prod_{r=1}^{n} \mathcal{P}^v\{E_{r+1}|E_r\},$$

and which is valid for any sequence $E_1 \supset E_2 \ldots E_r \supset E_n$ corresponding to the time sequence $t_1 < t_2 < \ldots t_r < t_n$ and a set of conditional probabilities $\mathcal{P}\{E_{r+1}|E_r\}, r = 1, 2 \ldots n - 1$. This can be written more explicitly as:

$$\mathcal{P}^v\{E_{r+1}, t_{r+1}\} = \mathcal{P}^v\{E_{r+1}, t_r\}\mathcal{P}^v\{E_r, t_r\} \tag{4.245}$$

where E_r corresponds to $t_r \in \mathbb{R}$ and E_{r+1} to $t_{r+1} > t_r$. This relation indicates that the velocity distributions at any time t during the steady flow of the discrete fluid can be established from that at any other time or if the initial one is known. Choosing a closed time interval $[t, s] \in \mathbb{R}$ and a point $\tau \in (t, s)$, the transition probability concerning the velocity field will satisfy the Chapman-Kolmogorov relation, i.e.:

$$\underline{P}\{t, s\} = \int\limits_{v \subset X} \underline{P}\{t, s\} d\underline{P}\{\tau, s\} \ .$$

It can be written in form of a matrix differential equation as:

$$\underline{P}\{t + s\} = \underline{P}\{t\}\underline{P}\{s\} \tag{4.246}$$

which holds for any small time interval $[t, s]$ during the steady flow of the discrete fluid, indicating again (as in the case of solids), the semi-group property of the system. As before, it can be represented by a one-parameter semi-group of linear transformations $T_t : t \geq 0$, $t \in \mathbb{R}$. These operators are defined on the function space $C(\mathcal{V}) \subset X$ by:

$$T_t[f(w)] = \int\limits_{v \subset X} f(u)P(t, w, du) \ ; \ f \in C(\mathcal{V}) \ . \tag{4.247}$$

T_t satisfies the following conditions:

$$T_t T_s = T_{t+s} \text{ for all } t, s > 0 \text{ in } \mathcal{V} \subset X. \tag{4.248}$$

The linear semi-group $\{T_t \ ; \ t \geq 0\}$ is contracting, if $\| T_t f \| \leq \| f \|$, $f \in C(\mathcal{V})$. For considerations of the Markov process of a single fluid element $\alpha \in M$; $(M \Rightarrow \mathcal{A}_t)$, the weak-limit of the class of operators is important. Thus, as shown in Chapt. 2 this limit is given by:

$$\text{w-lim}_{t \downarrow t_0} T_t[f(w)] = f(w) \text{ for every } f \in C(\mathcal{V}) \ , \tag{4.249}$$

since bounded observables or events induce a weak topology on $\mathcal{V}$ only, which is closely related to the Markov process. It is to be noted that the outcome in the velocity space $\mathcal{V}$ is a vector valued quantity so that in general a multidimensional random process should be considered for the description of the flow. However for the condition of a uniform (steady) flow this can be simplified to a scalar process in which the velocities are characterized by $v = |v|$ only. Returning to relation (4.246) the transition matrix expressed by $P_{ij}(t)$ represents the transition of a fluid element α from a state i at time t_r to a state j at time t_{r+1}. Similar to the form shown for solids, one also has in adopting the notion of a time-homogeneous process the following limits:

$$\lim_{t \to 0^+} \frac{1 - P_{ii}(t)}{t} = -\lambda_{ii} < \infty \text{ and } \lim_{t \to 0^+} \frac{P_{ij}(t)}{t} = \lambda_{ij},$$

where the λ_{ij}'s are again the elements of the transition intensity matrix $\underline{Q}$. In this case the evolution of the probability distribution follow Kolmogorov's matrix differential equation valid for the velocity space $\mathcal{V} \subset X$, i.e.:

$$\frac{d\underline{P}^v(t)}{dt} = \underline{Q}^v \underline{P}^v(t). \tag{4.250}$$

Thus during steady flow, and by the exclusion of viscous effects, turbulence etc., the model of a Markov process will be valid, but is rather of the Poisson type. This can readily be recognized by considering the steady flow to be associated with independent events in the terminology of probability theory. Such a process is characterized by a number of events occuring within time t and is a Markov process with the invariant measure $\mathcal{P}^v$. Correspondingly one has the transition intensities given by:

$$\lambda_{ij} = \begin{cases} \lambda & \text{for } j = i+1 \\ 0 & \text{for } j \neq i+1 \end{cases} ; \ (i = 0, 1, 2 \ldots). \tag{4.251}$$

This means that only direct transitions from a state $i \to j = i+1; (i = 0, 1, 2 \ldots)$ can take place. However, it is possible to consider also n-step transitions to occur, which are then based on recurrence relations that are usually required in simulation techniques (see also Chapt. 2). By the use of equation (4.250) a system of differential equations will be formed, the solution of which is given by:

$$\left. \begin{aligned} P_{ij}(t) &= \frac{(\lambda t)^{j-i}}{(j-i)!} e^{-\lambda t} \text{ for } j \geq i \\ &= 0 \qquad\qquad \text{for } j < i, \end{aligned} \right\} \tag{4.252}$$

representing a Poisson process. Evidently the notion of independent events can also be considered for an ensemble of "interacting" fluid elements within a specific meso domain, which requires however the use of a common transition function. The process then becomes one belonging to the class of "interacting Markov processes". Thus, in the simplest way from the stand point of classical lattice statistical mechanics, one can introduce a "common transition probability" function for two neighbouring configurations i, j of the particles of the fluid. This approach has been suggested by Preston [287] and in terms of coupled Markov chains by Spitzer [286], Holley [332] among others. By considering the configuration space $\mathcal{C} \subset X$ and a common transition function, which is valid for a specific configurational meso domain, where $M_t \Rightarrow A_t$ of the fluid body, a countably finite set of all possible configurations on that domain with independent Markov processes on it, gives therefore:

$$\sum_{i \in \mathcal{A}_t} P_t(i,j) = \sum_{j \in \mathcal{A}_t} P_t(i,j) = 1 \; ; \; (i,j) \in \mathcal{C} \subset X, t \geq 0 \tag{4.253}$$

where it is assumed that $P_t(i,j)$ is an invariant measure on $M_t \subset \mathcal{C}$. Hence in the case of a uniform (steady) flow condition of the fluid, where the motion of a single element of the structure is described by (4.252) the probability distribution pertaining to an ensemble of elements $\alpha = 1 \ldots N$, is expressed by:

$$\mathcal{P}^m(A) = e^{-m\lambda} \prod_{i=1}^{r} \frac{\lambda^{N_i}}{N_i!} \; ; \; N_i \in M_t \Rightarrow \mathcal{A}_t \subset \mathcal{C}^m \; ; \; i = 1 \ldots r \tag{4.254}$$

in which the Borel set A is an element of the sub-σ-algebra $\mathcal{F}^n$ of $\mathcal{F}$, $\{N_i\}$ defined by $A \in \mathcal{A}_t$, $\mathcal{P}^m$ designates an invariant measure on the product space $\mathcal{C}^m \otimes \mathcal{C}^m$ and where the fluid elements $\alpha = 1 \ldots N$ undergo a one-step transition from the state $i = 1, 2 \ldots r$. For a more detailed description of the steady flow, it is however necessary to introduce the notion of a "Gibbsian" random field. Thus following postulate P.1 of Chapt. 2 and recognizing the fluid elements forming a Gibbsian ensemble, the random field will be characterized by the triple $[\mathcal{C}, \mathcal{F}^e, \mathcal{P}]$ for $M \Rightarrow \mathcal{A}_t = \{\alpha\}^N$, on the domain $M \subset \mathbb{R}^3$ and the interaction potential $^{\alpha\beta}\phi(R) = \phi(^\alpha\mathbf{r}, {}^\beta\mathbf{r}): \mathcal{C} \times \mathcal{C} \to \mathbb{R}$. The latter satisfies the following properties:

$$\left.\begin{array}{l}
\text{(i) } \phi(^\alpha\mathbf{r}, {}^\beta\mathbf{r}) = \phi(^\beta\mathbf{r}, {}^\alpha\mathbf{r}); \text{ (symmetry)} \\[4pt]
\text{(ii) } \phi(^\alpha\mathbf{r}, {}^\beta\mathbf{r}) = \phi(0, {}^\beta\mathbf{r} - {}^\alpha\mathbf{r}); \text{ (homogeneity)} \\[4pt]
\text{(iii) } \phi(^\alpha\mathbf{r}, {}^\beta\mathbf{r}) = 0, \text{ if } |{}^\beta\mathbf{r} - {}^\alpha\mathbf{r}| > \delta \text{ (nearest neighbour property)}
\end{array}\right\} \tag{4.255}$$

For a more specific definition of this random field and by following Spitzer [286], one also requires a boundary value function f on ∂M (boundary of the domain M). This function represents an arbitrary map $f: \partial M \to \{0,1\}$. For convenience one can extend this description to a map $\overline{f}: \overline{M} \to \{0,1\}$, $\overline{M} = M \cup \partial M$ so that:

$$\overline{f}(^\alpha\mathbf{r}) = \left\{\begin{array}{ll}
f(^\alpha\mathbf{r}) & \text{for } {}^\alpha\mathbf{r} \in M \\[4pt]
\partial f(^\alpha\mathbf{r}) & \text{for } {}^\alpha\mathbf{r} \in \partial M.
\end{array}\right\} \tag{4.256}$$

Hence for a given volume of the domain M and an interaction potential satisfying (4.255), a boundary value function f defined above, the Gibbsian random field will be specified by:

$$P(^\alpha\mathbf{r}) = Z^{-1} \exp[-\frac{1}{2} \sum_{{}^\alpha\mathbf{r} \in M} \sum_{{}^\beta\mathbf{r} \in M} \overline{f}(^\alpha\mathbf{r}) \overline{f}(^\beta\mathbf{r}) \phi(^\alpha\mathbf{r}, {}^\beta\mathbf{r})]. \tag{4.257}$$

where $Z = Z(M)$ is the equivalent to the partition function of the canonical ensemble. Due to the equivalence of this Gibbsian random field and a Markov random field (see for instance Dobrushin [285], Spitzer [286]) one can represent the flow of discrete fluids in terms of Markov theory [23], [330].

(iii) Flow through a fully saturated porous medium:

Among the various types of porous media, two classes of fluid-saturated solids
are of considerable interest in engineering applications. The first class is represented by a "Kaolin-Water mixture", (soil) indicated in Chapt. 2 by the
Scanning-Electron-Micrograph of Fig. 5. This material consists of an assembly
of clay particles forming the skeleton or the solid-phase of the microstructure,
while the available pore space is entirely filled with water. The second group of
porous media is characterized by solid particles that are embedded in a fluid
matrix as shown in Fig. 6 (Chapt. 2). If such materials are subjected to an
external load as for instance in a simple confined compression test, the fluid or
β-phase is discharged and the solid or α-phase essentially carries the applied
load. Thus, during a relatively short period of time the microstructure of the
porous medium undergoes a structural change that is closely related to the
"transport of the β-phase". It is the main purpose of the present analysis to
use the stochastic mechanics approach in the description of the transport of
the β-phase with the restriction imposed by the randomly arranged α-phase of
the medium.

Most formulations of transport phenomena in porous media have been based
on continuum mechanics models with the acceptance of empirical relations such
as Darcy and the generalized Fick laws. A common feature of these theories is
the adoption of an "appropriate averaging" procedure in order to achieve the
transition from the microscopic description to the required macroscopic one
(see for instance [333–335]). This transition, as has been shown repeatedly in
this volume, can also be achieved by probabilistic arguments that are more in
line with experimental observations.

In general, for the formulation of random transport phenomena, the basic
principles and definitions of stochastic mechanics will apply. In the present case,
concerned with the transport of the β-phase of the porous medium, a micro
element of that phase or $\beta \in \mathcal{B}$ has a configuration specified by:

$$\mathbf{x}(\beta,t) \in \mathcal{C} \text{ or briefly } {}^{\beta}\mathbf{x}_t \in \mathcal{C} \subset X \tag{4.258}$$

and a motion: $\{{}^{\beta}\mathbf{x}(t); -\infty < t < \infty\}$ in the discrete and time continuous case,
where $\mathbf{x}$ refers to the current position vector of the β-elements with respect to
the Euclidean frame. In stochastic mechanics these motions, as shown previ-
ously, are regarded as stochastic processes designated here by $\{{}^{\beta}\mathbf{x}(t)\}$ or ${}^{\beta}\mathbf{x}_t$ for
each element β. Similarly the state-space $Z \equiv X$ is formed by the state-vectors
${}^{\beta}\mathbf{z} \in Z$, and is defined by a set of r-parameters representing the "states
$i = 1 \ldots r$" of an element β of the structure. The "event structure" is again
characterized by $E_n \subset X$, where:

$$E_n \triangleq \{\mathbf{z}_n^{(i)} < {}^{\beta}\mathbf{z}_n^{(i)} < \mathbf{z}_n^{(i)} + \Delta\mathbf{z}_n^{(i)}\}; \cup_n E_n = Z \equiv X$$

$$E_n \cap E_n = \emptyset, \ (k \in \mathbb{R}^1, i = 1 \ldots r)$$

resulting in a σ-algebra $\mathcal{F}_n$ as a sub-σ-algebra of $\mathcal{F}$ in $X = \mathbb{R}^n$.

The elements E_n of $\mathcal{F}_n$ together with $\mathcal{F}$ then define a measurable space $[X, \mathcal{F}]$. An appropriate measure on these sets as shown earlier, is then:

$$0 \leq \mathcal{P}\{E_n\} \leq 1 \ ; \ \mathcal{P}\{E_n\} = 0, \ \text{if } E_n = \emptyset \text{ and } \mathcal{P}\{X\} = 1.$$

In continuum models for the flow of the β-phase, the overall diffusive flux appearing in the mass balance equations is considered to be composed of a molecular diffusion and a mechanical dispersion. The former is commonly represented by Fick's law, whilst the mechanical dispersion is often characterized by assuming that its magnitude increases proportional to the Darcy velocity. This involves the application of a "dispersivity tensor" in the mass balance relations. In the stochastic formulation an equivalent result can be obtained by the use of the theory of Markov processes. From a molecular fluid dynamics point of view, one may assume that the β-phase behaves in a Newtonian manner and that it consists of a collection of indistinguishable particles (i.e., molecules $\beta_1, \beta_2 \ldots$). Hence, in an "unconstrained motion" of single particles $\beta_1, \beta_2 \ldots$ or the collective motion of an ensemble of them within a given meso domain, the random endomorphism will apply and can be represented by a set of function $\{\mathbf{x}(\beta, t) \equiv {}^\beta\mathbf{x}_t\}_{\beta=1}^N = M_t \subset \mathcal{C}$ or in the configuration space contained in X. One obtains therefore as shown before by means of the geometrical transformation a Poisson process.

Of greater interest however, is the constrained flow of the discrete fluid represented by the earlier deterministic differential eqn. (4.239), i.e.:

$$\frac{\partial^\beta \mathbf{x}(t)}{\partial t} = \mathbf{v}({}^\beta\mathbf{x}, t)$$

which has to be replaced by a set of non-linear stochastic differential equations to account for the dispersion and possible small (molecular) diffusion of the elements of the β-phase. It is evident that the description of the constrained flow will differ from the foregoing one, even though the theory of Markov processes can be used. This type of flow within an arbitrary meso-domain of the fluid body is indicated schematically in Fig. 41(a,b). The first class of porous media (Kaolin-Water compound) has a structure that consists of interconnected clusters of clay particles (α-phase), which restricts the relative motion of the β-phase with respect to the α-phase. The second group of porous media illustrated by the idealized Al-Resin structure of Fig. 41(b) permits a somewhat greater relative motion due to the availability of a larger "pore space".

Since the velocity random field ($\mathbf{v} \in \mathcal{V} \subset X$) is "unobservable", it is convenient for the further description of this flow to introduce formally a general "vector functional", which contains in its argument some of the controlling factors of the random phenomenon as well as "observable quantities" or:

$$\mathbf{f(v)} \Rightarrow \mathbf{f}[\mathbf{x}(t), t] \Rightarrow \mathcal{F}_\epsilon^t \{{}^\alpha c, {}^\beta c, \epsilon, \dot{\epsilon}, q, \dot{q} \ldots\} \tag{4.259}$$

in which ${}^\alpha c, {}^\beta c$ denote the volume fractions of the α and β phases, respectively, $\epsilon, \dot{\epsilon}$ the strains and strain rate, $q, \dot{q}$ the discharge and discharge-rate of the fluid phase. It is assumed that a direct relation between the elements of the "control space Θ" (see Fig. 42 stochastic state-space representation) or ${}^\alpha c, {}^\beta c \in \Theta$ and that of the "observation space $\mathcal{E}$", $\epsilon, q \in \mathcal{E}$ can be established. This will be more fully discussed subsequently. In this sense one can introduce a "system or velocity functional" $\mathbf{f} \in \mathcal{V} \subset X$ to represent the velocity field, which contains in its argument the random position vector of the fluid elements $\beta = 1 \ldots N$, the time t and a set of control parameters $\{\theta_t\}$.

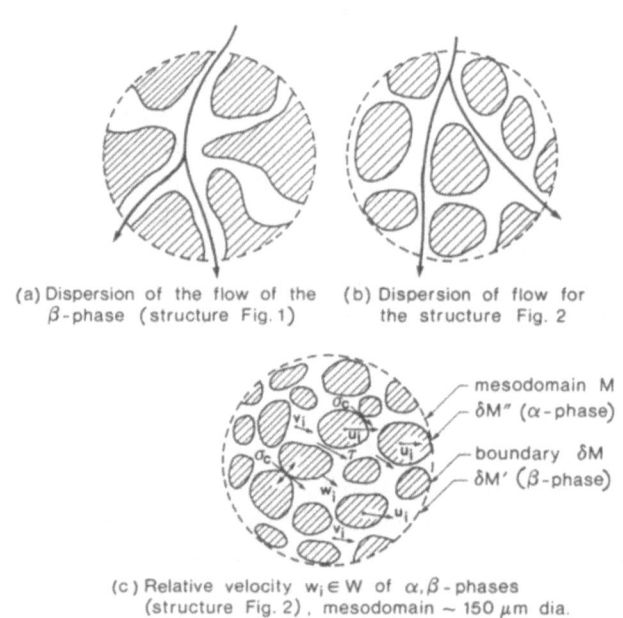

(a) Dispersion of the flow of the β-phase (structure Fig. 1)

(b) Dispersion of flow for the structure Fig. 2

mesodomain M
$\delta M''$ (α-phase)
boundary δM
$\delta M'$ (β-phase)

(c) Relative velocity $w_i \in W$ of α, β-phases (structure Fig. 2), mesodomain ~ 150 μm dia.

Fig. 41. Meso-domain structures and flow characteristics of the β-phase (150 $\mu m \cdot$ dia).

$$\sigma = \sigma({}^\alpha \sigma_c, {}^\beta \tau) : \quad {}^\alpha \sigma_c = \text{surface stress}$$
$${}^\beta \tau = \text{viscous stress}$$
$$u_i - \text{displacement of } \alpha\text{-phase, } (\dot{u}_i\text{-velocity})$$
$$v_i - \text{velocity of } \beta\text{-phase}$$
$$w_i - \text{relative velocity } w_i \in W \text{ of } \alpha, \beta\text{-phases}$$

Thus the evolution of the transport of the β-phase will be characterized by the following non-linear stochastic differential equation:

$$d\mathbf{x}(t) = \mathbf{f}[\mathbf{x}(t), \boldsymbol{\theta}_t, t]dt \; ; \;\; t \geq t_0, (t, t_0) \in T. \tag{4.260}$$

Recognizing that the set of parameters $\boldsymbol{\theta}_t$ implicitly stated above, character-ize a "disturbance function $\mathbf{g}(\mathbf{x}, t)$" on the constrained flow of the β-phase or $\mathbf{f}[\mathbf{x}(t), t]$, one can, on the assumption that the disturbances are of the order of Gaussian noise, rewrite (4.260) to give:

$$\frac{d\mathbf{x}(t)}{dt} = \mathbf{f}[\mathbf{x}(t), t] + \underline{g}[\mathbf{x}(t), t]\boldsymbol{\theta}_t \; ; \;\; t \geq t_0, (t, t_0) \in T \;, \tag{4.261}$$

with the initial condition: $\mathbf{x}(0) = \mathbf{x}_0$ at $t = 0 = t_0$ assumed to be independent of the noise process $(\boldsymbol{\theta}_t, t \geq t_0)$ and where $\underline{g}$ is a matrix function. It is well known that the process $\boldsymbol{\theta}_t$ is a delta-correlated random process and not integrable in the mean-square sense (see for instance [161]). It is therefore necessary to express the "n-dimensional random disturbance" by an n-dimensional Brow-nian motion or Wiener process. Thus, using the formal derivatives of $\boldsymbol{\theta}_t$, i.e., $\boldsymbol{\theta}_t \simeq d\mathbf{b}_t/dt$ where $\{\mathbf{b}_t, t \geq t_0\}$ is the vector process of independent Brownian motions, eqn. (4.261) can be written as follows:

$$d\mathbf{x}(t) = \mathbf{f}[\mathbf{x}(t), t]dt + \underline{g}[\mathbf{x}(t), t]d\mathbf{b}(t) \; ; \;\; t \geq t_0. \tag{4.262}$$

This relation is easily recognized as an Itô stochastic differential equation, the solution of which is to be interpreted in the sense of the integral equation, i.e.:

$$\mathbf{x}(t) = \mathbf{x}(t_0) + \int_{t_0}^{t} \mathbf{f}[\mathbf{x}(\tau), \tau]d\tau + \int_{t_0}^{t} \underline{g}[\mathbf{x}(\tau), \tau]d\mathbf{b}(\tau), \; (t_0 \leq \tau \leq t). \tag{4.263}$$

The existence, uniqueness and solution of this equation is comprehensively dealt with among others in refs. [48, 161]. On the basis of the above assumption, i.e., that the initial condition for the process $\mathbf{x}(t)$ is independent of the Brownian motion, the solution of $\mathbf{x}(t)$ is a Markov process, even a diffusion process in $\mathcal{V} \subset Z \equiv X$. By such a process is usually meant a continuous Markov process with a transition function $P\{s, x; t, y\}$ which in the one-dimensional case has to satisfy certain conditions. As already discussed (Chapt. 2) the variables (s, x) are the "forward" and (t, y) the "backward" variables. Hence the process is characterized by two functions, i.e., one function $a(s, x)$ representing the average trend of the evolution of the process during a small time interval and the other $\sigma^2(s, x)$ which determines the mean-square deviation of the process from its expected value. They are known as "drift" and "diffusion" coefficients, respectively. By introducing the probability density p or:

$$p\{s, x; t, y\} = \frac{\partial P\{s, x; t, y\}}{\partial y}$$

assumed to be continuous with respect to the times s and t, respectively, together with its first and second drivatives, it is a fundamental solution of the set of parabolic differential equations discussed earlier in Chapt. 3, i.e.:

$$\frac{\partial}{\partial s}p(s,x;t,y) = -a(s,x)\frac{\partial}{\partial x}p(s,x;t,y) - \frac{1}{2}\sigma^2(s,x)\frac{\partial^2}{\partial x^2}p(s,x;t,y)$$

$$\frac{\partial}{\partial t}p(s,x;t,y) = -\frac{\partial}{\partial y}[a(t,y)p(s,x;t,y)] + \frac{1}{2}\frac{\partial^2}{\partial y^2}[\sigma^2(t,y)p(s,x;t,y)]$$

in which the first relation is the Kolmogorov backward and the second Kolmogorov's forward or Fokker-Planck equation.

In the multi-dimensional case by recognizing the constrained flow of the β-phase as a diffusion process, the corresponding Itô-vector stochastic differential equation takes the form:

$$d\mathbf{x} = \mathbf{a}(\mathbf{x},t)dt + \underline{\sigma}(\mathbf{x},t)d\mathbf{w}(t) \tag{4.264}$$

in which the vector: $\mathbf{a} = [a_1(x,t), a_2(x,t)\ldots a_n(x,t)]^T$, $\mathbf{x} = [x_1, x_2 \ldots x_n]^T$, $\mathbf{w}(t)$ the vector of independent Brownian motion and $\underline{\sigma}(t)$ a matrix function or $\{\sigma_{jk}(t)\}_{jk\leq h}$. Assuming that the velocity functional in (4.259, 4.260) is a smooth function of the $(n+1)$-variables of the configuration vector $\mathbf{x} = [x_1, x_2 \ldots x_n]^T$ and time t, one obtains by using Itô's formula the following parabolic differential equation:

$$\left. \begin{array}{l} d\mathbf{f}[\mathbf{x}(t),t] = \Big[\dfrac{\partial \mathbf{f}}{\partial t}[\mathbf{x}(t),t] + \mathbf{a}(t)\boldsymbol{\nabla}_x \cdot \mathbf{f}[\mathbf{x}(t),t] \\[2mm] \qquad\qquad + \dfrac{1}{2}\displaystyle\sum_{j,k=1}^{n} \sigma_{jk}(t)\dfrac{\partial^2 \mathbf{f}}{\partial x_j \partial x_k}[\mathbf{x}(t),t]\Big] dt \\[4mm] \qquad\qquad + \displaystyle\sum_{j,k\leq n} \sigma_{jk}(t)\dfrac{\partial \mathbf{f}}{\partial x_j}[\mathbf{x}(t),t]d\mathbf{w}_j. \end{array} \right\} \tag{4.265}$$

By introducing the differential operator L_s, one obtains from the above relation the "backward Kolmogorov" equation in the form of:

$$L_s\mathbf{f} = \frac{\partial \mathbf{f}}{\partial s} + \left\{ \mathbf{a}(\boldsymbol{\nabla}_x \cdot \mathbf{f}) + \frac{1}{2}\sum_{j,k=1}^{n} a_{jk}\frac{\partial^2 \mathbf{f}}{\partial x_j \partial x_k} \right\} ;\ a_{jk} = (\underline{\sigma}\,\underline{\sigma}^T)_{jk}. \tag{4.266}$$

Denoting the bracket expression by the operator:

$$A_s = \left\{ \mathbf{a}\boldsymbol{\nabla}_x + \frac{1}{2}\sum_{j,k=1}^{n} a_{jk}\frac{\partial^2}{\partial x_j \partial x_k} \right\}$$

one obtains an abbreviated form of the backward equation, i.e.:

$$L_s \mathbf{f} = \frac{\partial \mathbf{f}}{\partial s} + A_s \mathbf{f} \tag{4.267}$$

Analogously, by using the forward variables (t, y) the "Kolmogorov forward" equation is then:

$$L_t \mathbf{f} = \frac{\partial \mathbf{f}}{\partial t} + \left\{ \nabla_Y (\mathbf{a} \cdot \mathbf{f}) - \frac{1}{2} \sum_{j,k=1}^{n} a_{jk} \frac{\partial^2 \mathbf{f}}{\partial y_j \partial y_k} \right\} \tag{4.268}$$

and by denoting the bracket term by A_t, one has:

$$L_t \mathbf{f} = \frac{\partial \mathbf{f}}{\partial t} + A_t \mathbf{f} \tag{4.269}$$

It is seen that the transport of the β-phase through the medium can be formulated by a Markov diffusion process in terms of the diffusion operators L_s, L_t, respectively and the velocity functional:

$$\mathbf{f(v)} = \mathbf{f}[\mathbf{x}(t), t].$$

Perhaps the most significant aspect of the stochastic transport theory is the consideration of the boundary value problem associated with the constrained flow of the β-phase. Thus considering Fig. 41(c) and the boundary ∂M of an arbitrary meso-domain M it may be noticed that:

$$\left. \begin{array}{l} \partial M = \partial M' \cup \partial M'' : \partial M' = \cup_{k=1}^{n} \partial M_k' \\ \partial M'' = \cup_{j=1}^{p} \partial M_j'' \end{array} \right\} \tag{4.270}$$

in which $\partial M'$ is that part of the boundary which will permit a "through flow" of the β-phase to any neighbouring domain of the medium, whilst $\partial M''$ is the section of the boundary, where the fluid can only be "absorbed" or "reflected". Hence assuming that $\mathbf{x}(t)$ is a solution of the dynamical system representing the constrained flow of the β-phase in form of the Itô stochastic differential equation (4.264), then for any domain M in $\mathbb{R}^n$ by placing a "perfectly absorbing" barrier on $\partial M''$ or the union of M'' of ∂M, the following probability will hold:

$$P\{\mathbf{x}(t) = \mathbf{x}(s) | \mathbf{x}(0) = \mathbf{x}_0 \in M\} = 1 \text{ for all } t > \tau \tag{4.271}$$

where τ refers to the "first exit time" of the process $\mathbf{x}(t)$ from the domains M. Consequently, the probability density will be zero, i.e.:

$$p\{s, x; t, y\} = 0 \text{ for all } \mathbf{x} \in \partial M'' \text{ and } \mathbf{y} \in M \tag{4.272}$$

so that no transition from a point $\mathbf{x} \in \partial M''$ to any point or configuration $\mathbf{y}$ inside the domain M can occur. The process $\mathbf{x}(t)$ is thus "absorbed". Moreover the probability density $p\{s, x; t, y\}$ will be the Green function of the backward Kolmogorov eqn. (4.266, 4.267) satisfying the following boundary conditions:

$$p\{s,x;t,y\} = \delta(\mathbf{x}-\mathbf{y}) \; ; \; t > s, \mathbf{x} \in \partial M''$$
$$p\{s,x;t,y\} \to \delta(\mathbf{x}-\mathbf{y}) \text{ as } s \uparrow t \; ; \; \mathbf{x},\mathbf{y} \in M.$$

$$\left.\right\} (4.273)$$

So far as the "reflection of elements" of the β-phase on $\partial M''$ is concerned, it may be assumed to occur instantaneously, so that after the return to the domain M, the d.e. (4.264) will describe again the flow. However, the reflection is restricted to the boundary region and thus requires some modification of the functions $a(s,x)$ and $\sigma^2(s,x)$. This is more conveniently dealt with by a singular perturbation technique (see for instance [336–338]). The boundary value problem for the constrained flow of the β-phase can be summarized in terms of the probability densities. They take the form of the diffusion equations (4.267, 4.268) i.e.:

$$\frac{\partial p}{\partial s} + A_s p = 0 \quad \text{for the backward equation}$$
$$\frac{\partial p}{\partial t} + A_t p = 0 \quad \text{for the forward equation}$$

$$\left.\right\} (4.274)$$

which conform to the previously mentioned Dirichlet problem. In particular by regarding the Markov process as a time-homogeneous one, the system of equations is strictly parabolic and the operators A_s, A_t are then elliptic and adjoint. Thus denoting by $\mathbf{u}(\mathbf{x}, s)$ and $\mathbf{u}(\mathbf{y}, t)$ the boundary value functions of the Kolmogorov equations, respectively, the following conditions in terms of these functions can be stated:

(i) Absorption or reflection of the β-phase with small (molecular) diffusion will occur, when:

$$\lim_{s \uparrow t} \mathbf{u}(\mathbf{x}, s) = 0, \; s < t \text{ on } \partial M'' \subset \partial M.$$

(ii) Through flow of the β-phase and the condition for a subclass of boundaries value problems of the parabolic type will occur, when:

$$\lim_{t \downarrow s} \mathbf{u}(\mathbf{y}, t) = 1, \; t > s \text{ on } \partial M' \subset \partial M.$$

In order to establish the relation of the above stochastic theory to the experimental observations on test samples of the porous media with the structures indicated in Figs. 5, 6, the connection between "unobservable" and "observable" quantities should be considered. Thus with reference to the medium of Fig. 6 (Al-Resin compound) material specimen have been subjected to a constant load in a simple "confined compression" test, resulting in the measurements of the volumetric strain "ϵ_v" as well as the determination of the distributions of the volume fractions "$^\alpha c, ^\beta c \in \{\theta_t\}$" of the α, β-phase, respectively.

These fractions as elements of the set $\{\theta_t\}$ were obtained from observations of cross-sections of test samples in a scanning-electron microscope (SEM) after careful microtoming and the subsequent evaluation (distribution functions) by means of quantitative stereology [339]. It is convenient to use for the distinction between observable and unobservable quantities, the "stochastic state-space" representation, already discussed in section 2.3 (Chapt. 2) of this volume. In the present case this space is indicated by Fig. 42. The velocity space $\mathcal{V}$ has been chosen as a subspace of the state-space $Z \subset X$. Thus in the simple case for a specimen to be subjected to a uni-axial confined compression test only, the main "observable" quantities are the volumetric strain and the discharge q as elements of $\mathcal{E} \subset X$. Evidently a more complete description of the material response requires also the consideration of the motion of the α-phase. In that case it is necessary to use a "relative velocity space", $W \subset Z \equiv X$, (Fig. 42) the elements w of which describe the motion of the α-phase with respect to the fluid or β-phase, i.e., $\mathbf{w} = {}^{\beta}\mathbf{v} - {}^{\alpha}\dot{\mathbf{u}}$. Similarly, a more general formulation requires the use not only of the strain ϵ or the strain rate $\dot{\epsilon}$, but also of the discharge vector $\mathbf{q}$ or its rate $\dot{\mathbf{q}}$ as elements of the "observation space $\mathcal{E} \subset X$". The latter is connected to the internal variable (stress) space Σ with $\boldsymbol{\sigma} = ({}^{\alpha}\boldsymbol{\sigma}_c, {}^{\beta}\boldsymbol{\tau}) \in \Sigma$ via the material operator M or its inverse. The quantities ${}^{\alpha}\boldsymbol{\sigma}_c$ denote the surface (contact) stress between the elements of the α-phase and ${}^{\beta}\boldsymbol{\tau}$ the viscous stress in the β-phase.

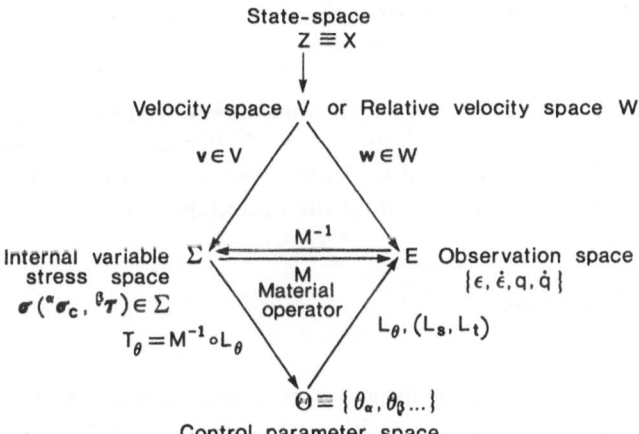

Fig. 42. Stochastic state-space representation (porous media).

Many models and examples of the above mentioned stresses ${}^{\alpha}\sigma_c, {}^{\beta}\tau$ have been given in a rather extensive literature, but will not be pursued here further (see for instance [341, 342]. So far as the material operator M is concerned, it is to be noted that its determination follows from the discussion given in section 4.1. Its validity is however restricted to the Newtonian behaviour of the fluid, the structure of which has been assumed to consist of undistinguishable particles. Evidently for more complex or non-Newtonian flows of discrete fluids some modifications of this operator will be required. It has been suggested more recently that for considerations of such flows a class of "multi-particle constitutive" relations should be used (see [341], [342]). The set of control parameters $\{\theta_t\}$ is connected to the observation space $\mathcal{E}$, (for the transport of the β-phase only) by a diffusion operator L_θ, i.e., L_s, L_t and the Σ-space by the "composition-operator: $T_\theta = M^{-1} \circ L_\theta$". In order to reduce the three dimensional analysis to the simple case of uni-axial loading of a test sample in a confined compression test, one may consider that the mean velocity of the flow of the β-phase is given by:

$$E\{{}^{\beta}\mathbf{v}\} = {}^{\beta}\overline{\mathbf{v}} = \int_{\mathcal{V} \subset X} {}^{\beta}\mathbf{v} d\mathcal{P}^{\mathbf{v}} , \qquad (4.275)$$

and hence a "reduced" form of the vector functional $\mathcal{F}_\epsilon^t$ (eqn. 4.259), so that it is represented simply by $\mathcal{F}_\epsilon^t\{{}^{\alpha}c\}$. Since the velocity space $\mathcal{V}$ is regarded as a Hilbert space, one has the following inner product:

$$< \mathbf{f}(\mathbf{v}), \ \mathcal{P}^{\mathbf{v}} >= \int_{\mathcal{V} \subset \mathbb{R}^n} \mathbf{f}(\mathbf{v}) d\mathcal{P}^{\mathbf{v}} \qquad (4.276)$$

where $\mathcal{P}^{\mathbf{v}} \in \mathcal{M}(X = \mathbb{R}^n)$ is a set of measures and $\mathbf{f}(\mathbf{v}) \in \mathcal{B}_0(\mathbb{R}^n)$ is a class of bounded functions on $\mathcal{B}_0$ or the space of all real-valued Borel measurable functions on $\mathbb{R}^n$. In the one-dimensional case $\mathbf{v}$ becomes $v_t \in \mathcal{V}$ and hence the flow process will be characterized by the probability density:

$$p(\overline{\mathbf{v}}, t) = \int_{t_0}^{t_1} p(\mathbf{v}, t) dt \qquad (4.277)$$

in which the upper limit of the integral indicates a certain time instance $t_1 = t_{cr}$ (critical time) at which no measurable outflow occurs. At this point the α-phase under the effect of the external load and constraint boundary conditions reaches a maximum of "compaction" so that only a minimum discharge or no noticeable outflow of the β-phase can take place (Fig. 43). Hence from the time instant onwards ($t_1 = t_{cr}$), the porous structure exhibits an approximately steady-state response with a considerable longer duration of time than that associated with relation (4.277). It is evident, that in this case the dominant control parameter is the volume fraction of the α-phase $(^{\alpha}c)$, so that the flow during $t_0 = 0$ and $t = t_1$ will only be affected by the unobservable $\sigma = \sigma(^{\alpha}\sigma_c, {}^{\beta}\tau)$ and the induced volumetric strain ϵ_v. Hence by considering the stochastic flow process v_t in form of a "two-component process" (see Chapt. 3), relation (4.277) can also be written in the form of:

$$p(\overline{\sigma}, \overline{\epsilon}_v) = \int_{t_0}^{t_1} p(\sigma, \epsilon) dt. \tag{4.278}$$

By adopting a phenomenological model for the interaction between the α and β-phase, respectively together with the required balance equations leads to a constitutive relation for the 1-dimensional case as follows:

$$\frac{d\overline{\epsilon}_v}{dt} = \kappa \cdot {}^{\alpha}c \frac{d}{dt} \left(\frac{\overline{\sigma}}{{}^{\alpha}c} \right). \tag{4.279}$$

Realizing from the topological features of the velocity space as a subspace of Z or X, that there exists a one-to-one correspondence between $\overline{\epsilon}_v$ and $^{\alpha}\overline{c}$ and that the latter parameter also controls the internal variable $\overline{\sigma}$, then by noting the mapping between $\overline{\epsilon}_v$ and $^{\alpha}\overline{c}$ as well as the operator T_θ, one obtains for the relation (4.268) the following expression:

$$p(T_\theta {}^{\alpha}\overline{c}, {}^{\alpha}\overline{c}) = \int_{t_0}^{t_1} p(T_\theta {}^{\alpha}c, {}^{\alpha}c) dt. \tag{4.280}$$

Hence the evolution of the probability density for the transport of the β-phase only in the one-dimensional case of the flow, $(v_t \in \mathcal{V})$ will be given by:

$$\frac{d}{dt} p(T_\theta {}^{\alpha}\overline{c}, {}^{\alpha}\overline{c}) = L_\theta p(T_\theta {}^{\alpha}\overline{c}, {}^{\alpha}\overline{c}) , \tag{4.281}$$

which can be reduced to:

$$\frac{d}{dt} p({}^{\alpha}\overline{c}) = L_\theta p({}^{\alpha}\overline{c}), \tag{4.282}$$

and which is an equivalent form to the previously given diffusion equation. For a constant applied stress evidently the conditions: $\overline{\sigma} = T_\theta {}^{\alpha}\overline{c}$ and $\overline{\epsilon}_v = \epsilon_v(^{\alpha}c)$ will hold. Hence for the case of the one-dimensional flow of the β-phase at any given time within the parameter set $(t_0, t_1) \in T$, it will be specified by the reduced functional $\mathcal{F}_{\epsilon}^t(^{\alpha}c)$ assumed to be linear and bounded on a dense set of the control space Θ. However this representation has been restricted to an "ϵ_v-projection" of $v_t \in \mathcal{V} = \Sigma \otimes \mathcal{E}$.

Since experimentally the time evolution and the probability distributions of $^{\alpha}c$ can be monitored [340] the evaluation of the latter can be specified by using:

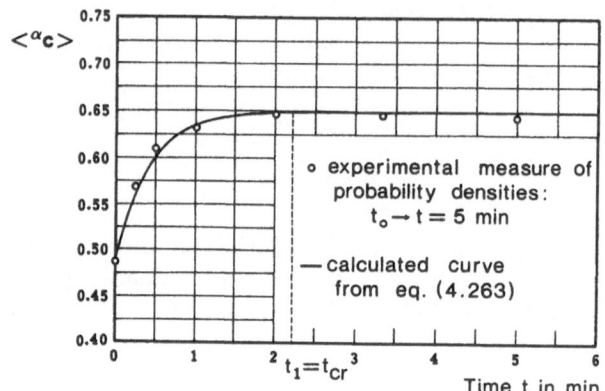

Fig. 43. Average values of the control parameter $^\alpha c$ versus time.

$$^\alpha c(t) = \int\limits_{t_0=0}^{t_1} K(t - t_1)(^\alpha \dot c)d\tau \qquad (4.283)$$

in which the Kernel function can be obtained from:

$$^\alpha c(t) = {}^\alpha c_0 + Ae^{-Bt} \qquad (4.284)$$

where the constants $^\alpha c_0$ and A, B are evaluated from the initial condition and the rate of change of the experimental values of $^\alpha c$ immediately after the initial condition ($t = 0^+$). The constant $^\alpha c_0$ can be established by considering the curve of $^\alpha c$ to approach the "steady-state" after the time $t_1 = t_{cr}$ has been reached. For the determination of the distributions of the control parameter $^\alpha c(t)$, SEM microscopy with the subsequent application of quantitative stereology to the obtained micrographs of cross-sections of material specimen have been employed. It has been found convenient for the purpose of these tests to use an Aluminium-Resin composite consisting of ethylene-glycol-ether as the β-phase in which (spherical) aluminium particles of $\sim 5\mu m$ dia. were inserted as the α-phase. The numerical evaluation of eqn. (4.284) shows good agreement with the experimentally determined results in Fig. 43 for an applied stress of $\sigma = 140 kPa$. The upper limit t_1 ($\sim$ no flow condition of the β-phase) was reached at $t_r \simeq 2 \cdot 2$ minutes.

It may be concluded from the above given examples and their numerical evaluations, that the stochastic mechanics approach can be readily extended to a wider class of discrete media. The use of probabilistic concepts in the representation of the material behaviour involves however the tools of modern analysis and stochastic calculus. The introduced three spatial-measuring scales and for dynamic problems the corresponding time-scales, lie at the foundation of the stochastic mechanics theory. Due to the limitation of the size of this volume it was not possible to include a discussion on the associated experimental work. The latter is important for the evaluation of random field quantities and the morphological analysis of the given microstructure. However one method of optical processing often used in this Laboratory has been discussed in ref. [343]. More recently a review of other experimental procedures in the study of discrete media has also been given in ref. [344].

References

[1] L. KRÜGER, L. J. DASTON and M. HEIDELBERGER (Eds.) : *The Probabilistic Revolution, Vol. I, Ideas in History,* MIT Press, Cambridge, Mass. (1987).

[2] L. KRÜGER, G. GIGERENZER and M. S. MORGAN (Eds.) : *The Probalistic Revolution, Vol. II, Ideas in the Sciences,* MIT Press, Cambridge, Mass. (1987).

[3] A. KAMLAH : *The Decline of the Laplacian Theory of Probability, a study of Stumpf, Kries and Meinong* in Vol. I of ref. [1], pp. 90-116.

[4] R. VON MISES : *Wahrscheinlichkeitsrechnung Deuticke,* Leipzig (1928).

[5] H. REICHENBACH : *The Theory of Probability,* Univ. of California Press (1949).

[6] A. N. KOLMOGOROV : *Grundbegriffe der Wahrscheinlichkeitsrechnung,* Springer, Berlin (1933).

[7] E. KNOBLOCH : *Émile Borel as a Probabilist* in Vol. I of ref. [1], pp. 215-233.

[8] D. HILBERT : *Mathematische Probleme. Archiv für Mathematik und Physik* (1901) and in *Gesammelte Abhandlungen,* Vol. III, Berlin (1935).

[9] E. BOREL : *Traité du calcul des probabilités et de ses applications,* (4 vols.), Paris, Gauthier-Villars (1925-39).

[10] A. N. KOLMOGOROV : *Foundations of the Theory of Probability,* Chelsea, New York (1950).

[11] E. BOREL : " *Les probabilités dénombrables,*" English translation by J. Barone and A. Novikov In: *A History of the Axiomatic Formulation of Probability, Arch. for History of Exact Sciences,* 18 (1977-78).

[12] L. KRÜGER : *Overview of Part VI (Physics)* in Vol. II of ref. [2].

[13] T. VON PLATO : *Probabilistic Physics the Classical Way* in Vol. II of ref. [2].

[14] W. GIBBS : *Elementary Principles of Statistical Mechanics,* Dover Publ., New york (1960).

[15] R. C. TOLMAN : *The Principles of Statistical Mechanics,* Univ. of Oxford Press, London (1938).

[16] S. G. BRUSH : *The Kind of Motion we call Heat,* 2 vols., North Holland, Amsterdam (1976).

[17] S. P. GUDDER : *Stochastic Methods in Quantum Mechanics,* North Holland, New York (1979).

[18] J. VON NEUMANN : *"Proof of the quasi-ergodic hypothesis,"* Proc. N.A.S. **18**, 70 (1932).

[19] G. D. BIRKHOFF : *"Proof of the ergodic theorem,"* Proc. N.A.S. **17**, (1931).

[20] A. I. KHINTCHINE : *Mathematical Foundations of Statistical Mechanics,* Dover Publ., New York (1949).

[21] K. YOSIDA and S. KAKUTANI : *"Operator-theoretical treatment of Markov processes and mean ergodic theorem,"* Annals of Math., **42**, Nov. 1 (1941).

[22] D. R. AXELRAD : *Micromechanics of Solids,* PWN-Elsevier Scientific Publ. Co., Amsterdam (1978).

[23] D. R. AXELRAD : *Foundations of the Probabilistic Mechanics of Discrete Media,* Foundations and Philosophy of Science Series, Pergamon Press, Ed. M. Bunge, Oxford (1984).

[24] K. YOSIDA : *Functional Analysis,* Springer, Berlin (1978).

[25] K. KURATOWSKI and A. MOSTOWSKI : *Set Theory,* North Holland, Amsterdam (1976).

[26] A. N. KOLMOGOROV and S. V. FOMIN : *Functional Analysis,* Vols. I and II, Graylock, Rochester, Albany (1961).

[27] W. RUDIN : *Functional Analysis,* McGraw Hill Co., New York (1973).

[28] D. H. GRIFFEL : *Applied Functional Analysis,* Ellis Horwood Ltd. Publ., Chichester, England (1981).

[29] J. L. KELLEY : *General Topology,* van Nostrand, New York (1955).

[30] Y. CHOQUET-BRUHAT, C. DEWITT-MORETTE and M. DILLARD-BLEICK : *Analysis, Manifolds and Physics,* North-Holland, Amsterdam (1977).

[31] H. H. SCHAEFFER : *Topological Vector Spaces,* MacMillan Series in Advanced Mathematics and Theor. Physics, MacMillan, New York (1966).

[32] G. F. SIMMONS : *Introduction to Topology and Modern Analysis,* McGraw Hill Co., New York (1963).

[33] N. BOURBAKI : *Eléménts de Mathématique, Vol. V, Espaces vectorials topologiques,* Hermann, Paris (1953).

[34] N. BOURBAKI : *Eléménts de Mathématique, Vol. III, Topologie général,* Hermann, Paris (1958).

[35] K. KURATOWSKI : *Topology, I, II,* Academic Press-PWN, New York and Warsaw (1966-1968).

[36] P. D. PANAGIOTOPOULOS : *Inequality Problems in Mechanics and Applications,* Birkhäuser Boston Inc. (1985).

[37] N. YA. VILENKIN *et al.* : *Functional Analysis,* Walters-Noordhoff Publ., Groningen (Holland) (1972).

[38] E. HILLE and PHILLIPS : *Functional Analysis and Semi-groups,* Publ. Am. Math. Soc. (1948).

[39] P. L. BUTZER and H. BERENS : *Semi-groups of Operators and Applications,* Springer, New York (1967).

[40] R. T. ROCKAFELLER : *Convex Analysis*, Princeton Univ. Press (1972).

[41] A. RÉNYI : *Foundations of Probability*, Holden-Day, San Francisco (1970).

[42] Yu. V. PROHOROV and YU. A. ROZANOV : *Probability Theory*, Springer, New York (1969).

[43] A. A. BOROWKOW : *Wahrscheinlichkeitstheorie*, Birkhäuser, Basel (1976).

[44] A. T. BHARUCHA-REID : *Elements of the Theory of Markov Processes and its Applications*, McGraw Hill, New York (1960).

[45] T. L. DOOB : *Stochastic Processes*, J. Wiley, New York (1961).

[46] J. NEVEU : *Mathematical Foundations of the Calculus of Probability*, Holden-Day, San Francisco (1965).

[47] P. BRÉMAUD : *Point Processes and Queues*, Martingale Dynamics, Springer-Verlag, New York (1981).

[48] R. J. ELLIOT : *Stochastic Calculus and Applications*, Springer-Verlag, New York (1982).

[49] D. R. AXELRAD : *"On the Transient Behaviour of Structured Solids,"* In *Trends in Applications of Pure Mathematics to Mechanics*, Eds. E. Kröner and K. Kirchgässner, Lecture Notes in Physics, 249, Springer-Verlag, Berlin (1985).

[50] T. L. DOOB : *Classical Potential Theory and its Probabilistic Counter Part*, Springer-Verlag, Berlin (1983).

[51] O. HANŠ : *"Generalized Random Variables,"* Trans. 1st Prague Conference on Information Theory and Random Processes (1956).

[52] H. UMEGAKI and A. T. BHARUCHA-REID : *"Banach space-valued random variables and tensor products of Banach spaces,"* J. Math. Anal. Appl., 31, (1970).

[53] P. R. HALMOS : *Measure Theory*, van Nostrand, New York (1950).

[54] S. SAKS : *Theory of the Integral*, PWN Polish Scientific Publ., Warsaw (1937).

[55] I. M. GEL'FAND and N. YA. VILENKIN : *Generalized Functions*, Vols. I-IV, Academic Press, New York (1964).

[56] E. MOURIER : *"Random elements in linear spaces,"* Proc. 5th Berkeley Symp. Math. Statistics and Probability, Vol. II (1965).

[57] Yu. V. PROHOROV : *"The method of characteristic functionals,"* Proc. 4th Berkeley Symp. Math. Statistics and Probability, Vol. II (1960).

[58] D. A. KAPPOS : *Probability Algebras and Stochastic Spaces*, Academic Press, New York (1969).

[59] P. BILLINGSLEY : *Convergence of Probability Measures*, Wiley, New York (1968).

[60] K. ITÔ and H. P. McKEAN : *Diffusion Processes and their Sample Paths*, Springer-Verlag, Berlin (1965).

[61] E. NELSON : *Dynamical Theories of Brownian Motion*, Princeton Univ. Press, Princeton, New Jersey (1967).

[62] E. EBERLEIN and M. S. TAQQU, Eds. : *Dependence in Probability and Statistics. A survey of recent results.* Birkhäuser Verlag, Basel (1986).

[63] M. FRÉCHET : *Généralités sur les Probabilités,* Gauthier-Villars, Paris (1950).

[64] W. FELLER : *An Introduction to Probability Theory and its Applications. Vols. I, II,* Wiley, New York (1966).

[65] B. V. GNEDENKO : *Theory of Probability,* Chelsea, New York (1963).

[66] M. PELIGRAD : *Recent advances in the central limit theorem and its weak invariance principle for mixing sequences of random variables (A survey)* in ref. [62], pp. 193-223 (1986).

[67] R. BRADLEY : *Basic properties of strong mixing conditions,* in ref. [62], pp. 165-192 (1986).

[68] I. A. IBRAGIMOV : *"Some limit theorems for stationary processes,"* Theor. Prob. Appl., $\underline{7}$, pp. 349-382 (1962).

[69] I. A. IBRAGIMOV : *"A note on the central limit theorem for dependent random variables,"* Theor. Prob. Appl., $\underline{20}$, pp. 135-141 (1975).

[70] P. ERDÖS and M. KAC : *"On certain limit theorems of the theory of probability,"* Bull. Am. Math. Soc., $\underline{52}$, pp. 292-302 (1946).

[71] P. ERDÖS and M. KAC : *"On the number of positive sums of independent random variables,"* Bull. Am. Math. Soc., $\underline{53}$, pp. 1011-1020 (1947).

[72] M. DONSKER : *"An invariance principle for certain probability limit theorems,"* Memoirs, AMS $\underline{6}$ (1957).

[73] W. PHILLIP : *Invariance principles for independent and weakly dependent random variables,* In ref. [62], pp. 224-266 (1986).

[74] A. M. YAGLOM : *An Introduction to the theory of stationary random functions,* Prentice Hall, Eaglewood Hills, New Jersey (1965).

[75] E. PARZEN : *Modern Probability Theory and its Application,* Wiley, New York (1960).

[76] J. MIKUSINSKI : *The Bochner Integral,* Birkhäuser Verlag, Basel (1978).

[77] N. WIENER : *Non-linear Problems in Random Theory,* MIT Press, Massachusetts (1958).

[78] M. LOÈVE : *Probability Theory,* von Nostrand, New York (1962).

[79] A. MOSTOWSKI : *Constructible sets with applications,* North-Holland, Amsterdam (1969).

[80] A. A. MARKOV : *Calculus of Probability,* 4th ed., Moscow (1924) [in Russian].

[81] E. DYNKIN : *Markov Processes,* Vols. I and II, Academic Press, New York (1965).

[82] R. M. BLUMENTHAL and R. K. GETOOR : *Markov Processes and Potential Theory,* Academic Press, New York (1968).

[83] M. ROSENBLATT : *Markov Processes. Structure and Asymptotic Behaviour,* Springer-Verlag, Berlin (1971).

[84] D. REVUZ : *Markov Chains,* North-Holland, Amsterdam (1975).

[85] R. KINDERMANN and J. L. SNELL : *"Markov Random Fields and their Applications,"* Contemporary Maths. Vol. I, Am. Math. Soc., Providence, R. I. (1980).

[86] R. L. DOBRUSHIN : *"Description of a random field by means of conditional probabilities and the conditions governing its regularity,"* Theory of Prob. and its Appl., 10, pp. 193-213 (1969).

[87] M. B. AVRIENTSEV : *"On a method of describing discrete parameter random fields,"* Problemy Peredachi Informatsii, 6, pp. 100-109 (1970).

[88] Yu. A. ROZANOV : *"On Gaussian fields with given conditional distributions,"* Theory of Prob. and its Appl., 12, pp. 381-391 (1967).

[89] Z. HASHIN : *J. Appl. Phys.,* 50, 481 (1983).

[90] M. J. BERAN : *Statistical Continuum Theories,* Interscience Publ. John Wiley, New York (1968).

[91] I. A. KUNIN : *Elastic Media with Microstructure,* Vol. II, Springer-Verlag, Berlin (1983).

[92] M. KAC : *Phys. Fluids,* Vol. 2, 8 (1959).

[93] J. K. PERCUS and G. O. WILLIAMS : *"The Intrinsic Interface,"* In *Fluid Interfacial Phenomena,* Ed. C. A. Croxton, John Wiley & Sons, New York (1986).

[94] M. KAC and J. LOGAN : *"Fluctuations,"* In *Fluctuation Phenomena,* Eds. E. W. Montroll and J. L. Lebowitz, North-Holland Publ. Co., Amsterdam (1979).

[95] H. J. HERRMANN and S. ROUX (Eds.) : *Statistical Methods for the Fracture of Disordered Media,* North-Holland Publ. Co., Amsterdam (1990).

[96] J. C. CHARMET, S. ROUX and E. GUYON (Eds.) : *Disorder and Fracture,* Plenum Press, New York (1990).

[97] J. P. HULIN, A. M. CAZABAT, E. GUYON and F. CARMONA (Eds.) : *Hydrodynamics of Dispersed Media,* North-Holland Publ. Co., Amsterdam (1990).

[98] C. A. CROXTON : *Statistical Mechanics of the Liquid State,* J. Wiley, Chichester (1980).

[99] D. N. ZUBAREV : *Non-equilibrium Statistical Thermodynamics,* Cons. Bureau-New York, London (1974).

[100] R. BALESCU : *Equilibrium and Non-equilibrium Statistical Mechanics,* J. Wiley, New York (1975).

[101] M. ISHII : *Thermo-fluid Dynamics Theory of Two-Phase Flow,* Collection de la direction des études et recherches d'électricité de France, Eyrolles, France (1975).

[102] I. PRIGOGINE : *Non-equilibrium Statistical Mechanics,* Wiley / Interscience, New York (1962).

[103] J. O. HIRSCHFELDER, C. F. CURTISS and R. B. BIRD : *Molecular Theory of Gases and Liquids,* John Wiley, New York (1954).

[104] J. P. BOON and S. YIP : *Molecular Hydrodynamics*, McGraw-Hill Inc., New York (1980).

[105] J. P. HANSEN and J. R. McDONALD : *Theory of Simple Liquids*, Academic Press, London (1976).

[106] A. K. MacPHERSON : *Atomic Mechanics of Solids*, North-Holland, Amsterdam (1990).

[107] N. METROPOLIS, A. W. ROSENBLUTH, M. N. ROSENBLUTH, A. H. TELLER and E. TELLER : *"Equation of state calculations by fast computing machines,"* J. Chem. Phys., 21, pp. 1087-1092 (1953).

[108] K. YOSIDA and S. KAKUTANI : *"Operator-theoretical treatment of Markov processes and mean ergodic theorem,"* Annals of Math., 42, No. 1 (1941).

[109] K. YOSIDA : *"The Markov process with a stable distribution,"* Proc. Imp. Acad. Japan, 16] (3) (1940).

[110] S. KAKUTANI : *"Ergodic theorem and the Markov proces with a stable distribution,"* Proc. Imp. Acad. Japan, 16 (3) (1940).

[111] S. HESS and W. LOOSE : *Molecular Dynamics: Test of Microscopic Models for the Material Properties of Matter*, In *Constitutive Laws and Microstructure*, Eds. D. R. Axelrad and W. Muschik, Springer-Verlag, Berlin (1986).

[112] K. BINDER : In *Monte-Carlo Methods in Statistical Physics*, Ed. K. Binder, Springer-Verlag, Berlin (1979, 1986).

[113] S. W. LOVESEY : *Dynamics of Solids and Liquids by Neutron Scattering*, Springer, Heidelberg (1977).

[114] J. C. DORE, G. WALFORD and D. I. PAGE : *Mol. Physics*, 29, 565 (1975).

[115] J. P. HANSEN : *Correlation functions and their relationships with experiments*, In *Microscopic Structure and Dynamics of Liquids*, Eds. J. Dupuy and J. J. Dianoux, Plenum Press, New York (1977).

[116] P. A. EGELSTAFF : *An Introduction to the Liquid State*, Academic Press, New York (1967).

[117] J. YVON : *Correlations and Entropy in Classical Statistical Mechanics*, Pergamon Press, Oxford (1969).

[118] E. ISING : *"Beitrag zur Theorie des Ferromagnetismus,"* Zeit. für Physik, 31, pp. 253-258 (1925).

[119] C. ITZYKSON and J. M. DROUFFE : *Statistical Field Theory*, Vols. I, II, Cambridge Monographs on Math. Physics, Cambridge University Press, New York (1989).

[120] E. W. MONTROLL and B. J. WEST : *On an enriched collection of stochastic processes*, In *Fluctuation Phenomena*, Eds. E. W. Montroll and J. L. Lebowtiz, Studies in Statistical Mechanics, North-Holland Publ. Co., Amsterdam (1979).

[121] S. R. BROADBENT and J. M. HAMMERSLEY : *Proc. Cambridge Philos. Soc.*, 53, 629 (1957).

[122] C. DOMB and M. S. GREEN (Eds.) : *Phase Transitions and Critical Phenomena*, Vols. 1-6, Academic Press, London (1972).

[123] D. STAUFFER : *Introduction to Percolation Theory*, Taylor and Francis, London (1985).

[124] N. CHRISTOFIDES : *Graph theory: an algorithmic approach*, Academic Press, New York (1975).

[125] F. HARARY : *Graph Theory*, Addison-Wesley, Mass. (1969).

[126] A. KOSSAKOWSKI : *"Time evolution in isolated and non-isolated quantum mechanical systems,"* Rev. Math. Phys., $\underline{3}$, 247 (1973).

[127] G. MACKEY : *The Mathematical Foundation of Quantum Mechanics*, Benjamin, New York (1963).

[128] W. MUSCHIK : *Aspects of Non-equilibrium thermodynamics – Six Lectures on Fundamentals and Methods*, World Scientific Series in Theoretical and Applied Mechanics, London (1990).

[129] A. T. BHARUCHA-REID : *Random Integral Equations*, Academic Press, New York (1972).

[130] K. ITÔ : *"On stochastic differential equations,"* Mem. Am. Math. Soc., $\underline{4}$ (1951).

[131] L. ARNOLD : *Stochastische Differentialgleichungen Theorie u. Anwendung*, Oldenburg, Müncheu (1974).

[132] O. HANŠ : *"Random operator equations,"* Proc. 4th Berkeley Symp. on Math. Stat. and Prob., Vol. II, pp. 185-202 (1960).

[133] A. ŠPAČEK : *"Zufällige Gleichungen,"* Czechoslovak Math. J., $\underline{5}$, pp. 462-466 (1955).

[134] H. P. McKEAN : *Stochastic Integrals*, Academic Press, New York (1969).

[135] J. J. MOREAU : *"La notion de sur-potentiel et les liaisons unilatérales en élastoplastique,"* C. R. Acad. de Paris, $\underline{267A}$, pp. 964-957 (1968).

[136] J. J. MOREAU : *"Evolution problems associated with a moving set in a Hilbert space,"* J. Diff. Equs., $\underline{26}$, pp. 347-374 (1977).

[137] I. I. GIHMAN and A. V. SKOROHOD : *The Theory of Stochastic Processes II*, Springer-Verlag, Berlin (1975).

[138] R. S. LIPSTER and A. N. SHIRYAYEV : *Statistics of Random Processes I, II: I. General Theory, II. Applications*, Springer-Verlag, Berlin (1978).

[139] C. DELLACHERIE : *Capacités et Processes Stochastiques*, Springer-Verlag, Berlin (1967).

[140] K. L. CHUNG : *Markov Chains with Stationary Transition Probabilities*, Springer, Berlin (1987).

[141] J. VON NEUMANN : *Functional Operators*, Vol. I, Ann. of Math. Studies, Princeton Univ. Press (1950).

[142] V. DOLEZAL : *Monotone Operators and Applications in Control Theory*, Elsevier Publ. Co., Amsterdam (1979).

[143] A. LASOTA and M. C. MACKEY : *Probabilistic Properties of Deterministic Systems*, Cambridge Univ. Press, Cambridge (1985).

[144] G. ADOMIAN : *Stochastic Systems,* Math. in Science Series 169, Academic Press, New York (1983).

[145] S. R. ADKE and S. M. MANJUNATH : *Finite Markov Processes,* John Wiley & Sons, New York (1985).

[146] W. GUZ : *"Markovian processes in classical and quantum statistical mechanics,"* Rep. Math. Phys., 7, pp. 205-214 (1975).

[147] M. L. SILVERSTEIN : *Symmetric Markov Processes,* Lecture Notes in Mathematics, 426, Springer-Verlag, Berlin (1974).

[148] M. SHARPE : *General Theory of Markov Processes,* Academic Press, New York (1988).

[149] E. HILLE and R. S. PHILLIPS : *Functional Analysis and Semi-groups,* Colloqu. Publ. Am. Math. Soc. (1957).

[150] A. KOLMOGOROV : *"Zur Theorie der Markoffschen Ketten,"* Math. Annalen, 112, pp. 155-160 (1935).

[151] A. EINSTEIN : *Ann. Physik,* 17, 19, 549, 371 (1905/6).

[152] K. J. ÅSTRÖM : *Introduction to Stochastic Control theory,* Academic Press, New York (1970).

[153] W. HORSTHEMKE and R. LEFEVER : *Noise Induced Transitions: Theory and Application in Physics, Chemistry and Biology,* Springer-Verlag, Berlin (1984).

[154] K. ITÔ : *"Stochastic integrals,"* Proc. Imp. Acad. Tokoyo, 90, pp. 519-524 (1944).

[155] K. ITÔ and M. NISIO : *"On the stationary solutions of stochastic differential equations,"* J. Math, Kyoto Univ., 4, 1 (1-7a) (1964).

[156] D. R. AXELRAD : *Stochastic analysis of structural changes in solids,* In *Constitutive Laws and Microstructure,* Eds. D. R. Axelrad and W. Muschik, Springer-Verlag, Berlin (1988).

[157] H. K. KUSHNER and D. S. CLARK : *Stochastic Approximation Methods for Constrained and Unconstrained Systems,* Springer, New York (1978).

[158] G. C. GOODWIN and R. L. PAYNE : *Dynamic System Identification,* Academic Press, New York (1977).

[159] M. AOKI : *Optimization of Stochastic Systems,* Academic Press, New York (1967).

[160] H. F. CHEN : *Recursive Estimation and Control of Stochastic Systems,* John Wiley & Sons, New York (1985).

[161] A. H. JAZWINSKI : *Stochastic Processes and Filterring Theory,* Vol. 64, In *Mathematics in Science & Engineering,* Ed. R. Bellman, Academic Press, New York (1970).

[162] M. H. A. DAVIS : *"The representation of Martingales Jump Processes,"* SIAM, Journ. Control and Optimaization, Vol. 14, No. 4 (1976).

[163] W. H. FLEMING and R. W. RICHEL : *Deterministic and Stochastic Optimal Control,* Springer-Verlag, Berlin (1975).

[164] R. RICHEL : *"A minimum principle for controlled Jump Processes,"* Lecture Notes in Economics and Math. Systems, No. 107, Springer-Verlag, Berlin (1975).

[165] M. RUDEMO : *"State Estimation for Partially Observed Markov Chains,"* Journ. Math. Analysis and Applications, Vol. 44, pp. 581-611 (1973).

[166] D. R. AXELRAD and J. W. PROVAN : *"Deformation theory of elastic polycrystalline solids,"* Arch. Stos., 25, Warsaw (1973).

[167] M. A. PINSKY : *Lectures on Random Evolution,* World Scientific, New Jersey (1991).

[168] H. MARGENAU and N. R. KESTNER : *Theory of Intermolecular Forces,* Pergamon Press, Oxford (1971).

[169] A. D. BUCKINGHAM : *"Interatomic and Intermolecular Forces,"* In *Microscopic Structure and Dynamics of Liquids,* Eds. J. Dupuy and A. J. Dianoux, Nato-Adv. Study Inst., Series B: Physics, France (1977), Plenum Press, New York (1978).

[170] P. HOBZA and R. ZAHRADNIK : *Weak Intermolecular Interactions in Chemistry and Biology,* Elsevier Publ., Amsterdam (1980).

[171] P. HOBZA and R. ZAHRADNIK : *Intermolecular Complexes,* Elsevier Publ., Amsterdam (1988).

[172] P. SCHUSTER, G. ZUNDEL and C. SANDORFY : *The Hydrogen Bond,* Vols. I-III, North-Holland, Amsterdam (1976).

[173] M. BORN and Y. HUANG : *Dynamical Theory of Crystal Lattices,* Pergamon Press, Oxford (1954).

[174] A. MOTT and B. JONES : *Properties of Metals and Alloys,* Pergamon Press, Oxford (1936).

[175] D. R. AXELRAD, J. W. PROVAN and S. EL-HELBAWI : *"Dislocation effects in elastic structured solids,"* Arch. Mech. Stos., 25, pp. 801-810 (1973).

[176] J. W. PROVAN and D. A. BAMIRO : *"Elastic grain-boundary responses in copper and aluminium,"* Acta Metallurgica, 25, pp. 309-319 (1977).

[177] W. BOLLMANN : *Crystal Defects and Crystalline Interfaces,* Springer-Verlag, New York (1970).

[178] C. GOUX et al. : *Surface Science,* Vol. 31 (1972).

[179] D. R. AXELRAD and J. KALOUSEK : *"Measurements of Microdeformations by Holographic X-ray Diffraction,"* In *Exp. Mechanics in Research and Development,* Ed. J. I. Pindera, Univ.of Waterloo Press, Ontario, Canada (1973).

[180] G. HASSON, J. V. BOOS, I. HERBEUVAL, M. BISCONDI and C. GOUX : " *Theoretical and experimental determinations of grain-boundary structures and energies,"* Surface Sciences, 31, pp. 115-137 (1972).

[181] D. R. AXELRAD and S. BASU : *"Mechanical relaxation theory of fibrous structures,"* In Adv. in Mol. Rel. and Int. Proc., 11, pp. 165-190, Elsevier Publ. Amsterdam (1977).

[182] D. R. AXELRAD : *"On the time-dependent behaviour of bond paper,"* In *General Constitutive Relations for Wood and Wood Based Materials,* Eds. B. A. Jayne, J. A. Johnson and R. W. Perkins, National Science Foundation, Syracuse Univ. Pess, Syracuse, N. Y. (1978).

[183] D. R. AXELRAD, Y. M. HADDAD and D. ATACK : *"Stochastic deformation theory of a two-dimensional fibrous network,"* 11th An. Meeting Soc. Eng. Science Proc., Ed. G. J. Dvorak, Durham University, North Carolina (1974).

[184] D. R. AXELRAD : *"Theory of bond failure in hydrogen-bonded solids,"* In *Adv. in Molecular Relaxation and Interaction Processes,* 15, pp. 51-69, Elsevier Publ., Amsterdam (1979).

[185] T. KIHARA : *Intermolecular Forces,* Wiley, Chichester, England (1978).

[186] A. FREY-WYSSLING : *Biochem. Biophys. Acta,* 18, pp. 166-168 (1955).

[187] J. JONES : *Journ. of Polymer Science,* 32, p. 371 (1958).

[188] C. Y. LIANG and R. H. MARCHERSAULT : *"Hydrogen bonds in native cellulose,"* Journ. of Polymer Science, 37 (1959).

[189] W. C. HAMILTON and J. A. IBERS : *Hydrogen Bonding in Solids,* Benjamin Publ., New York (1968).

[190] G. C. PIMENTEL and A. L. McLLELLAN : *The Hydrogen Bond,* Freeman Publ., San Francisco (1960).

[191] R. A. JACOBSON, J. A. WUNDERLICH and W. N. LIPSCOMB : *"The crystal and molecular structure of cellulose,"* Acta Cryst., 14, pp. 568-607 (1961).

[192] A. FREY-WYSSLING : *Macromolecules in Cell Structures,* Harvard Univ. Press (1957).

[193] R. S. TONNESEN and O. ELLEFSEN : *Cellulose and Cellulose Derivatives,* John Wiley & Sons, New York (1971).

[194] B. PEDERSEN : *Acta Crystallogr.,* 30, p. 289 (1974).

[195] D. R. AXELRAD : *"Rheology of discrete media,"* Proc. 7th Int. Congress on Rheology, Eds. C. Klason and J. Kubat, Gothenburg, Sweden (1976).

[196] RI. PERALTA-FABI : *Experimental Investigation of Creep Behaviour of Bond Paper,* Ph. D. Disertation, McGill University, Canada (1978).

[197] E. W. MONTROLL and B. J. WEST : *"On an enriched collection of slochustic processes,"* In *Fluctuation Phenomena,* Eds. E. W. Montroll and J. L. Lebowitz, North-Holland Publ. Co., Amsterdam (1979).

re198] M. E. GURTIN : *"Thermodynamics and the cohesive zone in fracture,"* ZAMP, 30, pp. 991-1003 (1979).

[199] D. MAUGIS : *"Fracture Mechanics and Solid Adhesion,"* In *Disorder and Fracture,* Eds. J. C. Charmet *et al.,* Plenum Press, New York (1990).

[200] J. W. PROVAN and C. I. MBANUGO : *"Stochastic fatigue crack growth: An experimental study,"* Res. Mechanica, Vol. 2, pp. 53-72 (1981).

[201] H. GHONEM and J. W. PROVAN : *"Micromechanics theory of fatigue crack initiation and propagation,"* Eng. Fracture Mechanics, Vol. 13, No. 4, pp. 963-977 (1980).

[202] P. MEAKIN, GAUG LI, L. M. SANDER, HONG YAN, F. GUINEA, O. PLA and E. LOUIS : *"Simple Stochastic Models for Material Failure,"* In *Disorder and Fracture,* Eds. J. C. Charmet, S. Roux and E. Guyon, Plenum Press, New York (1990).

[203] D. MAUGIS : *"Subcritical crack growth, surface energy, fracture toughness, stick-slips and embrittlement,"* J. Mat. Sci., 20, p. 3041 (1985).

[204] A. H. NISSAN : *Trans. 6th Fundamental Research Symp.,* Brit. Paper and Board Ind. Fed., Fibre-Water Interactions, Oxford (1977).

[205] H. S. FRANK and W. Y. WEN : *Discussions – Faraday Soc.,* 24, pp. 133-140 (1957).

[206] N. T. J. BAILEY : *The Mathematical Theory of Epidemics,* Hafner Publ. Co., New York (1957).

[207] H. W. HASKEY : *"A general expression for the mean in a simple stochastic epidemic,"* Biometrika, 44 (1957).

[208] KIRKWOOD : *J. Chem. Physics,* 3, pg. 300 (1935).

[209] F. F. ABRAHAM : *Phys. Rep.,* 53, 93 (1979).

[210] R. EVANS : *Adv. Phys.,* 28, 143 (1979).

[211] J. K. PERCUS : *Studies in Statistical Mechanics – The Liquid State of Matter,* Eds. E. W. Montroll and J. L. Lebowitz, North-Holland, Amsterdam (1982).

[212] J. S. ROWLINSON and B. WILSON : *Molecular Theory of Capillarity,* Clarendon Press, Oxford (1982).

[213] TEMPERLEY and TRAVENA : *Liquids and their Properties,* Ellis Horwood Ltd., Chichester (1978).

[214] J. L. LEBOWITZ and J. K. PERCUS : *J. Math. Phys.,* 4, p. 176 (1963).

[215] A. J. M. YANG, P. D. FLEMING and J. H. GIBBS : *J. Chem. Phys.,* 64, 3732 (1976).

[216] G. VAN KAMPEN : *Phys. Rev.,* 135 (1964).

[217] B. F. McCOY and H. T. DAVIS : *"Free energy theory of inhomogeneous fluids,"* Phys. Rev., 20 (1979).

[218] J. BRICMONT, J. L. LEBOWITZ and C. E. PFISTER : *"On the local structure of the phase separation line in the two-dimensional Ising system,"* Journ. Stat. Physics, Vol. 26, No. 2 (1981).

[219] J. BRICMONT, J. R. FONTAINE and J. L. LEBOWITZ : *"Surface Tension, Percolation and Roughening,"* Journ. Stat. Physics, Vol. 29, No. 2 (1982).

[220] R. L. DOBRUSHIN : *Theory Prob. Appl.,* 17, 582 (1972) and 18, 252 (1973).

[221] J. K. PERCUS and G. O. WILLIAMS : *"The intrinsic interface,"* In *Fluid Interfacial Phenomena,* Ed. C. A. Croxton, John Wiley & Sons Ltd., London (1986).

[222] K. OELSCHLÄGER : *"A limit theorem for a one-dimensional many-particle system with gradient interaction,"* In *Stochastic Modelling in Biology,* pp. 141-149, Ed. P. Tatu, World Scientific, London (1990).

[223] E. PRESUTTI : *"Collective phenomena in stochastic particle systems,"* In *Stochastic Processes – Math. and Physics,* Lecture Notes in Mathematics, Eds. S. Albeverio, P. Blanchard and L. Streit, pp. 195-232, Springer-Verlag, Heidelberg, New York (1987).

[224] M. G. MÜRMANN : *"The hydrodynamic limit of a one-dimensional neighbour gradient system,"* Journ. Stat. Physics, 48, pp. 769-788 (1987).

[225] L. A. SANTALÓ : *"Integral geometry and geometric probability,"* Encyclopedia of Mathematics and its Applications, Ed. Gian-Carlo Rota, Vol. 1, Addison-Wesley Publ. Co., Reading, Mass. (1976).

[226] W. BLASCHKE : *Vorlesungen über Differentialgeometric,* Springer-Verlag, Berlin (1923).

[227] H. HADWIGER : *"Über Statische Flächen u. Längenmessung,"* Mitt. Naturforsch. Gesellsch., Bern, pp. 53-58 (1939).

[228] H. HADWIGER : *Vorlesungen über Inhalt, Oberflache und Isoperimetric,* Springer-Verrlag, Berlin (1957).

[229] D. G. KENDALL : *"Foundations of a theory of random sets,"* In *Stochastic Geometry,* Eds. E. F. Harding and D. G. Kendall, Wiley & Sons, Chichester, England (1974).

[230] K. KRICKEBERG : *"Hyperplane processes,"* Bull. Int. Ind. Statis., 46, pp.623-660 (1975).

[231] R. E. MILES : *"The random division of space,"* Suppl. Adv. Appl. Prob., pp. 243-266 (1972).

[232] R. E. MILES : *"A Synopsis of Poisson Flats in Euclidean Spaces,"* In *Stochastic Geometry,* Eds. Harding and Kendall, Wiley & Sons, Chichster (1974).

[233] R. E. MILES : *"Stereology for embedded aggregates of not necessarily convex particles,"* In Proc. 2nd Int. Workshop on Stereology and Stoch. Geometry, pp. 127-147, Aarhus (1983).

[234] G. MATHERON : *Random Sets and Integral Geometry,* John Wiley, New York (1975).

[235] J. SERRA : *Image Analysis and Mathematical Morphology,* Vols. I, II, Academic Press, New York (1982).

[236] D. STOYAN, W. S. KENDALL and J. MECKE : *Stochastic Geometry and its Applications,* Wiley Series in Probability and Math. Statistics, John Wiley & Sons Ltd., New York (1987).

[237] A. CAUCHY : *"Notes sur divers théorèmes relatifs à la rectification des courbes et à la quadrature des surfaces,"* C. R. Acad. Sci., Paris, 13 (1841).

[238] M. W. CROFTON : *"On the theory of local probability, applied to straight lines drawn at random in a plane, the method being used and extended to the proof of certain new theorems in the integral calculus,"* Phil. Trans. Roy. Soc., London, 158 (1868).

[239] G. CHOQUET : *Topology,* Academic Press, New York (1966).

[240] M. G. KENDALL and P. A. P. MORAN : *Geometric Probability,* Griffins Publ., London (1963).

[241] D. J. DALEY and D. VERE-JONES : *An Introduction to the Theory of Point Processes,* Springer Series in Statistics, Springer-Verlag, Berlin (1988).

[242] B. D. RIPLEY : *Spacial Statistics,* Wiley & Sons, New York (1981).

[243] O. KALLENBERG : *Random Measures,* Akademie Verlag, Berlin (1976) and 3rd ed., Academic Press, London (1963).

[244] O. KALLENBERG : *"An informal guide to the theory of conditioning in point processes,"* Int. Stat. Rev., 52, pp. 151-164 (1984).

[245] S. K. SRINIVASAN : *Stochastic Point Processes and their Applications,* Griffin Publ., London (1974).

[246] L. JANOSSY : *"On the absorption of a nuclear cascade,"* Proc. R. Irish Acad. Sc., Sect. A 53, (1950).

[247] O. MACCHI : *"The coincidence approach to stochastic point processes,"* Adv. Appl. Probab., 7, pp. 83-122 (1975).

[248] M. WESTCOTT : *"The probability generating functional,"* J. Austral. Math. Soc., 14, pp. 448-466 (1972).

[249] B. D. RIPLEY : *"The second-order analysis of spacial point processes,"* J. Appl. Prob., 13, pp. 255-266 (1976).

[250] B. D. RIPLEY : *Spacial Statistics,* John Wiley & Sons Ltd., New York (1981).

[251] J. NEYMAN and E. L. SCOTT : *"Statistical approach to problems of cosmology,"* J. R. Stat. Soc., B-20, (1958).

[252] D. L. SNYDER : *Random Point Processes,* John Wiley & Sons Ltd., New York/London (1975).

[253] K. KRICKEBERG : *"The Cox Process,"* Inst. Nazionale di Alta Matematicam Symp. Mat., 9, pp. 151-167 (1972).

[254] A. F. KARR : *Point Processes and their Statistical Reference,* Dekker, New York/Zürich (1985).

[255] E. CARTAN : *"Les systémes différentiels extérieurs et leurs applications géometriques,"* Act. Sci. Indust., 994 (1945).

[256] E. CARTAN : *Le principe de dualité et certaines intégrales multiples de l'espace tangentiel et de l'espace reglé,* (Oeuvres Complètes, Vol. 1) Gauthier-Villar, Paris (1952).

[257] E. WEIBEL : *Stereological Methods,* Vol. 2, Academic Press, New York (1980).

[258] R. E. MILES : *"The random division of space,"* Adv. Appl. Prob. (Special Suppl.), 265 pp. (1972).

[259] D. JEULIN : *"On image analysis and micromechanics,"* Revue Phys. Appl., 23, pp. 549-556 (1988).

[260] D. JEULIN : *"Morphological modelling of images by sequential random functions,"* Signal Processing, 16, pp. 403-431, North-Holland (1989).

[261] E. E. UNDERWOOD : *Quantitative Stereology*, Addison-Wesley Publ. Co., Reading, Mass. (1970).

[262] M. A. DELESSE : *"Procede mecanique pour determiner la composition des roches,"* C. R. Acad. Sci. (Paris), 25, pp. 544-545 (1947).

[263] D. R. AXELRAD : *Random Theory of Deformation of Structured Media*, Lect. Notes - Int. Centre Mech. Sci., Udine (Italy), Springer-Verlag, Vienna (1971).

[264] D. R. AXELRAD : *"Micromechanics of Discrete Systems,"* Arch. Mech. Stos., 28, 299 (Warsaw) (1973).

[265] R. J. KNOPS and E. W. WILKES : *"Theory of elastic stability,"* Handbuch der Physik, Bd. VI, a/3, Springer-Verlag, Berlin (1964).

[266] E. TONTI : *"On the formal structure of physical theories,"* Inst. di Matematica del Politec. Milano, (Italy) (1975).

[267] S. MIZOHATA : *The Theory of Partial Differential Equations*, Cambridge University Press (1973).

[268] S. BASU : *On a general deformation theory of structured solids*, Ph.D. Dissertation, McGill University, Montréal, Canada (1975).

[269] K. F. ALBRECHT, V. CHETVERIKOV, R. FUNKE, W. EBELING, W. MENDE and M. PESCHEL : *"Random phenomena in non-linear systems in connection with the Volterra approach,"* In Stochastic Phenomena and Chaotic Behaviour in Complex Systems, Ed. P. Schuster, Springer-Verlag, Berlin (1984).

[270] E. STEIN, G. ZHANG and J. A. KÖNIG : *"Micromechanical modelling and computation of shakedown with non-linear kinematic hardening including examples for 2D-problems,"* In Recent Developments in Micromechanics, Eds. D. R. Axelrad and W. Muschik, Springer-Verlag, Berlin (1991).

[271] A. T. BHARUCHA-REID : *"On random-operator equations in Banach space,"* Bull. Acad. Polon. Sci., Ser. Sci. Math., Astronom. Phys., vol. 7, pp. 561-564 (1959).

[272] D. KANNAN and A. T. BHARUCHA-REID : *"An operator valued stochastic integral,"* Proc. Japan Acad., 47, pp. 472-476 (1971).

[273] G. ADOMIAN : *"Random operator equations in mathematical physics,"* I. J. Math. Phys., 11, pp. 1069-1084 (1974).

[274] A. V. GRANATO : *"Internal friction studies of dislocation motion,"* Dislocation Dynamics, McGraw-Hill Co., New York (1966).

[275] E. KRÖNER : *Kontinuumtheorie der Versetzungen und Eigenspannungen*, Springer-Verlag, Berlin (1958).

[276] J. P. HIRTH and J. LOTHE : *Theory of dislocations*, McGraw Hill Co.,
New York (1968).

[277] D. R. AXELRAD and J. KALOUSEK : *"Measurements of microdeformations by holographic interferometry – X-ray diffraction,"* In *Exp. Mechanics*, S.M. 9, pp. 387-401, University of Waterloo, Canada (1973).

[278] J. KALOUSEK : *Experimental Investigations of the Deformation of Structured Media*, Ph.D. Dissertation, McGill University, Montréal, Canada
(1973).

[279] J. W. PROVAN and D. R. AXELRAD : *"The effective elastic response of
randomly oriented polycrystalline solids in tension,"* Arch. Mechs., 28, 3,
pp. 531-547 (Warsaw) (1976).

[280] J. MASON : *Methods of Functional Analysis for Applications in Solid Mechanics*, Elsevier Publ. Co., Amsterdam (1985).

[281] P. BILLINGSLEY : *Ergodic Theory and Information*, John Wiley & Sons,
New York (1965).

[282] D. R. AXELRAD and W. FRYDRYCHOWICZ : *"Stochastic theory of the
inelastic behaviour of multi-component high-temperature materials,"* Math.
Models and Methods in Applied Sciences, vol. 2, No. 1, pp. 37-52, World
scientific Publ. (1992).

[283] S. K. SRINIVASAN : *Stochastic point processes and their applications*,
Hafner Press Ltd. (1974).

[284] J. E. MOYAL : *"The general theory of stochastic population processes,"*
Acta Math., 108, pp. 1-31 (1962).

[285] R. L. DOBRUSHIN : *"Description of a random field by means of conditional probabilities and the conditions governing its regularity,"* Theory
Prob. Appl., 10 (1969).

[286] F. SPITZER : *"Markov random fields and Gibbs ensembles,"* Amer. Math.
Monthly, 78 (1971).

[287] C. J. PRESTON : *Gibbs States on Countable Sets*, Cambridge Univ. Press,
Cambridge (1974).

[288] C. J. PRESTON : *Random Fields*, Springer Lecture Notes in Mathematics,
vol. 534, Springer-Verlag, Berlin (1976).

[289] R. J. ADLER : *The Geometry of Random Fields*, John Wiley, New York
(1981).

[290] J. KAMPÉ de FERIÉT : *"Statistical fluid mechanics, two-dimensional
gravity waves,"* In *Partial Differential Equations and Continuum Mechanics*, Madison, Wisconsin, pp. 136-197 (1962).

[291] J. B. KELLER : *"Wave propagation in random media,"* Proc. Symp. Appl.
Math., vol. 13, pp. 227-246, Am. Math. Soc. Providence, R. I. (1962).

[292] U. FRISCH : *"Wave propagation in random media,"* In *Probabilistic Methods in Appl. Math.*, vol. I, Ed. Bharucha-Reid, Academic Press, New York
(1968).

[293] D. R. AXELRAD and M. OSTOJA-STARZEWSKI : *"Probabilistic approach to the wave propagation in structured solids,"* Proc. 9th U. S. Nat. Congr. of Appl. Mechanics, ASME Publ. (1982).

[294] M. OSTOJA-StARZEWSKI : *Microdynamics of structured solids,* Ph.D. Dissertation, McGill University, Montréal, Canada (1983).

[295] D. R. AXELRAD and M. OSTOJA-STARZEWSKI : *"Plane wave propagation in discrete solids by the Monte-Carlo simulation,"* In Mathematical Modelling in Science and Technology 4th Int. Conf. Proceedings, Zürich (1984), Pergamon Press, New York (1984).

[296] B. K. BISWAS : *Monte-Carlo Simulation of Wave-propagation in Polycrystalline Solids,* M. Eng. Thesis, McGill University, Montréal, Canada (1983).

[297] D. R. AXELRAD : *"Mechanical models of relaxation phenomena,* Advances in Molecular Relaxation Processes, 2, pp. 41-68, Elsevier Publ. Co., Amsterdam (1970).

[298] Y. M. HADDAD : *Response Behaviour of a 2D-Fibrous Network,* Ph.D. Dissertation, McGill University, Montréal, Canada (1975).

[299] D. R. AXELRAD, Y. M. HADDAD and D. ATACK : *"Stochastic deformation theory of a two dimensional fibrous network,"* 11th Am. Symp. Soc. of Eng. Sci. Proceedings, Ed. G. J. Dvorak, Durham, North Carolina, U.S.A. (1975).

[300] Yu. N. RABOTNOV : *Creep Problems in Structural Members,* North-Holland Publ. Co., Amsterdam (1969).

[301] D. R. AXELRAD and K. REZAI : *"Determination of surface displacements by holographic electro-optical processing,"* J. Appl. Optics, 21, 2001 (1982).

[302] K. REZAI : *Determination of Surface Displacements by Holographic-Electro-Optical Processing,* Ph.D. Dissertation, McGill University, Montréal, canada (1981).

[303] D. R. AXELRAD, K. REZAI and D. ATACK : *"Probabilistic mechanics of fibrous structures,"* J. Appl. Maths. and Physics (ZAMP), vol. 35 (1984).

[304] C. MACK : *"An exact formula for $Q_k(n)$, the probable number of K-aggregates in a random distribution of points,"* Phil. Mag. 7th Series, 39, 778 (1948).

[305] D. R. AXELRAD, M. R. KAMAL and W. FRYDRYCHOWICZ : *"Stochastic flow dynamics of melts,"* In *Recent Developments in Micromechanics,* Eds. D. R. Axelrad and W. Muschik, Springer-Verlag, Berlin (1991).

[306] D. R. AXELRAD, W. FRYDRYCHOWICZ and M. R. KAMAL : *"Stochastic analysis of polymer melts with a binary structure,"* Z. angew. Math. Phys. (ZAMP), 43 (1992).

[307] P. G. de GENNES : *J. Chem. Phys.,* 55, 572 (1971).

[308] P. G. de GENNES : *Scaling Concepts in Polymer Physics,* Cornell Univ. Press, Ithaca (1979).

[309] M. DOI and S. F. EDWARDS : *Chem. Soc., Faraday Trans.,* 2, 74 (1978).

[310] W. HESS : *"Generalized Rouse theory for entangled polymeric liquids,"* Macromolecules, 21, pp. 2620-2632 (1986).

[311] W. HESS : *"Self-diffusion and reptation in semi-dilute polymer solutions,"* Macromolecules, 19, pp. 1305-1404 (1986).

[312] T. A. KAVASSALIS and J. NOOLANDI : *"New view of entanglements in dense polymer systems,"* Phys. Rev. Lett., 59 (1987).

[313] T. A. KAVASSALIS and J. NOOLANDI : *"A new theory of entanglements and dynamics in dense polymer systems,"* Macromolecules, 21, pp. 2869-2879 (1988).

[314] M. RUBINSTEIN and E. HELFAND : *"Statistics of entanglement of polymers,"* J. Chem. Phys., 82 (1985).

[315] G. ALLEGRA and F. GANAZZOLI : *"Configurations and dynamics of real chains,"* J. Chem. Phys., 76 (1982).

[316] KAZUYOCHI IWATA : *"Topological theory of entanglement,"* J. Chem. Phys., 73 (1980).

[317] M. DOI and S. F. EDWARDS : *The Theory of Polymer Dynamics,* Clarendon Press, Oxford (1986).

[318] E. KRÖNER, D. CHASSAPIS and R. TAKSERMAN-KROZER : *"The physics of temporary polymer networks,"* In *Biological and Synthetic Polymer Networks,* Ed. O. Kramer, Elsevier Publ. Co., Amsterdam (1988).

[319] P. J. FLORY : *"The configuration of real polymer chains,"* J. Chem. Phys., 17 (1949).

[320] P. J. FLORY : *Spacial Configuration of Macromolecule Chains,* Nobel Foundation Reprint, John Wiley & Sons, New York (1975).

[321] J. HAFNER : *"Quantum theory of structures,"* In *The Structure of Binary Compounds,* Eds. F. R. de Boer and D. G. Pettifor, North-Holland, Amsterdam (1989).

[322] E. KRÖNER and TAKSERMAN-KROZER : *"Statistical Mechanics of temporary polymer networks,"* Rheol. Acta, Part I and II (1984).

[323] I. MÜLLER and K. WILMANSKI : *"Residual deformations and strain recovery in polymers with folded and oriented crystals,"* OFG – Report, T. U. Berlin (1986) [unpublished].

[324] W. HESS : *"Tracer diffusion in polymeric mixtures,"* Macromolecules, 20 (1987).

[325] B. J. MEISTER : *"Uniting molecular network theory and reptation theory to predict the rheological behaviour of entangled linear polymers,"* Macromolecules, 22, pp. 3611-3619 (1989).

[326] H. C. ÖTTINGER : *"Diffusion equation versus coupled Langevin equations approach to hydrodynamics of dilute polymer solutions,"* J. Rheology, 33 (1989).

[327] G. I. TAYLOR : *Proc. Roy. Soc. A,* 102, pp. 180-189 (1922) and 104, pp. 213-218 (1923).

[328] R. E. MEYER : *Introduction to Mathematical Fluid Dynamics*, Wiley-Interscience Publ., New York (1971).

[329] G. K. BATCHELOR: *"Developments in microhydrodynamics,"* 14th Congr. IUTAM, pp. 33-35, North-Holland, Amsterdam (1976).

[330] D. R. AXELRAD : *"Micromechanics of fluids,"* In *Mechanics of Structured Media*, Ed. A. P. S. Selvadurai, Carlton University, Canada, Elsevier Publ., Amsterdam (1982).

[331] D. R. AXELRAD : *"Markov theory in the mechanics of discrete fluids,"* Int. Journ. Eng. Sci., 20, No. 2, pp. 181-186 (1982).

[332] R. HOLLEY : *"Markovian interaction processes with finite range interactions,"* Annals of Math. Statistics, 43, No. 6, pp. 1961-1976 (1971).

[333] J. BAER : *Dynamics of fluids in porous media*, Elsevier Publ. Co., NewYork (1982).

[334] C. M. MARLE : *"On macroscopic equations governing multi-phase flow with diffusion and chemical reactions in porous media,"* Int. Journ. Eng. Sci., 20, No. 5 (1982).

[335] P. M. ADLER and H. BRENNER : *"Transport processes in spacially periodic capillary networks, I-III,"* In *Physico-chemicsl Hydrodynamics*, Vol. 5, No. 3/4, Pergamon Press, Oxford (1984).

[336] K. ITÔ : *"On stochastic differential equations,"* Mem. Amer. Math. Soc., 4 (1951).

[337] H. J. KUSHNER : *Stochastic stability and control*, Academic Press, New York (1967).

[338] A. N. SHIRYAYEV : *"On stochastic equations in the theory of conditional Markov processes,"* Theor. Probability Appl., 11 (1966).

[339] D. R. AXELRAD : *"Stochastic transport theory of porous media,"* Journ. Appl. Math. and Physics (ZAMP), 41, pp. 157-173, Birkhauser Verlag Basel (1991).

[340] D. R. AXELRAD : *"Stochastic analysis of the fluid flow in a fully saturated porous media,"* In *Trends in Applications of Mathematics to Mechanics*, Interaction of Mechanics & Mathematics Series, Eds. W. Schneider, H. Troger and F. Ziegler, Longman Scientific, England (1991).

[341] M. B. ALLEN : *"Transport equations in porous media,"* In *Multi-phase flow in porous media*, Eds. C. A. Brebbia and S. A. Orszaq, Lecture Notes in Eng., No. 34, Springer-Verlag, Berlin (1988).

[342] H. C. ÖTTINGER : *"A class of multiparticle constitutive equations,"* J. Rheology, 36 (1991).

[343] D. R. AXELRAD and K. REZAI : *"Experimental Procedures in Micromechanics Proc. SEM,"* Conference on Exp. Mechanics, Cambridge, MA (1989).

[344] D. R. AXELRAD : *Measurement of microdeformations in structural solids,"* In *Recent Developments in Micromechanics*, Eds. D. R. Axelrad and W. Muschik, Springer-Verlag, Berlin (1991).

Subject Index

Drift coefficient 118

Einstein 123
Endomorphism 114, 218, 232, 261, 294, 302
Ensemble
  canonical 77, 81, 300
  Gibbs 4, 77
  isothermal-isobaric 77
Epoch 107
Ergodic
  hypothesis 3
  theory 4, 5
Error(s)
  law of 3
  distribution of 3
Event 1, 2, 22, 30, 39, 48, 53, 73, 98
External field 67

Fatigue failure 171
Fibrous structures 154, 161, 172
  stochastic analysis of 264-276
Field variables 63
Filtration (of a space) 109
Fluctuations 186, 188
Fluid structure
  binary 182
  multi-phase 182
Fokker-Planck equation 119, 123, 290, 304
Fracture and bond failure 171
Fracture mechanics 230
Function(s)
  auto-correlation 46, 48, 85
  auto-covariance 51
  Borel 9, 27, 38
  characteristic 27, 36, 41, 51, 59
  Clausius' virial 76
  conditional density 53
  conditional probability 53

continuous 9, 40
correlation 50, 53, 84, 87
covariance 46
cross-correlation 46
density 38, 45
distance 8, 11, 18
distribution 22, 23, 27, 28, 38, 40, 41, 74
  Green's 132
  mean (value) 45-8
  measurable 31, 33, 38
  moment generating 27, 28
  pair-distribution 80
  partition 78, 88
  radial distribution 80, 86
  random 30, 37, 45, 50
  single-particle 79
Functional
  characteristic 36
  continuous 17, 34
  correlation 45, 46, 47, 48, 51, 58
  linear 17, 34, 36, 51, 58
  material 233
  Minkowski 11
  random 47, 58
  system 125
  variance 58
Functional central limit theorem 44

Generalised contact force 67
Generalised surface force 168
Geometric probability 189, 190
Gibbsian random field(s) 59, 300
Graph theory 57, 90
Grain boundary 63, 159, 160, 244

Hamiltonian 4, 79, 80, 163
Harmonic analysis 49
Helmholtz free energy 153, 178, 181
History

334

335